Computer Science Workbench

Series Editor: Tosiyasu L. Kunii

Springer Japan KK

Computer Science Workbench

N. Magnenat Thalmann, D. Thalmann: Image Synthesis. Theory and Practice. XV, 400 pp., 223 figs., including 80 in color. 1987

B.A. Barsky: Computer Graphics and Geometric Modeling Using Beta-splines. IX, 156 pp., 85 figs., including 31 in color. 1987

H. Kitagawa, T.L. Kunii: The Unnormalized Relational Data Model. For Office Form Processor Design. XIII, 164 pp., 78 figs. 1989

N. Magnenat Thalmann, D. Thalmann: Computer Animation. Theory and Practice. Second Revised Edition. XIII, 245 pp., 156 figs., including 73 in color. 1990

N. Magnenat Thalmann, D. Thalmann: Synthetic Actors in Computer-Generated 3D Films. X, 129 pp., 133 figs., including 83 in color. 1990

K. Fujimura: Motion Planning in Dynamic Environments. XIII, 178 pp., 85 figs. 1991

M. Suk, S.M. Bhandarkar: Three-Dimensional Object Recognition from Range Images. XXII, 308 pp., 107 figs. 1992

H. Ishikawa: Object-Oriented Database System. Design and Implementation for Advanced Applications. XVIII, 166 pp., 38 figs. 1993

S.Z. Li: Markov Random Field Modeling in Computer Vision. XVI, 260 pp., 71 figs. 1995

Z.-Q. Liu, S. Miyamoto: Soft Computing and Human-Centered Machines. 20, 346pp., 95 figs. 2000

Z.-Q. Liu, S. Miyamoto (Eds.)

Soft Computing and Human-Centered Machines

With 95 Figures

 Springer

Zhi-Qiang Liu
Professor
Department of Computer Science and Software Engineering
The University of Melbourne
Parkville, Vic 3052, Australia

Sadaaki Miyamoto
Professor
Institute of Engineering Mechanics and Systems
The University of Tsukuba
Ibaraki 305-8573, Japan

Library of Congress Cataloging-in-Publication Data

Soft computing and human-centered machines / Z.-Q. Liu, S. Miyamoto (eds.).
    p. cm. — (Computer science workbench)
  Includes bibliographical references.
  ISBN 978-4-431-67986-8   ISBN 978-4-431-67907-3 (eBook)
  DOI 10.1007/978-4-431-67907-3

    1. Soft computing. 2. Human-computer interaction. I. Liu, Z.-Q. (Zhi-Qiang), 1950- II.
Miyamoto, Sadaaki, 1950- III. Series.

QA76.9.S63 S615 2000
006.3—dc21

00-021951

Printed on acid-free paper

Typesetting: Camera-ready by the editors and authors

SPIN: 10755623

Series Preface

Computer Science Workbench is a monograph series which will provide you with an in-depth working knowledge of current developments in computer technology. Every volume in this series will deal with a topic of importance in computer science and elaborate on how you yourself can build systems related to the main theme. You will be able to develop a variety of systems, including computer software tools, computer graphics, computer animation, database management systems, and computer-aided design and manufacturing systems. Computer Science Workbench represents an important new contribution in the field of practical computer technology.

Tosiyasu L. Kunii

Preface

With the advent of digital computers some five decades ago and the widespread use of computer networks recently, we have gained enormous power in gathering information and manufacturing. Yet, this increase in computing power has not given us *freedom* in a real sense, we are increasingly enslaved by the very machine we built for gaining freedom and efficiency. *Making machines to serve mankind* is an essential issue we are facing. Building human-centered systems is an imperative task for scientists and engineers in the new millennium.

The topic of human-centered servant modules covers a vast area. In our projects we have focused our efforts on developing theories and techniques based on fuzzy theories. Chapters 2 to 12 in this book collectively deal with the theoretical, methodological, and applicational aspects of human-centered systems. Each chapter presents the most recent research results by the authors on a particular topic.

We begin this book with an introduction to new developments in the theory of fuzzy multisets. Chapter 2 discusses in detail the theory of multisets and fuzzy multisets. Since in any intelligent system, one of the most fundamental requirements is effective and efficient schemes for querying databases or knowledge bases according to multirelations, the fuzzy multiset theory offers a natural framework for developing such schemes. This chapter applies the fuzzy multiset theory to the design of a query language, fuzzy SQL, for fuzzy databases. Chapter 2 also points out that when we take fuzzy generalization into account, we must further investigate a number of theoretically interesting problems in multisets that will have significant practical consequences. Chapter 3 introduces the theory of rough sets which is very useful for knowledge classification and approximation.

Chapter 4 gives a comprehensive discussion of fuzzy cognitive maps (FCM). Human-centered machines must be able to carry out causal reasoning in uncertain environments. The fuzzy cognitive map offers a natural framework for representing knowledge and inference. After an introduction to the basic theory of FCM, this chapter presents some major recent developments and applications. Furthermore, this chapter also presents a new method for systematically analyzing FCMs which will have an important impact on building large-scale FCMs in many real-world applications, for instance, stock exchange analysis, decision making, and planning.

Unsupervised learning, also known as clustering and learning by competition, is a major learning paradigm that deals with initially unlabeled data and is most useful when labeled training samples are not available or are known little. This is particularly relevant to human-centered machines, because in many applications human-centered machines must be able to *autonomously* discover the regularities in the data and extract useful properties to fulfill the task requirement. Chapter 5 discusses off-line (batched) data clustering techniques and gives special attention to the recent results in the fuzzy c-means clustering algorithm and its variants. Chapter 6 is concerned with competitive learning methods with an emphasis on soft-competition schemes.

In most human-centered systems we have to combine fuzzy subsets. For instance, in decision making we need to combine the several criteria to form an overall decision function. Chapter 7 introduces the general framework for fusing fuzzy information and a set of aggregation operators, namely, ordered weighted aggregation (OWA).

Chapter 8 introduces a novel, *selective* competitive learning paradigm, fuzzy gated neural networks (FGNN), for dynamic feature space partitioning, which is robust and can achieve accurate classification. FGNN eliminates the need for training the synaptic weights to a set of nodes. This property makes it easy and fast to train and simple to implement, which is especially important in human-centered systems because they have to operate in near real-time with intensive interactivity.

Chapter 9 discusses one of the fundamental and difficult issues in human-machine interaction: the *Kansei information* that models human emotional data based on the analysis of facial expressions, gestures, and voice. Recently Kansei information has gained considerable attention, especially in Japan in the development of intelligent systems. Chapter 9 introduces the basic concept of Kansei information that is characterized by its subjectivity, ambiguity, vagueness, and situatedness which are in sharp contrast to the traditional concept of information. This chapter focuses on the Kansei information extracted from facial expressions using soft computing methods.

As we constantly use vague judgment in forming our own opinions or making decisions, we have to model vague judgment in order to develop systems that are able to interact and understand human instructions and commands. Chapter 10 studies vagueness in human judgment and decision making from perspectives of the behavioral decision theory, fuzzy measure theory, and fuzzy sets. Based on extensive surveys by questionnaires, the author suggests that it is more appropriate to use fuzzy sets for representing vagueness in judgment and decision making which were traditionally treated as random processes. The author also proposes a fuzzy rating method for measuring a certain type of vagueness.

Chapter 11 studies application of chaotic systems to the analysis of time series. The authors propose a new method for dealing with time series using a deterministic system which is based on the possibilistic framework instead of the traditional probabilistic framework. The underlying motivation for this study is that many real systems, such as biological, social, and economic systems, are highly non-stationary. It is almost impossible to change parameter values and to re-create the phenomena in these systems. This study will have important implications in the development of intelligent systems.

The development of human-centered systems is a major undertaking that necessitates close collaboration between researchers from different fields of science and engineering. This book includes original contributions by the authors. In particular, Chapters 2, 3, 5, 9, 10, and 11 represent some of the major outcomes from the research project *Soft Computing and Human-Centered Information Systems* during 1997–1999 at the Tsukuba Advanced Research Alliance (TARA), University of Tsukuba. Moreover, to make the material self-contained, we have included Chapter 12 which gives a concise, easy-to-understand introduction to fuzzy set theory.

Acknowledgments

The co-editor, Sadaaki Miyamoto, who is the project leader for the development of Soft Computing and Human-Centered Information Systems at TARA, would like to express his sincere appreciation to Professor Kazuo Toraichi at the University of Tsukuba for his encouragement in carrying out this project. Professor Toraichi is the supervisor of the projects in the Aspect of Information Management through the Multimedia at TARA.

Special thanks go to Professor Toshiyasu L. Kunii of Hosei University, Tokyo for his encouragement and continued support to the editors during the preparation of this book. The authors would like to extend their appreciation to the staff of Springer-Tokyo for their cooperation, patience, and professional advice. We would also like to thank Julia Liu for her efforts in proof reading some of the chapters.

Zhi-Qiang Liu
Sadaaki Miyamoto

December 1999

List of authors

Chapter 1
> Zhi-Qiang Liu
> Department of Computer Science and Software Engineering
> The University of Melbourne
> Email:zliu@cs.mu.oz.au

> Sadaaki Miyamoto
> Institute of Engineering Mechanics and Systems
> University of Tsukuba
> Email:miyamoto@esys.tsukuba.ac.jp

Chapter 2
> Sadaaki Miyamoto

Chapter 3
> Tetsuya Murai
> Division of Systems and Information Engineering
> Graduate School of Engineering
> Hokkaido University
> Email:murahiko@main.eng.hokudai.ac.jp

> Michinori Nakata
> Department of Information Science
> Chiba-Keizai College
> Email:nakata@chiba-kc.ac.jp

> Masaru Shimbo
> Division of Systems and Information Engineering
> Graduate School of Engineering
> Hokkaido University
> Email:shimbo@main.eng.hokudai.ac.jp

Chapter 4
> Zhi-Qiang Liu

Chapter 5
> Sadaaki Miyamoto

> Kazutaka Umayahara
> Institute of Engineering Mechanics and Systems
> University of Tsukuba
> Email:uma@esys.tsukuba.ac.jp

Chapter 6
> Zhi-Qiang Liu

> Michael Glickman
> Department of Computer Science and Software Engineering

The University of Melbourne
Email:mickg@cs.mu.oz.au

Yajun Zhang
Department of Computer Science and Software Engineering
The University of Melbourne
Email:zyj@cs.mu.oz.au

Chapter 7
Ronald R. Yager
Iona College
ryager@iona.edu

Chapter 8
Zhi-Qiang Liu

Venketachalam Chandrasekran
Department of Computer Science and Software Engineering
The University of Melbourne
Email:vc@cs.mu.oz.au

Chapter 9
Takehisa Onisawa
Institute of Engineering Mechanics and Systems
University of Tsukuba
Email:onisawa@esys.tsukuba.ac.jp

Chapter 10
Kazuhisa Takemura
Institute of Policy and Planning Sciences
University of Tsukuba
Email:takemura@shako.sk.tsukuba.ac.jp

Chapter 11
Tohru Ikeguchi
Department of Applied Electronics
Faculty of Industrial Science and Technology
Science University of Tokyo
Email:tohru@te.noda.sut.ac.jp

Tadashi Iokibe
System Development Section
Computer System Business Unit
Meidensha Corporation
Email:iokibe-t@honsha.meidensha.co.jp

Kazuyuki Aihara
Department of Mathematical Engineering and Information Physics
Graduate School of Engineering

The University of Tokyo
Email:aihara@sat.t.u-tokyo.ac.jp

Chapter 12
Sadaaki Miyamoto

Contents

1
Introduction

Zhi-Qiang Liu
Sadaaki Miyamoto

1.1 The Third Industrial Revolution: human-centered machines

Since the first industrial revolution over two and half centuries ago when for the first time in our human history we started to use automated mechanical machines for manufacturing, we have seen dramatic changes in our life styles and our ability to produce and to alter our environments. With the advent of electronic computing machines more than half a century ago and the invention of the semiconductor in 1947, we gained a power that forever changed the outlook of our society and speeded the technological progress at a rate that defies the wildest of our imaginations. Driving this second industrial revolution is the semiconductor microchip. The early stage of this industrial revolution was characterized by centralized computing machines and mechatronics. The economy was geared for mass production, mass market, and mass media. Thanks to the universal accessibility of powerful computers and the Internet, especially since the early 90's, we are now connected literally to our private homes. Currently there are over 100 million computers connected to the Internet world wide, a number that is growing rapidly and will continue to do so in the new millennium. The world is soon to be fully networked and almost every aspect of our life will be related to or influenced by the computer. It is becoming obvious as we are counting down to year 2000 that we will soon be able to do most things on the Web, for instance, education, entertainment, shopping, banking, investment, and so on. Indeed, we are living in a networked world that is information-based and decentralized. The Web symbolizes this era.

With our powerful technology, we have been able to gain ever-deeper insights into nature of things, to explore ideas and places once were considered not possible, and to change our living environment: we seem to have the ability to control our destiny. The Web and the decentralized world have also created a new phenomenon: information explosion and saturation; constant upgrading of computing devices that increases only speed and memory size. As a consequence, we have been inundated with information we cannot digest and showered with gadgets that are difficult to operate. Such a development trend has almost completely ignored the issues of **human-centeredness**. As a result, the technological advancement has yet shown its promise in improving productivity and quality. In a sense, we have created a *machine-centered* culture; for instance, *learning-to-use* the computer has become a major educational activity in most primary schools in the developed world. A fundamental problem associated with the current technological development is that the machines we built lack the required intelligence that enables *human-centered* functionality.

From the historic perspective, it is inevitable that the third industrial revolution will be characterized by intelligent systems. Through the last two industrial revolutions, we have built ourselves industrial muscles (engines and automated mechatronics) and computing machines. However, we have not been able to fully integrate these two powerful components into functional systems that can operate intelligently. In recent years, the demand for intelligent systems has been rapidly increasing and gaining importance in virtually all economical and social sectors. As we are intelligent beings, machines that we build must behave intelligently in order to interact and communicate with the human and to serve the human. That is, we must be able to make machines that are human-centered and are true **servants** to humans.

These servants are not simply robots we use on the assembly line; they must be able to interact with and understand their human masters and make decisions to best serve the human. Although we have seen spectacular progresses in the last three decades in robotics, particularly, in robot's sensors, motor, and navigation capability, the progress in robotic intelligence is rather disappointing. Most robots used in manufacturing are on fixed and pre-programmed trajectories to carry out repeated actions. The traditional approach to autonomous robot design is to program a plan for navigation. In reality, however, such a design will almost certain to fail, because it is impossible to include all information and rules in the design due to the unpredictability of the dynamics and uncertainties in the environment, for instance, unknown topology and unexpected moving objects, such as humans or vehicles. To make the matter worse, sensors, and actuators are intrinsically noisy. The so-called *autonomous* robots are still primitive beings that cannot safely operate in any real working and living environment, let alone to serve the disabled, elderly, or persons in need of assistance, because, lacking intelligence, these robots themselves are fundamentally handicapped and require constant help. This is primarily due to the lack of a proper framework for the development of truly intelligent robots that have the ability to make proper decisions based on the task specifications.

The autonomous robot can become intelligent and robust in its behavior, only if it can learn in addition to following the rules in the pre-coded knowledge base; that is, the ability to acquire new knowledge, to modify and update the existing knowledge base. Traditional machine learning and pattern recognition paradigms have found many applications. However, these learning paradigms were developed mostly without the consideration of embodiedness that is essential to making the learning process an integral part of the human-centered servants. It is inevitable that most servant modules will have to function in a free working space that changes constantly. In addition, as the servant module is roaming around or interacting with humans or other servants, its view of the environment also changes. It is the dynamic environment, constant interaction, and uncertainty that make the learning process difficult and complex.

Due to recent rapid growth in electronic commerce and the explosion of the Internet, web-mining and software intelligent agents have gained significant attention. However, web-mining has the following major problems:

1. lack of structure;

2. intrinsically distributive;

3. multimedia contents.

Software agents must possess the following important properties: autonomy, adaptivity, intelligence, agent awareness, mobility, and anthropomorphism and so on [12], virtually all that are required for making a human! However, most so-called intelligent software agents cannot deliver the promised increase in productivity, enhanced software application, and human-like behavior. One of the major problems with the current approach to developing software agents, apart from the hype, is that most developers still rely on techniques in traditional artificial intelligence. As a result, such "intelligent" agents show very little intelligence, if any.

Soft computing can make a difference. With soft computing, we are able to instill the reasoning and learning ability in the servant modules and agents so that they can function effectively and intelligently. With soft computing, it is now possible to extend a large variety of learning techniques to machines, which include dynamic cognitive learning, neural-fuzzy based learning, and genetic-evolutionary type learning paradigms.

For an increasingly aging society, providing servant modules that are simple, intelligent, pleasant and flexible to our needs is perhaps the single most important requirement. Yet, we are serving the machines that we use. In summary, to serve the human, a machine must possess the following basic characteristics:

- multiple sensors that can be tuned and are self-organizable in order to communicate with the human whom it is serving and are aware of its environment in which it is operating.

- learning capability that enables the machine to adapt to different users and different operating conditions.

- reasoning and decision making capability that enables the machine to perform the *necessary* tasks autonomously, flexibly, and effectively.

- self-recovery ability with which the machine is able to recover itself from events such as power failure and software or hardware upgrading.

- robustness so that the machine's performance will not drastically deteriorate in sudden change of environments and conditions.

Over the last half a century, research on cybernetics, artificial intelligence, neural networks, genetic algorithms, evolutionary programming and computation, and especially, the advent of fuzzy logic and fuzzy sets [14] have paved the way for the development of soft computing systems that are capable of serving humans. In an effort to promote the research on human-centered machines and servant modules, in this book we present some of the major developments and paradigms in this growing area.

Unlike traditional computing systems that are based on rigid, crisp computation, human-centered machines must have the capability to reason, to learn, and to react in the way that is consistent with the way humans carry out their daily routines. To be effective, human-centered machines must handle uncertainties in the context that is significantly different from conventional techniques used in signal processing and control systems. In such uncertain environments, reasoning do not follow the additive axiom required by most classic systems and cannot be effectively modeled by probability functions based on crisp sets.

In addition to being developed in advanced, rigorous mathematical and logical frameworks, soft computing theories take into account the natural attributes in learning and reasoning. Therefore, soft computing is effective in real-world applications and has demonstrated great potentials in the development of functional, robust human-centered systems.

1.2   Soft Computing: a unifying framework for intelligent systems

Soft Computing is a unifying framework that combines techniques in neural networks, fuzzy theory, genetic algorithms, and artificial intelligence to develop intelligent systems that are able to reason and learn in dynamic, imprecise, and uncertain environments [15, 7].

Methods in Soft Computing

There are three well-known methods in soft computing: artificial neural networks, fuzzy sets, and genetic-evolutionary algorithms. A common feature in these methods is that they all attempt to imitate human-biological systems.

1. Artificial neural networks have learning schemes inherited from biological neural systems [1].

2. Fuzzy theory is based on the fact that human reasoning logic is not crisp and admits degrees [14].

3. Genetic algorithms and evolutionary computation are for optimization and learning; and are based on the basic principles in genetics and evolution [5, 6, 4].

Another common feature is that they are *soft* counterparts of traditional techniques:

1. Artificial neural networks are similar to traditional pattern recognition methods which are largely based on probabilistic techniques. Whereas traditional pattern recognition methods use linear models, most neural networks use nonlinear models.

2. Fuzzy theory has its counterparts in at least three different conventional areas of study: classical logic, probability theory, and control theory. Traditional methods are usually analytical. It is often argued that the models in the traditional methods are *objective*, whereas the models in fuzzy theory are *subjective*. However, the fact is that in conventional methods, we routinely make assumptions about certain models, e.g., the ubiquitous Gaussian assumption and independence assumption, which are themselves subjective and often result in non-robust behaviors. There is no absolute objectivity.

3. Genetic algorithms and evolutionary computation are similar to the conventional methods of mathematical programming as far as solving combinatorial optimization problems is concerned. However, genetic algorithms and evolutionary computation are easier to use, because they are mostly *data-driven*, whereas mathematical programming needs to carefully model the problem, which, in most real applications, is complex and difficult.

There are other methods that will also have a great potential in soft computing.

1. Rough sets proposed by Pawlak [11] approximate crisp sets or fuzzy sets [3] using the categorical information. Since categories or classes are a fundamental structure in understanding of objects, rough set theory offers a functional and descriptive framework for representing and analyzing human knowledge. Indeed, it has been used for knowledge discovery [16].

2. Multisets [8] which are sometimes called "bags" [9] frequently arise
   in information processing (e.g., in relational databases), and hence
   are referred to as one of the basic data structures [9]. Since multisets
   have a theoretically apparent and weak structure, they have not been
   studied in detail. The introduction of fuzziness makes the theory of
   multisets more sophisticated; and theoretically nontrivial problems
   arise in fuzzy multisets [13].

3. Chaotic systems have received considerable attention in physics and
   other areas. As we will see later, we can also apply chaos theory to
   soft computing.

References

[1] J.C. Bezdek and S.K. Pal, "Fuzzy models for pattern recognition back-
    ground, significance, and key points," in *Fuzzy Models for Pattern
    Recognition*, J.C. Bezdek and S.K. Pal, (Eds.), IEEE Press, pp.1-27,
    1992.

[2] V. Cherkassky and F. Mulier, *Learning From Data*, John Whiley &
    Sons, Inc., 1998.

[3] D. Dubois and H. Prade, "Rough fuzzy sets and fuzzy rough sets".
    Int. J. General Systems, Vol.17, pp.191-209, 1990.

[4] D.B. Fogel, *Evolutionary Computation: toward a new philosophy of
    machine intelligence* IEEE Press, Piscataway, NJ, 1995.

[5] D.E. Goldberg, *Genetic algorithms in Search, Optimization, and Ma-
    chine Learning*, Addison-Wesley, Reading, MA, 1989.

[6] J.H. Holland, *Adaptation in Natural and Artificial Systems*. MIT
    Press, MA, 1992.

[7] J.S.R. Jang, C.T. Sun, and E. Mizutani, *Neural-fuzzy and Soft comput-
    ing: a computational approach to learning and machine intelligence*.
    Prentice Hall, 1997.

[8] D.E. Knuth, *The Art of Computer Programming, Vol.2: Seminumeri-
    cal Algorithms*. Addison-Wesley, Reading, Massachusetts, 1969.

[9] Z. Manna and R. Waldinger, *The Logical Basis for Computer Program-
    ming, Vol.1: Deductive Reasoning*, Addison-Wesley, Reading, Mas-
    sachusetts, 1985.

[10] T.M. Mitchell, *Machine Learning*. WCB/Mcgraw-Hill, 1997.

[11] Z. Pawlak, *Rough Sets*. Kluwer Academic Publishers, Dordrecht, 1991.

[12] Y. Shoham, "What we talk about when we talk about software agents," *IEEE Intelligent Systems*, pp.28-31, March/April 1999.

[13] R.R. Yager, "On the theory of bags," *Int. J. General Systems*, Vol.13, pp.23-37, 1986.

[14] L.A. Zadeh, "Fuzzy sets," *Information and Control*. Vol.8, pp.338-353, 1965.

[15] L.A. Zadeh, "Fuzzy logic, neural networks and soft computing," Course Announcement, the University of California at Berkeley, Nov. 1992.

[16] W. Ziarko, ed., *Rough Sets, Fuzzy Sets, and Knowledge Discovery*, Springer, London, 1994.

2

Multisets and Fuzzy Multisets

Sadaaki Miyamoto

2.1 Introduction

Multisets, also called *bags*, are relatively unknown to researchers, partly because it has weaker mathematical properties and has been considered that applications are relatively limited.

Knuth [7] made an earlier, brief reference to multisets; In their book, Manna and Waldinger [14] devoted a chapter to multisets which they called *bags*. However, to be consistent with the current literature, throughout this chapter, we use *multiset*. Blizard [2] considered an axiomatic treatment of multisets.

Yager was the first to study fuzzy multisets and developed the basic operations [23]. In this chapter, we will modify and develop some new operations that replace Yager's. Based on the operations introduced by Yager, several researchers have carried out studies on other aspects fuzzy multisets [10, 11, 19, 20, 21, 24].

This chapter shows that, when we take fuzzy generalizations into account, we need to further investigate the theory of multisets, as there is a number of theoretically interesting problems that have important impact on the development of intelligent systems. We show that multisets are potentially very useful in information management such as relational database systems. Furthermore, we show that the origin of multisets comes from a simple sequential processing of a data set.

We illustrate the application of multisets by a query language, fuzzy structured query language (SQL), for fuzzy database systems.

2.2 Multisets

Multiset: a collection of repeated symbols

Multisets are a collection of objects in which an object can be repeated. We begin with a simple example.

Example 1. Let $X = \{a, b, c, d\}$ be an example of a universal set in which multisets are considered. An intuitive introduction of a multiset is

$$M = \{a, a, b, c, c, c\}.$$

M means that we observe two copies of the same object a, one of b, and three of c. We do not observe d in M. Accordingly we can use an alternative expression of the same multiset:

$$M = \{2/a, 1/b, 3/c\}.$$

Notice that the latter representation is somewhat similar to the representation of fuzzy set: $A = \sum_i \mu_i/x_i$. Moreover since the order of the symbols in M is of no concern, the multiset may be expressed as

$$M = \{b, a, c, a, c, c\}.$$

or

$$
\begin{aligned}
M &= \{2/a, 3/c, 1/b\} \\
  &= \{2/a, 3/c, 1/b, 0/d\}.
\end{aligned}
$$

The object of no copy may either be written or omitted as above.

Generally, let $X = \{x_1, \ldots, x_n\}$ be a finite universal set and assume that multisets are considered in this set. (This assumption of finite universal set is for simplicity. We can discuss the case of infinite universal sets in the same manner.) Let $\mathbf{N} = \{0, 1, 2, \ldots\}$ be the set of natural numbers. A (crisp) multiset, M, is characterized by a function

$$C_M \colon X \to \mathbf{N}. \tag{2.1}$$

When $C_M(x_i) = n_i$, it means that there are n_i copies of x_i in M. The function $C_M(\cdot)$ is called the count function, and hence it may also be written as $Count_M(\cdot)$.

For the above example,

$$C_M(a) = 2, \quad C_M(b) = 1, \quad C_M(c) = 3, \quad C_M(d) = 0.$$

We thus write

$$M = \{C_M(x)/x : x \in X\}. \tag{2.2}$$

We also write

$$M = \{k_1/x_1, \ldots, k_n/x_n\}$$

or

$$M = \{\overbrace{x_1, .., x_1}^{k_1}, \ldots, \overbrace{x_n, .., x_n}^{k_n}\}$$

as in the above example. The collection of all multisets of a universal set X is denoted by $\mathcal{M}(X)$, while the collection of all ordinary sets of X is denoted by 2^X.

Remark. The symbol $\{\cdot\}$ is used for both sets and multisets. It therefore shows a multiset in general in this chapter.

We call each $x_i \in X$ an object. Thus a particular object may be observed many times in a multiset. In contrast, if we want to indicate an element in a multiset, we call it a copy or an occurrence of the object. Hence two occurrences of the object are indistinguishable in a multiset.

Basic operations for multisets

We introduce basic relations and operations for multisets. The inclusion and equality between two multisets L and M are defined to be

$$L \subseteq M \iff C_L(x) \le C_M(x), \quad \text{for all } x \in X, \tag{2.3}$$
$$L = M \iff C_L(x) = C_M(x), \quad \text{for all } x \in X. \tag{2.4}$$

The union and intersection are given by

$$C_{L \cup M}(x) = C_L(x) \vee C_M(x), \quad \text{for all } x \in X, \tag{2.5}$$
$$C_{L \cap M}(x) = C_L(x) \wedge C_M(x), \quad \text{for all } x \in X \tag{2.6}$$

where $\vee$ and $\wedge$ imply max and min, respectively, as in most literature.

Readers may have noticed that there is resemblance between the count function $C_M(\cdot)$ and the membership function $\mu_A(\cdot)$. Indeed, the unions and the intersections have the same operations of maximum and minimum for both.

By an argument similar to that for fuzzy sets, we can drive the following propositions.

Proposition 1. The set $\mathcal{M}(X)$ is a distributive lattice for which the order is defined by the inclusion, and the union and intersection are the least upper bound (lub) and greatest lower bound (glb) operations, respectively.

Proof: This proof has another purpose of providing background knowledge of a lattice.

Let us recall that a lattice [13] is defined as follows. A partial order $\preceq$ is a binary relation that satisfies

Reflexivity: $x \preceq x$ for all x.

Antisymmetry: if $x \preceq y$ and $y \preceq x$, then $x = y$.

Transitivity: if $x \preceq y$ and $y \preceq z$, then $x \preceq z$.

A set with a partial order is called a poset.

Let X be a poset. An element $u \in X$ is called the least upper bound for x and y when $x \preceq u$, $y \preceq u$, and moreover if for any $u' \in X$ such that

$x \preceq u'$ and $y \preceq u'$, $u \preceq u'$ holds. The least upper bound is denoted by $u = \text{lub}(x, y)$. An element $\ell \in X$ is called the greatest lower bound for x and y when $\ell \preceq x$, $\ell \preceq y$, and moreover if for any $\ell' \in X$ such that $\ell' \preceq x$ and $\ell' \preceq y$, $\ell' \preceq \ell$ holds. The greatest lower bound is denoted by $\ell = \text{glb}(x, y)$. It is not difficult to see that $\text{lub}(x, y)$ and $\text{glb}(x, y)$ are unique, if they exist.

A lattice X is a poset such that for any $x, y \in X$, $\text{lub}(x, y)$ and $\text{glb}(x, y)$ exist.

Let us turn our attention to the present set. It is evident to see that $\mathcal{M}(X)$ is a poset with the inclusion $\subseteq$ for multisets. Moreover $\cup$ and $\cap$ respectively are the lub and glb operations, since for any $V \in \mathcal{M}(X)$ such that $L \subseteq V$ and $M \subseteq V$,

$$C_{L \cup M}(x) = C_L(x) \vee C_M(x) \leq C_V(x)$$

and hence $L \cup M \subseteq V$; the fact that $\cap$ defines the glb is proved likewise.

When a lattice satisfies the distributive law, it is called a distributive lattice. For the multisets the distributive law

$$(L \cup M) \cap N = (L \cap N) \cup (M \cap N) \tag{2.7}$$
$$(L \cap M) \cup N = (L \cup N) \cap (M \cup N) \tag{2.8}$$

should be proved. From the definitions it is sufficient to check if

$$(\ell \vee m) \wedge n = (\ell \wedge n) \vee (m \wedge n), \text{ and}$$
$$(\ell \wedge m) \vee n = (\ell \wedge n) \vee (m \wedge n)$$

hold for three natural numbers ℓ, m, n, which is an easy exercise.

An operation of ν-cut, denoted by $\mathcal{P}_\nu \colon \mathcal{M}(X) \to 2^X$, that projects a multiset into an ordinary set is introduced: for an arbitrary multiset $L \in \mathcal{M}(X)$

$$x \in \mathcal{P}_\nu(L) \iff C_L(x) \geq \nu, \tag{2.9}$$

where ν is an arbitrary natural number Notice that the ν-cut corresponds to the α-cut for fuzzy sets. In particular, $\mathcal{P}_1(\cdot)$ is also written as $\mathcal{P}(\cdot)$.

We can prove the next two propositions.

Proposition 2. For arbitrary two multisets $L, M \in \mathcal{M}(X)$,

$$L \subseteq M \iff \mathcal{P}_\nu(L) \subseteq \mathcal{P}_\nu(M) \quad \text{for all } \nu \in \{1, 2, \dots\}, \tag{2.10}$$
$$L = M \iff \mathcal{P}_\nu(L) = \mathcal{P}_\nu(M) \quad \text{for all } \nu \in \{1, 2, \dots\}. \tag{2.11}$$

Proposition 3. For arbitrary two multisets $L, M \in \mathcal{M}(X)$,

$$\mathcal{P}_\nu(L \cup M) = \mathcal{P}_\nu(L) \cup \mathcal{P}_\nu(M) \quad \text{for all } \nu \in \{1, 2, \dots\}, \tag{2.12}$$
$$\mathcal{P}_\nu(L \cap M) = \mathcal{P}_\nu(L) \cap \mathcal{P}_\nu(M) \quad \text{for all } \nu \in \{1, 2, \dots\}. \tag{2.13}$$

An ordinary set can be regarded as a particular kind of multiset. More precisely, a map of embedding $\mathcal{E}\colon 2^X \to \mathcal{M}(X)$ is defined, i.e., for an arbitrary ordinary set A of X,

$$C_{\mathcal{E}(A)}(x) = 1 \iff x \in A, \qquad C_{\mathcal{E}(A)}(x) = 0 \iff x \notin A. \qquad (2.14)$$

No transformation occurs in $\mathcal{E}(A)$. For example, if $A = \{a, b, c\}$ then $\mathcal{E}(A) = \{a, b, c\}$. The only difference is that the former is an ordinary set while the latter is a multiset. The following proposition seems almost trivial, but it justifies the operations of union and intersection for multisets, and moreover it will be useful in considering fuzzy multisets.

Proposition 4. For an arbitrary ordinary set A of X,

$$\begin{aligned}
\mathcal{P}(\mathcal{E}(A)) &= A, & (2.15) \\
\mathcal{E}(A \cup B) &= \mathcal{E}(A) \cup \mathcal{E}(B), & (2.16) \\
\mathcal{E}(A \cap B) &= \mathcal{E}(A) \cap \mathcal{E}(B). & (2.17)
\end{aligned}$$

There is another operation of *addition* for multisets, which is defined by

$$C_{L \oplus M}(x) = C_L(x) + C_M(x), \quad \text{for all } x \in X. \qquad (2.18)$$

The addition is characteristic to multisets. To see this, let us add an element to a set A and a multiset M. The addition of an element x to a set A is obviously $\{x\} \cup A$, whereas the addition to a multiset M should be $\{x\} \oplus M$, since the number of symbols of x in M should be increased by one.

Example 2. For $X = \{a, b, c, d\}$, let $M = \{2/a, 1/b, 3/c\}$ and

$$\begin{aligned}
L &= \{1/a, 2/b, 2/d\} \\
&= \{a, b, b, d, d\}.
\end{aligned}$$

Then,

$$\begin{aligned}
L \cup M &= \{2/a, 2/b, 3/c, 2/d\} \\
&= \{a, a, b, b, c, c, c, d, d\}, \\
L \cap M &= \{1/a, 1/b\} \\
&= \{a, b\}, \\
L \oplus M &= \{3/a, 3/b, 3/c, 2/d\} \\
&= \{a, a, a, b, b, b, c, c, c, d, d\}.
\end{aligned}$$

Remark. The symbol $\oplus$ in this chapter *does not* mean the addition of *fuzzy numbers*, unlike the frequent usage of this symbol elsewhere.

Other operations

For an ordinary set A, we have the complement A^C. For multisets, however, the complement does not exist, instead, we can define a relative complementation, in other words, the operation of difference. Namely, for two multisets L and M, the difference $L - M$ is defined by

$$C_{L-M}(x) = \{C_L(x) - C_M(x)\} \vee 0, \quad \text{for all } x \in X. \qquad (2.19)$$

Notice that

$$\mathcal{E}(A^C) = \mathcal{E}(X) - \mathcal{E}(A).$$

The reason why we cannot introduce the complementation for multisets is obvious: the complement of an ordinary set is defined in the closed world of the universal set, whereas the world is *open* for multisets. Therefore we can consider the *relative* operation of difference.

Let X and Y be two universal sets. Assume $L \subseteq X$ and $M \subseteq Y$. Then the Cartesian product $L \times M$, which is the multiset of $X \times Y$, is defined as follows.

$$C_{L \times M}(x, y) = C_L(x) \cdot C_M(y), \quad \text{for all } (x, y) \in X \times Y. \qquad (2.20)$$

Thus, for $L_1, L_2 \in \mathcal{M}(X)$ and $M \in \mathcal{M}(Y)$, we have

$$(L_1 \oplus L_2) \times M = (L_1 \times M) \oplus (L_2 \times M). \qquad (2.21)$$

The number of elements in a multiset that is sometimes called cardinality is denoted by $|\cdot|$, i.e.,

$$|M| = \sum_{x \in X} C_M(x). \qquad (2.22)$$

Example 3 Let $X = \{a, b, c\}$ and $Y = \{x, y, z\}$. Assume

$$
\begin{aligned}
L_1 &= \{a, b, c, c, c\}, \\
L_2 &= \{b, b, c\}, \\
M &= \{x, z, z\}.
\end{aligned}
$$

Then

$$
\begin{aligned}
L_1 - L_2 &= \{a, c, c\}, \\
L_2 \times M &= \{(b, x), (b, x), (c, x), (b, z), (b, z), (c, z), (b, z), (b, z), (c, z)\}, \\
|L_1| &= 5, \quad |L_2| = 3, \quad |M| = 3.
\end{aligned}
$$

Origin of multisets

We introduce an interpretation of an origin of multisets and the above operations.

Example 4. Let us consider an example where

$$X = \{a_1, a_2, \ldots, b_1, b_2, \ldots, c_1, c_2, \ldots\}$$

and

$$A = \{a_{i_1}, a_{i_2}, b_{j_1}\},$$
$$B = \{a_{p_1}, b_{q_1}, b_{q_2}\}.$$

Although we distinguish between the symbols a and b, we do not have the knowledge of the indices within a particular symbol. That is, we do not know i_1, i_2, and p_1.

Suppose we want to perform a set operation, say, the union. Since we cannot precisely obtain $A \cup B$, we can only estimate or calculate bounds for $A \cup B$. The theory of multisets is related to the calculation of the upper and lower bounds.

For the above example i_1 (or i_2) and p_1 may or may not be identical for the symbol a, and j_1 and q_1 (or q_2) may or may not be identical for the symbol b. Suppose $i_r \neq p_1$ and $j_1 \neq q_r$ ($r = 1, 2$), then we have the upper bound

$$A \cup B = \{a_{i_1}, a_{i_2}, a_{p_1}, b_{j_1}, b_{q_1}, b_{q_2}\}, \tag{2.23}$$

whereas the lower bound is obtained if $i_r = p_1$ and $j_1 = q_{r'}$ (for some $r, r' \in \{1, 2\}$); that is,

$$A \cup B = \{a_{i_1}, a_{i_2}, b_{q_1}, b_{q_2}\}. \tag{2.24}$$

Now, since we do not know i_1, i_2, j_1, j_2, *etc.*, we omit the indices, we obtain the multisets $\hat{A}$ and $\hat{B}$ as follows,

$$\hat{A} = \{a, a, b\}, \qquad \hat{B} = \{a, b, b\}.$$

The upper bound (2.23) corresponds to the addition,

$$\hat{A} \oplus \hat{B} = \{a, a, a, b, b, b\}$$

and the lower bound to the multiset union,

$$\hat{A} \cup \hat{B} = \{a, a, b, b\}.$$

We thus interpret a multiset as having indistinguishable elements, whence the multiset addition and union provide the upper and lower bounds for the union of the underlying distinguishable objects: the union means maximum matching of symbols, whereas the addition implies no match. Notice that the multiset intersection provides the upper bound for $A \cap B$ by the maximum matching; for the above example

$$A \cap B = \{a_{p_1}, b_{q_1}\}$$

and

$$\hat{A} \cap \hat{B} = \{a, b\}.$$

Notice that the lower bound is trivial, since no match means $A \cap B = \emptyset$.

The definition of the Cartesian product is justified from this interpretation. For example, let $C = \{a_{i_1}, a_{i_2}\}$ and $D = \{a_{p_1}, b_{q_1}\}$. Even if we do not have knowledge about the indices, the Cartesian product is unique:

$$C \times D = \{(a_{i_1}, a_{p_1}), (a_{i_2}, a_{p_1}), (a_{i_1}, b_{q_1}), (a_{i_2}, b_{q_1})\}.$$

Hence, we have

$$\hat{C} \times \hat{D} = \{(a, a), (a, a), (a, b), (a, b)\}.$$

from $\hat{C} = \{a, a\}$ and $\hat{D} = \{a, b\}$.

Multiset-valued image

Another interpretation concerning the origin of multisets comes from an image. Let us consider again a simple example.

Example 5.   Assume two universal sets $X = \{a, b, c, d, e\}$ and $Y = \{x, y, z\}$. Consider a mapping $f \colon X \to Y$ given by

$$f(a) = f(b) = f(c) = x, \qquad f(d) = f(e) = y,$$

and a set $A = \{a, b, d, e\}$ of X. Suppose that the set A is realized as a sequential file in an information processing system and consider a procedure that inputs an element in A one by one, applies f to the element, and outputs the result sequentially; the output file is denoted by B:

$$Input\ A \to [\,f\,] \to Output\ B.$$

For the sequence a, b, d, e, the output is x, x, y, y. Since A is a set, even if the order of the elements in the input file is altered, say a, d, e, b, it is considered to be the same input. For the latter we have x, y, y, x as the output. Thus, since x, x, y, y and x, y, y, x should be regarded as the same output, we have a multiset

$$B = \{x, x, y, y\} = \{2/x, 2/y\}.$$

This transformation can be extended to mapping a multiset into another multiset, for example,

$$M = \{x, x, \ldots, z, z, z\} \xrightarrow{f} \{f(x), f(x), \ldots, f(z), f(z), f(z)\}. \qquad (2.25)$$

The present transformation of multisets should be distinguished from the ordinary image $f(A)$ of a set A, it therefore is denoted by $f[M]$ hereafter.

Note that

$$f[M] = \bigoplus_{x \in M} \{f(x)\}, \qquad (2.26)$$

$$f(A) = \bigcup_{x \in A} \{f(x)\}. \qquad (2.27)$$

The equation (2.26) is regarded as the definition of $f[M]$.

The above mentioned interpretation using the indices $a_{i_1}, a_{i_2}, \ldots$ can be represented by a mapping that ignores the indices, for example,

$$X = \{a_1, a_2, \ldots, b_1, b_2, \ldots, c_1, c_2, \ldots\} \text{ and } Y = \{a, b, c, \ldots\}.$$

Assume

$$f(a_{i_1}) = f(a_{i_2}) = \cdots = a, \qquad f(b_{j_1}) = f(b_{j_2}) = \cdots = b,$$

and so on. In effect, f deletes the indices. We can see that $f[\cdot]$ provides the required multisets. Readers may consider Example 4 using this function.

We can prove the following proposition.

Proposition 5. For multisets L and M of X and a function $f : X \to Y$,

$$
\begin{aligned}
f[L \oplus M] &= f[L] \oplus f[M], & (2.28) \\
f[L \cup M] &\supseteq f[L] \cup f[M], & (2.29) \\
f[L \cap M] &\subseteq f[L] \cap f[M]. & (2.30)
\end{aligned}
$$

Proof

The first equation (2.28) is obvious from (2.26) and the associativity of $\oplus$.

For the second and third relations, notice the following.

$$
\begin{aligned}
C_{f[L] \cup f[M]}(y) &= C_{f[L]}(y) \vee C_{f[M]}(y) \\
&= \left(\sum_{x \in f^{-1}(y)} C_L(x) \right) \vee \left(\sum_{x \in f^{-1}(y)} C_M(x) \right) \\
&\leq \sum_{x \in f^{-1}(y)} (C_L(x) \vee C_M(x)) \\
&= C_{f[L \cup M]}(y); \\
C_{f[L] \cap f[M]}(y) &= C_{f[L]}(y) \wedge C_{f[M]}(y) \\
&= \left(\sum_{x \in f^{-1}(y)} C_L(x) \right) \wedge \left(\sum_{x \in f^{-1}(y)} C_M(x) \right) \\
&\geq \sum_{x \in f^{-1}(y)} (C_L(x) \wedge C_M(x)) \\
&= C_{f[L \cap M]}(y).
\end{aligned}
$$

The above relations should be contrasted with the ordinary image for ordinary sets:

$$
\begin{aligned}
f(A \cup B) &= f(A) \cup f(B), \\
f(A \cap B) &\subseteq f(A) \cap f(B).
\end{aligned}
$$

The image by a multivariable function can be defined in the same way. Let X_i, $i = 1, \ldots, p$ and Y be universal sets and L_i be multisets of X_i, $i = 1, \ldots, p$. Assume $f : X_1 \times \cdots \times X_p \to Y$. Then,

$$
f[L_1, \cdots, L_p] = \bigoplus_{(x_1, \ldots, x_p) \in L_1 \times \cdots \times L_p} \{ f(x_1, \ldots, x_p) \}, \tag{2.31}
$$

where $(x_1, \ldots, x_p)$ take all copies, $L_1 \times \cdots \times L_p$. Hence the number of elements in $f[L_1, \cdots, L_p]$ is $|L_1| \cdot |L_2| \cdots |L_p|$.

Assume $p = 2$ for simplicity. Then the following relations hold, for which the proof is omitted.

$$
\begin{aligned}
f[L_1 \oplus L_2, M] &= f[L_1, M] \oplus f[L_2, M], & (2.32) \\
f[L_1 \cup L_2, M] &\supseteq f[L_1, M] \cup f[L_2, M], & (2.33) \\
f[L_1 \cap L_2, M] &\subseteq f[L_1, M] \cap f[L_2, M]. & (2.34)
\end{aligned}
$$

2.3   Fuzzy Multisets

Yager introduced fuzzy multisets and called them *fuzzy bags* [23]. He defines a fuzzy multiset of X by a crisp multiset of $X \times I$, where $I = [0, 1]$ is the unit interval.

When

$$A = \{C_A(x, \mu)/(x, \mu) : x \in X, \mu \in I\}$$

is a fuzzy multiset, it means that the number of x with the membership degree μ is $C_A(x, \mu)$.

For example, let

$$A = \{1/(a, 0.7), 2/(a, 0.3), 1/(b, 0.5)\}.$$

This means that there are three a's with the degrees of 0.7, 0.3, and 0.3, and one b with the degree of 0.5. Hence we can write

$$A = \{(a, 0.7), (a, 0.3), (a, 0.3), (b, 0.5)\}.$$

The class of fuzzy multisets is defined as the class of crisp multisets of $X \times I$. The collection of all fuzzy multisets of X is denoted by $\mathcal{FM}(X)$ whereas the collection of all fuzzy sets of X is $\mathcal{F}(X)$.

The definitions of the inclusion, equality, union, and intersection of fuzzy bags introduced by Yager are those of crisp bags of $X \times I$.

However, there are some drawbacks in these definitions. First, the fuzzy bag operations are incompatible with those of ordinary fuzzy sets. To see this, let us extend the embedding operator $\mathcal{E}(\cdot)$ onto fuzzy multisets, that is, $\mathcal{E}(\cdot)$ maps an ordinary fuzzy set into the fuzzy bag of the same content. Assume $A = \sum_i \mu_i/x_i$. Then, define

$$\mathcal{E}(A) = \{1/(x_i, \mu_i)\}, \tag{2.35}$$

where $\mu_i \neq 0$.

Consider an example in which $E = 0.1/a + 0.6/b$ and $F = 0.5/b$. Then, $\mathcal{E}(E) = \{1/(a, 0.1), 1/(b, 0.6)\}$ and $\mathcal{E}(F) = \{1/(b, 0.5)\}$. From the operation by Yager [23], we have

$$\mathcal{E}(E) \cup \mathcal{E}(F) = \{1/(a, 0.1), 1/(b, 0.6), 1/(b, 0.5)\},$$

for $E \cup F = E$.

In general, the compatibility between the fuzzy multiset relations and operations and those of ordinary fuzzy sets are expressed as

$$\mathcal{E}(A) \subseteq \mathcal{E}(B) \iff A \subseteq B, \tag{2.36}$$
$$\mathcal{E}(A) \cup \mathcal{E}(B) = \mathcal{E}(A \cup B), \tag{2.37}$$
$$\mathcal{E}(A) \cap \mathcal{E}(B) = \mathcal{E}(A \cap B), \tag{2.38}$$

where A and B are ordinary fuzzy sets. The above example shows that these relations do not hold.

Second, it is difficult to obtain the operation of α-cut from the definitions by Yager [23]. Notice that we wish to define the α-cut so that the resulting crisp multisets have another set of compatibility relations, for example, suppose that A and B are two fuzzy multisets and that we have appropriately defined the α-cut A_α. We wish to obtain the following relations (cf. [5, 15]).

$$A \subseteq B \iff A_\alpha \subseteq B_\alpha, \ \forall \alpha \in (0, 1], \qquad (2.39)$$

$$(A \cup B)_\alpha = A_\alpha \cup B_\alpha, \qquad (2.40)$$

$$(A \cap B)_\alpha = A_\alpha \cap B_\alpha. \qquad (2.41)$$

With these motivations in mind, we introduce a standard form: First, we collect the membership degrees for each object $x \in X$:

$$A = \{n_i/(x_i, \mu_i)\}_{i=1,\ldots,n} \qquad (2.42)$$

$$= \{\{\mu_i', \ldots, \nu_i'\}/x_i\}_{i=1,\ldots,n}. \qquad (2.43)$$

For this example, we have

$$A = \{\{0.3, 0.7, 0.3\}/a, \{0.5\}/b\}.$$

Then, we arrange these degrees $\{\mu_i', \ldots, \nu_i'\}$ into the decreasing order: $\mu_i'' \geq \ldots \geq \nu_i''$. The resulting sequence is called the grade sequence for x_i. We write, instead of $\mu_i'', \ldots, \nu_i''$, $\mu_A^1(x), \ldots, \mu_A^p(x)$ for $x \in X$ ($\mu_A^1(x) \geq \ldots \geq \mu_A^p(x)$) and assume that p is independent of x by appending appropriate numbers of zero degrees for simplicity. Thus,

$$A = \{(\mu_A^1(x), \ldots, \mu_A^p(x))/x \ : \ x \in X\}. \qquad (2.44)$$

For example,

$$A = \{\{0.3, 0.7, 0.3\}/a, \{0.5\}/b\}$$

$$= \{(0.7, 0.3, 0.3)/a, (0.5, 0, 0)/b, (0, 0, 0)/c\}$$

in this form.

2.3.1 Basic Operations of Fuzzy Multisets

Now, we can define the basic relations and operations as follows, assuming that A and B are two fuzzy multisets of X.

(I) [inclusion]

$$A \subseteq B \iff \mu_A^j(x) \leq \mu_B^j(x), \ j = 1, 2, \ldots, p, \ \forall x \in X.$$

(II) [equality]

$$A = B \iff \mu_A^j(x) = \mu_B^j(x), \ j = 1, 2, \ldots, p, \ \forall x \in X.$$

(III) [union]

$$\mu_{A \cup B}^j(x) = \mu_A^j(x) \vee \mu_B^j(x), \ j = 1, 2, \ldots, p, \ \forall x \in X.$$

(IV) [intersection]

$$\mu_{A \cap B}^j(x) = \mu_A^j(x) \wedge \mu_B^j(x), \ j = 1, 2, \ldots, p, \ \forall x \in X.$$

(V) [α-cut]
For arbitrarily given $\alpha \in (0, 1]$, the number of copies of x in A_α, the α-cut of A, is defined by

$$
\begin{aligned}
C_{A_\alpha}(x) = 0 &\iff \mu_A^1(x) < \alpha, \\
C_{A_\alpha}(x) = k &\iff \mu_A^k(x) \geq \alpha \text{ and } \mu_A^{k+1}(x) < \alpha, \\
&\qquad (k < p), \\
C_{A_\alpha}(x) = p &\iff \mu_A^p(x) \geq \alpha.
\end{aligned}
$$

(VI) [Cartesian product]
Assume $X = \{x_1, \ldots, x_m\}$ and $Y = \{y_1, \ldots, y_n\}$ are two universal sets and assume

$$
\begin{aligned}
A &= \{(x_i, \mu_i) : x_i \in X\} \\
C &= \{(y_j, \nu_j) : y_j \in Y\}
\end{aligned}
$$

are two fuzzy multisets. Then the Cartesian product $A \times C$ is defined:

$$A \times C = \{((x_i, y_j), \mu_i \wedge \nu_j) : (x_i, y_j) \in X \times Y\}.$$

Example 6. Let $X = \{a, b, c\}$ and $Y = \{x, y\}$. Assume

$$
\begin{aligned}
A &= \{(a, 0.3), (a, 0.5), (a, 0.7), (b, 0.5)\}, \\
B &= \{(a, 0.9), (a, 0.2), (c, 0.4)\}, \\
C &= \{(x, 0.5), (y, 0.8), (y, 0.1)\}.
\end{aligned}
$$

We then have

$$
\begin{aligned}
A \cup B &= \{(0.9, 0.5, 0.3)/a, (0.5)/b, (0.4)/c\}, \\
A \cap B &= \{(0.7, 0.2)/a\}, \\
A \oplus B &= \{(a, 0.3), (a, 0.5), (a, 0.7), (a, 0.9), (a, 0.2), (c, 0.4), (b, 0.5)\}, \\
B \times C &= \{((a, x), 0.5), ((a, x), 0.2), ((c, x), 0.4), ((a, y), 0.8), ((a, y), 0.2), \\
&\qquad ((c, y), 0.4), ((a, y), 0.1), ((a, y), 0.1), ((c, y), 0.1)\}, \\
A_{0.4} &= \{a, a, b\}.
\end{aligned}
$$

We have the following propositions, of which the proofs are omitted.

Proposition 6. For ordinary fuzzy sets A and B and the above introduced relations and operations (I–IV), the relations (2.36), (2.37), and (2.38) hold. Moreover, for the Cartesian product,

$$\mathcal{P}(A \times C) \;=\; \mathcal{P}(A) \times \mathcal{P}(C).$$

Proposition 7. For fuzzy multisets A and B and the above introduced relations and operations (I–IV), the relations (2.39), (2.40), and (2.41) hold. Moreover, for the Cartesian product,

$$(A \times C)_\alpha = A_\alpha \times C_\alpha, \quad \forall \alpha \in [0, 1].$$

Proposition 8. Fuzzy multisets satisfy

(a) [the commutative law]

$$A \cup B = B \cup A, \quad A \cap B = B \cap A.$$

(b) [the associative law]

$$A \cup (B \cup C) \;=\; (A \cup B) \cup C,$$
$$A \cap (B \cap C) \;=\; (A \cap B) \cap C.$$

(c) [the distributive law]

$$A \cup (B \cap C) \;=\; (A \cup B) \cap (A \cup C),$$
$$A \cap (B \cup C) \;=\; (A \cap B) \cup (A \cap C).$$

Thus, the class of all fuzzy multisets of X forms a distributive lattice [13].

We show the proof of the distributive law in order to see how the α-cut works. Let us note that the last law holds if and only if

$$[A \cap (B \cup C)]_\alpha = [(A \cap B) \cup (A \cap C)]_\alpha \tag{2.45}$$

is valid for all $\alpha \in (0, 1]$, in view of (2.39). To show (2.45), we see that the following holds.

$$
\begin{aligned}
[A \cap (B \cup C)]_\alpha &= A_\alpha \cap (B \cup C)_\alpha \\
&= A_\alpha \cap (B_\alpha \cup C_\alpha) \\
&= (A_\alpha \cap B_\alpha) \cup (A_\alpha \cap C_\alpha) \\
&= (A \cap B)_\alpha \cup (A \cap C)_\alpha \\
&= [(A \cap B) \cup (A \cap C)]_\alpha,
\end{aligned}
$$

Notice the relations (2.40) and (2.41), and that the distributive law is valid for crisp multisets [14].

Addition

The addition $\oplus$ introduced by Yager [23] is given as follows. For $A = \{m_{ik}/(x_i, \mu_k)\}$ and $B = \{n_{ik}/(x_i, \mu_k)\}$,

$$A \oplus B = \{(m_{ik} + n_{ik})/(x_i, \mu_k)\}. \tag{2.46}$$

Proposition 9. The commutativity of the addition with the α-cut holds:

$$(A \oplus B)_\alpha = A_\alpha \oplus B_\alpha. \tag{2.47}$$

Since definitions of other operations, such as t-norms and conorms, are straightforward, we omit the details.

ν-cut for fuzzy multiset

The ν-cut for crisp multisets that corresponds to the α-cut for ordinary multisets can be extended for fuzzy multisets.

The ν-cut now is defined to be a transformation $\mathcal{P}_\nu(\cdot)$ from $\mathcal{FM}(X)$ into $\mathcal{F}(X)$; for arbitrarily given fuzzy multiset $A \in \mathcal{FM}(X)$,

$$\mu_{\mathcal{P}_\nu(A)}(x) = \mu_A^\nu(x), \quad x \in X. \tag{2.48}$$

$\mathcal{P}_\nu(A)$ is the νth member of the grade sequence. It is easy to see that the next proposition holds.

Proposition 10. For arbitrarily given $A \in \mathcal{FM}(X)$,

$$\{\mathcal{P}_\nu(A)\}_\alpha = \mathcal{P}_\nu(A_\alpha) \tag{2.49}$$

for all $\alpha \in [0, 1]$ and $\nu \in \{1, 2, \dots\}$.
The ν-cut and the α-cut are thus commutative.

2.4   Infinite Fuzzy Multisets

2.4.1   Infinite Sequence of Memberships and Computability

Assume again that $X = \{x_1, \dots, x_n\}$ is a finite universal set and multisets are considered in this set. [1]
Our problem here is to extend (2.43) to an infinite collection,

$$A = \{\{\mu_i', \dots, \nu_i', \dots\}/x_i\}_{i=1,\dots,n}, \tag{2.50}$$

---

[1] The universal set can be countably infinite. For simplicity we assume finiteness but the argument is the same for the infinite universe.

and to define an appropriate class of the collection of memberships $\{\mu'_i, \ldots, \nu'_i, \ldots\}$. It is difficult to deal with such an infinite collection without ordering the memberships, since the basic relations and operations are defined by ordered sequences of memberships.

Our starting point is that α-cuts for the infinite fuzzy multisets should provide well-defined crisp multisets. If we suppose that the collection is ordered from the first, namely, it is given by a monotone decreasing sequence of memberships $(\mu^1, \ldots, \mu^n, \ldots)$ $(\mu^1 \geq \ldots \geq \mu^n \geq \ldots)$ that approaches to zero $(\mu^n \to 0)$, its α-cuts are well-defined crisp multisets. This class is, however, too narrow, since it does not generalize the finite fuzzy multisets in which $\{\mu'_i, \ldots, \nu'_i\}$ is not ordered. Thus, the sequence must include any sequences of the form: $[\mu'_i, \ldots, \nu'_i, 0, \ldots, 0, \ldots]$ for which only finite members are nonzero but they are not sorted in general. Here we handle a class of sequences denoted by $[\nu_1, \ldots, \nu_n, \ldots]$ to distinguish them from the ordered sequences $(\mu^1, \ldots, \mu^n, \ldots)$.

The solution turns out to be an infinite sequence that approaches to zero $(\nu_n \to 0)$, but is not necessarily monotone.

In the rest of this section, we discuss the justification of this statement. Note that this class is exact, since if a sequence ν_n does not converge to zero, an α-cut for it will provide a crisp multiset of infinite count which is not well-defined.

Let the class of infinite sequences of nonnegative numbers that approaches to zero be S and an arbitrary element of S be $[\nu_1, \ldots, \nu_n, \ldots] \in S$, where $\lim_{n \to \infty} \nu_n = 0$, but the sequence is not necessarily monotone.

Now, let us consider an infinite fuzzy multiset A of X given by

$$A = \{[\nu_1, \ldots, \nu_n, \ldots]/x : x \in X\}. \tag{2.51}$$

We should consider if this class provides well-defined fuzzy multisets that are compatible with those in the previous section.

The fundamental idea is to sort $[\nu_1, \ldots, \nu_n, \ldots]$ into monotone nonincreasing sequence $(\mu^1, \ldots, \mu^m, \ldots)$ $(\mu^1 \geq \ldots \geq \mu^m \geq \ldots)$. The problem is therefore whether we can sort infinite sequences.

A trivial solution is to *assume* that the first sequence $[\nu_1, \ldots, \nu_n, \ldots]$ can always be ordered. Then we have no problem at all. A better idea is to adopt the following assumption of *computability for sorting.*

Recall that the statement that the sequence $\nu_1, \ldots, \nu_n, \ldots$ of nonnegative numbers approaches to zero ($\lim_{n \to \infty} \nu_n = 0$):

For arbitrary $\epsilon > 0$, there exists a positive integer N such that for every $n > N$, $\nu_n < \epsilon$.

Assumption C (computability for sorting):

For arbitrarily given $\epsilon > 0$, the positive integer $N(\epsilon)$ that satisfies the above property is computable by some effective procedure.

This assumption covers a broad class of such sequences: it is difficult to discover a sequence $\nu_1, \ldots, \nu_n, \ldots$ that approaches to zero that does not satisfy Assumption **C**. [2]

Using this assumption, we can reorder and construct the infinite grade sequence $(\mu^1, \ldots, \mu^p, \ldots)$ by the following procedure.

(a) (find μ^1) Take the first nonzero element $\nu_i(> 0)$ and Let $\epsilon = \nu_i$. Calculate $N(\epsilon)$ and from the finite subsequence $\nu_1, \ldots, \nu_{N(\epsilon)}$ find the maximum element

$$\nu_k = \max\{\nu_1, \ldots, \nu_{N(\epsilon)}\}.$$

Let $\mu^1 = \nu_k$. Set $\nu_k = 0$.

(b) (find μ^j) Assume that $\mu^1, \ldots, \mu^{j-1}$ have been found. It is obvious that step (a) can be used for finding the next element μ^j in the grade sequence.

2.4.2 Operations for Infinite Fuzzy Multisets

Thus, by the reordering we can define the basic relations and operations for infinite fuzzy multisets. As the infinite sequence of grades is well-defined, we use the symbols $\mu_A^1(x), \ldots$ etc. instead of $\nu_1, \ldots, \nu_n, \ldots$.

(I) [inclusion]

$$A \subseteq B \Leftrightarrow \mu_A^j(x) \leq \mu_B^j(x), \; j = 1, 2, \ldots, p, \ldots, \; \forall x \in X.$$

(II) [equality]

$$A = B \Leftrightarrow \mu_A^j(x) = \mu_B^j(x), \; j = 1, 2, \ldots, p, \ldots, \; \forall x \in X.$$

(III) [union]

$$\mu_{A \cup B}^j(x) = \mu_A^j(x) \vee \mu_B^j(x), \; j = 1, 2, \ldots, p, \ldots, \; \forall x \in X.$$

(IV) [intersection]

$$\mu_{A \cap B}^j(x) = \mu_A^j(x) \wedge \mu_B^j(x), \; j = 1, 2, \ldots, p, \ldots, \; \forall x \in X.$$

(V) [α-cut] For arbitrarily given $\alpha \in (0, 1]$, the number of copies of x in A_α, the α - cut of A, is defined by

$$C_{A_\alpha}(x) = 0 \quad \Longleftrightarrow \quad \mu_A^1(x) < \alpha,$$
$$C_{A_\alpha}(x) = k \quad \Longleftrightarrow \quad \mu_A^k(x) \geq \alpha \text{ and } \mu_A^{k+1}(x) < \alpha.$$

---

[2]It is possible to construct such a sequence by using probabilistic factors, but it is difficult to construct without the use of randomness.

Notice that the α-cut of an infinite fuzzy multiset always exists, which is obvious from the assumption.

When addition of two infinite fuzzy multisets should be considered, the computability assumption is useful. For the addition $A \oplus B$, first construct $\nu_1, \dots, \nu_n, \dots$ by letting

$$\nu_{2k-1} = \mu_A^k(x),$$
$$\nu_{2k} = \mu_B^k(x),$$
$$(k = 1, 2, \dots).$$

Note that $\nu_1, \dots, \nu_n, \dots$ approaches to zero and Assumption C holds for this sequence. (Use the computability for A and B for the proof.) Hence the grade sequence for $A \oplus B$ is calculated.

The propositions parallel to Propositions 7, 8, and 9 with the replacement of finite fuzzy multisets by infinite fuzzy multisets hold. It is easy to prove these propositions. We omit the details.

A useful operation for an infinite fuzzy multiset A is *fuzzy α-cut* $F_\alpha(A)$, which maps an infinite fuzzy multiset into a finite fuzzy multiset. [3] It is defined as follows. Let p be a positive number such that

$$\mu_A^k(x) \geq \alpha \quad \text{for } k \leq p,$$
$$\mu_A^k(x) < \alpha \quad \text{for } k > p.$$

Then,

$$\mu_{F_\alpha(A)}^k(x) = \mu_A^k(x), \quad k \leq p,$$
$$\mu_{F_\alpha(A)}^k(x) = 0, \quad k > p.$$

The fuzzy α-cut for ordinary fuzzy sets was introduced by Radecki [18].

Remark. Since the present approach is constructive, operations such as the union of an infinite number of fuzzy multisets: $\bigcup_i^{\infty} A_i$ (called infinite union here) is not considered. Even for finite fuzzy multisets, the infinite union may not be a well-defined fuzzy multiset. In contrast, the infinite intersection $\bigcap_i^{\infty} A_i$ is well-defined, although we must use the continuity of real numbers.

Thus, the set of finite and infinite fuzzy multisets is not a complete lattice. One simple way to guarantee the existence of the infinite union is to assume an *a priori* upper bound, which is the maximum fuzzy multiset

---

[3] In applications we should frequently assume that finite observations $F_\alpha(A)$ instead of the true A should be used, and therefore the fuzzy α-cut is necessary.

for a specified class. For such a bounded class, however, the addition should be excluded from the operations of the lattice.

Images of fuzzy multisets

We can extend (2.26) and (2.27) to fuzzy multisets by defining

$$f[A] = \bigoplus_{x \in A} \{f(x)\}, \qquad (2.52)$$

$$f(A) = \bigcup_{x \in A} \{f(x)\}. \qquad (2.53)$$

for a fuzzy multiset $A \in \mathcal{FM}(X)$. We assume that $\mathcal{FM}(X)$ includes infinite fuzzy multisets.

In order to obtain explicit representations, let $X = \{x_1, \ldots, x_m\}$ and $Y = \{y_1, \ldots, y_n\}$ be two universal sets. Let $f: X \to Y$ and assume $A = \{(x_i, \mu_i)\}$, then

$$f[A] = \{(f(x_i), \mu_i)\}.$$

In contrast, $f(A)$ requires the form of the grade sequence. We have

$$f(A) = \{(\max_{x \in f^{-1}(y_j)} \mu_A^1(x), \ldots, \max_{x \in f^{-1}(y_j)} \mu_A^p(x))/y_j, \quad j = 1, \ldots, n\}.$$

Example 7. Assume two universal sets $X = \{a, b, c, d, e\}$ and $Y = \{x, y, z\}$. Consider the mapping $f: X \to Y$ in Example 5:

$$f(a) = f(b) = f(c) = x, \qquad f(d) = f(e) = y,$$

and

$$A = \{(a, 0.1), (a, 0.5), (b, 0.7), (d, 0.9), (e, 0.8)\}.$$

We have

$$f[A] = \{(x, 0.1), (x, 0.5), (x, 0.7), (y, 0.9), (y, 0.8)\},$$
$$f(A) = \{(x, 0.5), (x, 0.7), (y, 0.9)\}.$$

We can see that the following proposition holds.

Proposition 11. For an arbitrarily given $A, B \in \mathcal{FM}(X)$,

$$f[A \oplus B]_\alpha = f[A]_\alpha \oplus f[B]_\alpha$$
$$f[A \cup B]_\alpha = f[A]_\alpha \cup f[B]_\alpha,$$

for all $\alpha \in [0, 1]$.

Using this proposition and Proposition 5, we obtain

Proposition 12. For an arbitrarily given $A, B \in \mathcal{FM}(X)$,

$$
\begin{aligned}
f[A \oplus B] &= f[A] \oplus f[B], \\
f[A \cup B] &\supseteq f[A] \cup f[B], \\
f[A \cap B] &\subseteq f[A] \cap f[B], \\
f(A \cup B) &= f(A) \cup f(B), \\
f(A \cap B) &\subseteq f(A) \cap f(B).
\end{aligned}
$$

We can consider narrower classes of fuzzy sets of X and infinite fuzzy multisets of Y: for a fuzzy set C and a fuzzy multiset D in these classes we assume that the following cardinalities are finite:

$$
|C| = \sum_{x \in X} \mu_C(x) < \infty,
$$

$$
|D| = \sum_{y \in Y} \sum_{k=1}^{\infty} \mu_D^k(y) < \infty.
$$

From Proposition 12 and

$$
|A| = |f[A]|,
$$

the following proposition holds.

Proposition 13. For fuzzy sets A and B of X in the above class,

$$
\begin{aligned}
|A \oplus B| &= |f[A] \oplus f[B]|, & (2.54) \\
|A \cup B| &\geq |f[A] \cup f[B]|, & (2.55) \\
|A \cap B| &\leq |f[A] \cap f[B]|. & (2.56)
\end{aligned}
$$

Notice that the respective bounds of the right hand sides are exact, i.e., they cannot be improved.

The proofs are omitted since they are straightforward.

2.5   Another Fuzzification

We can consider another type of fuzzification of the crisp multiset. A brief discussion is sufficient for understanding this.

This fuzzification is done by allowing that the *Count* function takes the value of fuzzy numbers. Let $\mathcal{FN}(\boldsymbol{R}^+)$ be the collection of all strictly non-negative fuzzy numbers. [4] A multiset M is then characterized by a function $C_M(\cdot)\colon X \to \mathcal{FN}(\boldsymbol{R}^+)$.

It is straightforward to extend the basic operations. In the following definitions, the right hand sides use the ordering and the operations of fuzzy numbers.

(I) [inclusion]

$$M \subseteq N \iff C_M(x) \leq C_N(x) \quad \text{for all } x \in X.$$

(II) [equality]

$$M = N \iff C_M(x) = C_N(x) \quad \text{for all } x \in X.$$

(III) [union]

$$C_{M \cup N}(x) = \max\{C_M(x), C_N(x)\}.$$

(IV) [intersection]

$$C_{M \cap N}(x) = \min\{C_M(x), C_N(x)\}.$$

(V) [addition]

$$C_{M \oplus N}(x) = C_M(x) + C_N(x).$$

In this way we can handle the case when the count of an element of X is *about k*.

2.6 Application to Query Language for Fuzzy Database

Let us consider an application of fuzzy multisets in the Section 2.3 to a class of relational databases. The motivation for fuzzy multisets in fuzzy database is apparent. In crisp relational databases, the derived records from SELECT in SQL form a multiset [4]. This naturally leads to fuzzy multisets as the results of applying SELECT in fuzzy SQL [3].

---

[4] A strictly nonnegative fuzzy number N means that $\mu_N(x) = 0$ for all $x < 0$.

2.6.1   Fuzzy Multirelations

SQL allows multiple copies of the same element. Consider a simple relation scheme $R(C,Y)$ which consists of the following tuples:

$$R = \{(C_1, a), (C_2, a), (C_1, b), (C_3, a), (C_2, b)\}.$$

This relation is an ordinary subset of $C \times Y$. From a query SELECT ALL Y FROM **R**, we obtain $\{a, a, b, a, b\}$, which is a multiset.

Consider a simple fuzzy relation scheme $FR(C,Y,M)$ in which the third attribute M implies the degree of membership. Assume that FR is the following set of tuples:

$$FR \;=\; \{(C_1, a, 0.2), (C_2, a, 0.1), (C_1, b, 0.5),$$
$$(C_3, a, 0.3), (C_2, b, 0.5)\}.$$

Applying

SELECT ALL Y FROM **FR**

we have

$$\{(a, 0.2), (a, 0.1), (b, 0.5), (a, 0.3), (b, 0.5)\}.$$

Thus we have three different memberships: 0.2, 0.1, 0.3 for a, and two equal memberships: 0.5, 0.5 for b.

2.6.2   Functions in Fuzzy SQL

Such derived relations or fuzzy relations should be called multirelations or fuzzy multirelations. We have more operations than those for ordinary sets and fuzzy sets. They should be included as functions for an extension of a query language for manipulating multirelations. Some of them are as follows.

(1) SELECT ALL:

As shown above, it is a function that derives a (fuzzy) multiset from an ordinary set or another (fuzzy) multiset.

(2) ADDITION:

It simply collects all tuples from two multirelations of the same attributes.

(3) UNION:

It is defined as the union of fuzzy multisets.

(4) INTERSECT:

It is defined as the intersection of fuzzy multisets.

(5) Reduction to an ordinary (fuzzy) relation:

Some functions such as SELECT DISTINCT request reduction of fuzzy multirelations to ordinary fuzzy relation. The reduction is done by taking out the copy of the maximum membership degree among the multiple copies of the same element of the (ordinary) relation, which corresponds to the extension principle. Notice that sorting membership degrees is necessary in order to perform the reduction. Possibility distributions should be handled strictly in the reduction. That is, two possibility distributions $\Pi = \sum_i \mu_i/x_i$ and $\Pi' = \sum_i \nu_i/x_i$ are strictly equal if and only if $\mu_i = \nu_i$, $\forall i$. Two tuples are not equal and should not be reduced to one if there exists a pair of unequal possibility distributions corresponding to the same attribute.

(6) Reduction to a crisp multirelation:

Another reduction is to apply an α-cut to a fuzzy multirelation and obtain a crisp multirelation.

(7) Selection or join from two multirelations:

Assume that we have p tuples $(a, \mu), \ldots, (a, \mu')$ in a scheme $\mathbf{FR}(A,M)$ and q tuples $(b, \nu), \ldots, (b, \nu')$ in another scheme $\mathbf{FS}(B,N)$ where M and N are memberships. Consider a query

"SELECT A, B FROM $\mathbf{FR}$, $\mathbf{FS}$ WHERE $\mathbf{FR}$.A $\mathcal{O}$ $\mathbf{FS}$.B"

in which $\mathcal{O}$ is an operator such as $=, \neq, >$, etc. Then, we have pq copies of $(a, b, \mu \wedge \nu), \ldots, (a, b, \mu' \wedge \nu')$ in the resulting multirelation. This function is naturally extended to the case when the attribute values are possibility distributions.

A problem is whether to use the data structure in the form of the collection of plain tuples like

$$\{\ldots (x, \mu), \ldots (x, \mu') \ldots (x', \nu), \ldots (x', \nu') \ldots\}$$

or in the standard form of the grade sequence such as

$$\{\ldots (x, (\mu, \ldots, \mu')) \ldots (y, (\phi, \ldots, \phi'))\}.$$

In the former form x and x' may be the same element, whereas x and y are different in the latter form. The latter form saves space, but requires calculations for the data transformation from the first to the second form. Notice also that the second form is useful when UNION and INTERSECT are performed.

2.7   Conclusion

The theory of multisets and fuzzy multisets have been studied. There are two fuzzifications of crisp multisets; each of them have appropriate applications. Fuzzy multisets are theoretically interesting and useful in many practical applications, in particular, in the development of human-centered machines. Although we have not discussed t-norms and conorms, we can study their properties in relation to the infinite fuzzy multisets and subclasses of the collections of all fuzzy multisets.

As an application, we have presented the fuzzy SQL. There is room for further study in this area, since there are many related topics in crisp relational databases and multisets [8, 12].

Recently, Banâtre and D. Le Métayer have proposed a programming language on the basis of multisets [1]. The use of fuzzy multisets for such purposes should also be considered.

As shown in this chapter, multisets are naturally derived from sequential processing of information items, we therefore expect further developments in systems of multisets, and in particular, fuzzy multisets.

2.8 REFERENCES

[1] J.-P. Banâtre and D. Le Métayer, "Programming by multiset transformation," *Comm. ACM*, Vol. 36, No. 1, pp. 98–111, 1993.

[2] W. D. Blizard, "Multiset theory," *Notre Dame Journal of Formal logic*, Vol. 30, No. 1, pp. 36–66, 1989.

[3] P. Bosc, "Some approaches for processing SQLf nested queries," *International Journal of Intelligent Systems*, Vol. 11, No. 9, 613–632, 1996.

[4] J. Celko, *Joe Celko's SQL for Smarties: Advanced SQL Programming*, Morgan Kaufmann, 1995.

[5] D. Dubois and H. Prade, *Possibility Theory: An Approach to Computerized Processing of Uncertainty*, Plenum, New York, 1988.

[6] K. S. Kim and S. Miyamoto, Application of fuzzy multisets to fuzzy database systems, *Proc. of 1996 Asian Fuzzy Systems Symposium*, Dec. 11–14, 1996, Kenting, Taiwan, R.O.C. pp. 115–120, 1996.

[7] D. E. Knuth, *The Art of Computer Programming, Vol.2 / Seminumerical Algorithms*, Addison-Wesley, Reading, Massachusetts, 1969.

[8] A. Klausner and N. Goodman, Multirelations – semantics and languages, *Proc. VLDB85*, Aug. 21-23, Stockholm, pp. 251–258, 1985.

[9] R. Kruse, J. Gebhardt, and F. Klawonn, *Foundations of Fuzzy Systems*, Wiley, Chichester, 1994.

[10] B. Li, W. Peizhang, and L. Xihui, Fuzzy bags with set-valued statistics, *Comput. Math. Applic.*, Vol. 15, pp. 811–818, 1988.

[11] B. Li, Fuzzy bags and applications, *Fuzzy Sets and Systems*, Vol. 34, pp. 61–71, 1990.

[12] L. Libkin and L. Wong, Query languages for bags and aggregate functions, *J. Comput. Sys. Sci.*, Vol. 55, pp. 241–272, 1997.

[13] S. MacLane and G. Birkoff, *Algebra, 2nd ed*, Macmillan, New York, 1979.

[14] Z. Manna and R. Waldinger, *The Logical Basis for Computer Programming, Vol. 1: Deductive Reasoning*, Addison-Wesley, Reading, Massachusetts, 1985.

[15] S. Miyamoto, *Fuzzy Sets in Information Retrieval and Cluster Analysis*, Kluwer Academic Publishers, Dordrecht, 1990.

[16] Z. Pawlak, *Rough Sets*, Kluwer, Dordrecht, 1991.

[17] Z. Pawlak, Hard and soft sets, In: W. Ziarko (ed.), *Rough Sets, Fuzzy Sets and Knowledge Discovery, Proceedings of the International Workshop on Rough Sets and Knowledge Discovery (RSKD'93)*, Banff, Alberta, Canada, October 12-15, Springer-Verlag, Berlin, pp. 130–135, 1993.

[18] T. Radecki, Level fuzzy sets, *J. of Cybernetics*, Vol. 7, pp. 189–198, 1977.

[19] A. Rebai, Canonical fuzzy bags and bag fuzzy measures as a basis for MADM with mixed non cardinal data, *European J. of Operational Res.*, Vol. 78, pp. 34–48, 1994.

[20] A. Rebai and J.-M. Martel, A fuzzy bag approach to choosing the "best" multiattributed potential actions in a multiple judgement and non cardinal data context, *Fuzzy Sets and Systems*, Vol. 87, pp. 159–166, 1997.

[21] A. Ramer and C.-C. Wang, Fuzzy multisets, *Proc. of 1996 Asian Fuzzy Systems Symposium*, Dec. 11–14, 1996, Kenting, Taiwan, R.O.C., pp. 429–434, 1996.

[22] J. D. Ullman, *Principles of Database and Knowledge-Base Systems: Vol. 1*, Computer Science Press, Rockville, Maryland, 1988.

[23] R. R. Yager, On the theory of bags, *Int. J. General Systems*, Vol. 13, pp. 23–37, 1986.

[24] R. R. Yager, Cardinality of fuzzy sets via bags, *Mathl. Modelling*, Vol. 9, pp. 441–446, 1987.

3

Modal Logic, Rough Sets, and Fuzzy Sets

Tetsuya Murai
Michinori Nakata
Masaru Shimbo

3.1   Introduction

The theory of *rough sets* proposed by Pawlak [9, 10] is a mathematical theory for dealing with *knowledge* in terms of *classification* and *approximation*. According to Pawlak,

> We simply assume here that knowledge is based on the ability to classify objects, and by object we mean anything we can think of, for example, real things, states, abstract concepts, processes, moments of time, etc.
>
> Thus knowledge, in our approach, is necessarily connected with the variety of classification patterns related to specific parts of the real or abstract world, called here the *universe of discourse* (in short *the universe*). (Pawlak [10], p.2)

A family of subsets $\mathcal{X}$ from the universe U is referred to as (*abstract*) *knowledge about* U when it forms a classification[1] (partition) of U:

$$(1) \quad \forall X \in \mathcal{X} (X \neq \emptyset),$$
$$(2) \quad \cup \mathcal{X} = \cup \{X \mid X \in \mathcal{X}\} = U,$$
$$(3) \quad \forall X, Y \in \mathcal{X} (X \cap Y \neq \emptyset \Rightarrow X = Y).$$

For any classification $\mathcal{X}$ of U, a binary relation R on U is defined by

$$xRy \Leftrightarrow \exists X \in \mathcal{X} (x, y \in X).$$

The relation R obviously satisfies the following three conditions:

Reflexivity:	$\forall x \in U (xRx)$,
Symmetry:	$\forall x, y \in U (xRy \Rightarrow yRx)$,
Transitivity:	$\forall x, y, z \in U ((xRy \text{ and } yRz) \Rightarrow xRz)$.

---

[1] In his book [10], strictly speaking, Pawlak first refers to *any* family of subsets as abstract knowledge; and then he confines himself to the case that knowledge is given as a classification.

Therefore, it is an *equivalence relation* on U. Note that any equivalence relation also satisfies the following two important properties:

Seriality: $\forall x \in U \, \exists y \in U \, (xRy),$

Euclidness: $\forall x, y, z \in U \, ((xRy \text{ and } xRz) \Rightarrow yRz).$

Conversely, given an equivalence relation R on U, it is well-known that the *quotient set* of U with respect to R is defined by

$$U/R = \{[x]_R \mid x \in U\},$$

where $[x]_R$ denotes an *R-equivalence class* containing an element x:

$$[x]_R = \{y \in U \mid xRy\}.$$

The quotient set can be easily shown to be a partition of U, and we can therefore use the two notions of classifications and equivalence relations interchangeably. In what follows, we shall mainly use equivalence relations, which are also referred to as *knowledge*.

There are two kinds of subsets under given knowledge R:

1. One kind contains those subsets that can be precisely explained in terms of R in the sense that they are a union of one or more R-equivalence classes, which play a part in 'building blocks' for subsets. Such a subset is said to be *R-definable*, or *R-exact*, in U.

2. Another kind contains those subsets that cannot be explained in terms of R, so they are said to be *R-rough*, *R-undefinable* or *R-inexact*, in U.

For convenience, the empty set $\emptyset$ is also regarded as being R-definable. By a *trivially* R-definable subset, we mean it is either U or $\emptyset$, because they can be R-definable for any equivalence relation R on U.

In order to capture the concept of R-roughness in an exact way, the notion of approximation is introduced. For any subset X of U, *lower* and *upper* approximations of X are defined using R-equivalence classes as 'building blocks.' The *lower R-approximation* of X in U, denoted $\underline{R}(X)$, is the union of R-equivalence classes that are contained in X:

$$\underline{R}(X) \overset{\text{def}}{=} \cup \{Y \in U/R \mid Y \subseteq X\} = \{x \in U \mid [x]_R \subseteq X\}.$$

On the other hand, the *upper R-approximation* of X in U, denoted $\overline{R}(X)$, is the union of R-equivalence classes that have non-empty intersection with X:

$$\overline{R}(X) \overset{\text{def}}{=} \cup \{Y \in U/R \mid Y \cap X \neq \emptyset\} = \{x \in U \mid [x]_R \cap X \neq \emptyset\}.$$

Then, we can easily show the following characterization of R-definability and roughness in terms of approximation: for any subset X in U,

$$X \text{ is } R\text{-definable} \Leftrightarrow \underline{R}(X) = X = \overline{R}(X),$$
$$X \text{ is } R\text{-rough} \Leftrightarrow \underline{R}(X) \subset X \subset \overline{R}(X).$$

Note that $\underline{R}(X) = X$ if and only if $X = \overline{R}(X)$. By approximation, we can distinguish three regions with respect to each subset X in U:

R-positive region of X: $POS_R(X) \overset{\text{def}}{=} \underline{R}(X)$,

R-negative region of X: $NEG_R(X) \overset{\text{def}}{=} U \setminus \overline{R}(X)$,

R-borderline region of X: $BN_R(X) \overset{\text{def}}{=} \overline{R}(X) \setminus \underline{R}(X)$.

Now we can understand that a fundamental unit of Pawlak's rough set theory is the pair of the universe and an equivalence relation on it:

$$<U, R>,$$

which is called a *Pawlak approximation space*. The pair is an example of the notion of *Kripke frames* in modal logic. In general, any pair of a set U and an arbitrary relation R on U can be a Kripke frame. Here, we can find our starting point to examine some relationships between rough set theory and modal logic, where it should be noted that we confine ourselves to modal propositional logic.

It should be stressed that this chapter will focus on *only one aspect* of rough set theory. The readers can find many other fruitful aspects in rough set theory that are not dealt with in modal logic (see Pawlak's book [10], for details).

3.2 Language for Modal Logic

Modal logic [2, 4, 5] is primarily the logic of *necessity* and *possibility*,[2] an extension of classical logic with two unary operators, $\Box$ and $\Diamond$, by which we can express a new kind of sentences like

$\Box p$: It is *necessary* that p,

$\Diamond p$: It is *possible* that p.

More precisely, given a countable[3] set of *atomic sentences*

$$\mathcal{P} = \{p_1, p_2, \cdots, p_n(, \cdots)\},$$

a *language* $\mathcal{L}_{ML}(\mathcal{P})$ for modal propositional logic is formed[4] from $\mathcal{P}$ as a set of *sentences* closed under the usual propositional operators such as $\top$ (the truth constant), $\bot$ (the falsity constant), $\neg$ (negation), $\wedge$ (conjunction), $\vee$ (disjunction), $\rightarrow$ (conditional), and $\leftrightarrow$ (biconditional) as well as the two *modal* operators $\Box$ (necessity) and $\Diamond$ (possibility):

---

[2]In the literature, there are other kinds of modalities with respect to *obligatoriness, knowledge, belief, tense, provability, programs*, and so on.

[3]Finite or countably infinite.

[4]Parentheses '(' and ')' are also needed. The most-outside parentheses are usually omitted.

atomic sentences: $p_i \in \mathcal{P} \Rightarrow p_i \in \mathcal{L}_{ML}(\mathcal{P})$,
zero-place operators: $\top, \bot \in \mathcal{L}_{ML}(\mathcal{P})$,
unary operators: $p \in \mathcal{L}_{ML}(\mathcal{P}) \Rightarrow \neg p, \Box p, \Diamond p \in \mathcal{L}_{ML}(\mathcal{P})$,
binary operators: $p, p' \in \mathcal{L}_{ML}(\mathcal{P}) \Rightarrow (p \wedge p'), (p \vee p'), (p \rightarrow p'), (p \leftrightarrow p') \in \mathcal{L}_{ML}(\mathcal{P})$.

A *schema* is defined as a set of sentences that have a common particular syntactic form. For example, the following set of sentences

$$\{\Box p \rightarrow p \mid p \in \mathcal{L}_{ML}(\mathcal{P})\}$$

is referred to as the schema **T** and is often represented simply by '$\Box p \rightarrow p$' unless confusion arises. Elements in **T** are called its *instances*. So the following three sentences

$$\Box p_1 \rightarrow p_1, \quad \Box(p_2 \wedge p_1) \rightarrow (p_2 \wedge p_1), \quad \text{and} \quad \Box(p_1 \rightarrow \Diamond p_3) \rightarrow (p_1 \rightarrow \Diamond p_3)$$

are instances of **T** since p_1, $(p_2 \wedge p_1)$, $(p_1 \rightarrow \Diamond p_3)$ are elements in $\mathcal{L}_{ML}(\mathcal{P})$, respectively.

For a sentence p, p^* is defined in a recursive way as

$$
\begin{aligned}
p_i^* &= \neg p_i \ (i = 1, 2, 3, \cdots) & (p \vee q)^* &= (p^*) \wedge (q^*) \\
\top^* &= \bot & (p \rightarrow q)^* &= \neg(p^*) \wedge (q^*) \\
\bot^* &= \top & (p \leftrightarrow q)^* &= (p^*) \leftrightarrow \neg(q^*) \\
(\neg p)^* &= \neg(p^*) & (\Box p)^* &= \Diamond(p^*) \\
(p \wedge q)^* &= (p^*) \vee (q^*) & (\Diamond p)^* &= \Box(p^*)
\end{aligned}
$$

Then the *dual* of a schema **S** is defined by

$$\mathbf{S}^* \overset{\text{def}}{=} \{\neg(p^*) \mid p \in \mathbf{S}\}.$$

Several schemas and their duals that are well-known in the literature are shown in Table 3.1.

TABLE 3.1. Schemas in Modal Logic

Df $\Diamond = \Diamond p \leftrightarrow \neg \Box \neg p$	**(Df** $\Diamond)^* = $**Df** $\Box = \Box p \leftrightarrow \neg \Diamond \neg p$
M $= \Box(p \wedge q) \rightarrow (\Box p \wedge \Box q)$	**M**$^* = $**M** $\Diamond = (\Diamond p \vee \Diamond q) \rightarrow \Diamond(p \vee q)$
C $= (\Box p \wedge \Box q) \rightarrow \Box(p \wedge q)$	**C**$^* = $**C** $\Diamond = \Diamond(p \vee q) \rightarrow (\Diamond p \vee \Diamond q)$
N $= \Box \top$	**N**$^* = $**N** $\Diamond = \neg \Diamond \bot$
P $\Box = \neg \Box \bot$	**(P** $\Box)^* = $**P** $= \Diamond \top$
K $= \Box(p \rightarrow q) \rightarrow (\Box p \rightarrow \Box q)$	**K**$^* = $**K** $\Diamond = (\neg \Diamond p \wedge \Diamond q) \rightarrow \Diamond(\neg p \wedge q)$
D $= \Box p \rightarrow \Diamond p$	**D**$^* = $**D** (autodual)
T $= \Box p \rightarrow p$	**T**$^* = $**T** $\Diamond = p \rightarrow \Diamond p$
B $= p \rightarrow \Box \Diamond p$	**B**$^* = $**B** $\Diamond = \Diamond \Box p \rightarrow p$
4 $= \Box p \rightarrow \Box \Box p$	**4**$^* = $**4** $\Diamond = \Diamond \Diamond p \rightarrow \Diamond p$
5 $= \Diamond p \rightarrow \Box \Diamond p$	**5**$^* = $**5** $\Diamond = \Diamond \Box p \rightarrow \Box p$

3.3   Kripke Semantics for Modal Logic

From the point of view of semantics, the modal operators *cannot* be truth-functional, because, if they are, then they must, strangely enough, be equal to one of the sentences shown in Table 3.2. So what we need is a *non-*

TABLE 3.2. Truth-functional zero-place and unary operators

p	$\neg p$	$\top$	$\bot$
T	F	T	F
F	T	T	F

truth-functional way of interpreting modal operators and, in fact, the so-called *possible-worlds*, or *Kripke, semantics* was developed in the late 1950's independently by S. Kanger, H. J. J. Hinttika, and S. A. Kripke, and it is now recognized as the most powerful apparatus for describing modal logic.

Definition 1 *A Kripke frame $\mathcal{F}$ is a tuple*

$$<U, R>,$$

where U is a non-empty set and R is a binary relation on U.

Elements in U are called *possible worlds, states of affairs,* or *points*. R is often called an *accessibility relation*, which describes the way each world is accessible from other worlds. So, for each x in U, we can define the set of possible worlds that are 'accessible' from x:

$$U_R(x) \stackrel{\text{def}}{=} \{y \in U \mid xRy\} \ .$$

Note that, when R is an equivalence relation on U, $U_R(x)$ is the R-equivalence class containing x:

$$U_R(x) = [x]_R.$$

A Kripke model is defined by a Kripke frame plus a valuation:

Definition 2 *Given a Kripke frame $\mathcal{F} = <U, R>$, a Kripke model $\mathcal{M}$ on $\mathcal{F}$ is a triple*

$$<U, R, V>,$$

where V is a valuation that assigns exactly one of the truth-values, T (true) and F (false), to each atomic sentence at each possible world, i.e.,

$$V : \mathcal{P} \times U \to \{T, F\} \ .$$

We can extend a valuation as *truth conditions* for any sentence in the following form: given a Kripke model $\mathcal{M} = <U, R, V>$, let

$$\mathcal{M}, x \models p$$

mean that a sentence p is true at a possible world x in $\mathcal{M}$. Then,

$$
\begin{aligned}
\mathcal{M}, x \models \mathsf{p}_i &\overset{\text{def}}{\Longleftrightarrow} V(\mathsf{p}_i, x) = T, \\
\mathcal{M}, x \models \top, & \\
\text{Not } (\mathcal{M}, x \models \bot), & \\
\mathcal{M}, x \models \neg p &\overset{\text{def}}{\Longleftrightarrow} \text{not } (\mathcal{M}, x \models p), \\
\mathcal{M}, x \models p \vee p' &\overset{\text{def}}{\Longleftrightarrow} \mathcal{M}, x \models p \text{ or } \mathcal{M}, x \models p', \\
\mathcal{M}, x \models p \wedge p' &\overset{\text{def}}{\Longleftrightarrow} \mathcal{M}, x \models p \text{ and } \mathcal{M}, x \models p', \\
\mathcal{M}, x \models p \rightarrow p' &\overset{\text{def}}{\Longleftrightarrow} \mathcal{M}, x \models p \Rightarrow \mathcal{M}, x \models p', \\
\mathcal{M}, x \models p \leftrightarrow p' &\overset{\text{def}}{\Longleftrightarrow} \mathcal{M}, x \models p \Leftrightarrow \mathcal{M}, x \models p'.
\end{aligned}
$$

Truth conditions of modal operators date from G. W. F. Leibniz, who stated that $\Box p$ holds when p is true at *all* possible worlds. We write them as the following definition:

Definition 3 *Let $\mathcal{M} = <U, R, V>$ be a Kripke model and let x be an arbitrary element in U, then*

$$
\begin{aligned}
\mathcal{M}, x \models \Box p &\overset{\text{def}}{\Longleftrightarrow} \forall y \in U \, (xRy \Rightarrow \mathcal{M}, y \models p), \\
\mathcal{M}, x \models \Diamond p &\overset{\text{def}}{\Longleftrightarrow} \exists y \in U \, (xRy \text{ and } \mathcal{M}, y \models p).
\end{aligned}
$$

3.4   Truth Sets and Generalized Lower and Upper Approximations

Given a Kripke model $\mathcal{M} = <U, R, V>$, for any sentence p in $\mathcal{L}_{\text{ML}}(\mathcal{P})$, its *truth set* $\|p\|^{\mathcal{M}}$ in $\mathcal{M}$ is defined as the set of possible worlds at which p is true in the model $\mathcal{M}$:

$$\|p\|^{\mathcal{M}} \overset{\text{def}}{=} \{x \in U \mid \mathcal{M}, x \models p\}.$$

For the truth sets we have the following properties:

$$
\begin{aligned}
\|\top\|^{\mathcal{M}} &= U, \\
\|\bot\|^{\mathcal{M}} &= \emptyset, \\
\|\neg p\|^{\mathcal{M}} &= (\|p\|^{\mathcal{M}})^{\mathrm{C}}, \\
\|p \wedge q\|^{\mathcal{M}} &= \|p\|^{\mathcal{M}} \cap \|q\|^{\mathcal{M}}, \\
\|p \vee q\|^{\mathcal{M}} &= \|p\|^{\mathcal{M}} \cup \|q\|^{\mathcal{M}}, \\
\|p \rightarrow q\|^{\mathcal{M}} &= (\|p\|^{\mathcal{M}})^{\mathrm{C}} \cup \|q\|^{\mathcal{M}}, \\
\|p \leftrightarrow q\|^{\mathcal{M}} &= ((\|p\|^{\mathcal{M}})^{\mathrm{C}} \cup \|q\|^{\mathcal{M}}) \cap ((\|q\|^{\mathcal{M}})^{\mathrm{C}} \cup \|p\|^{\mathcal{M}}).
\end{aligned}
$$

Note that, using both $U_R(x)$ and $\|p\|^{\mathcal{M}}$, the truth conditions for modal operators in Definition 3 can be rewritten in a simpler way:

Proposition 4 *Let* $\mathcal{M} = <U, R, V>$ *be a Kripke model, then*

$$\mathcal{M}, x \models \Box p \Leftrightarrow U_R(x) \subseteq \|p\|^{\mathcal{M}},$$
$$\mathcal{M}, x \models \Diamond p \Leftrightarrow U_R(x) \cap \|p\|^{\mathcal{M}} \neq \emptyset.$$

By Proposition 4, we have the following proposition:

Proposition 5 *Let* $\mathcal{M} = <U, R, V>$ *be a Kripke model, then*

$$\|\Box p\|^{\mathcal{M}} = \{x \in U \mid U_R(x) \subseteq \|p\|^{\mathcal{M}}\},$$
$$\|\Diamond p\|^{\mathcal{M}} = \{x \in U \mid U_R(x) \cap \|p\|^{\mathcal{M}} \neq \emptyset\}.$$

If we may apply this formulation into an arbitrary subset X in U, then we obtain the notion of *generalized lower and upper approximations* developed by Lin, Wang, Yao [7, 8, 14, 15] and others:

Definition 6 *Given a Kripke frame* $\mathcal{F} = <U, R>$, *for each subset* X *in* U,

$$\underline{R}(X) \overset{\text{def}}{=} \{x \in U \mid U_R(x) \subseteq X\},$$
$$\overline{R}(X) \overset{\text{def}}{=} \{x \in U \mid U_R(x) \cap X \neq \emptyset\}.$$

Properties of an accessibility relation characterize the generalized approximations as shown in Table 3.3. For example, we can easily show that

$$R \text{ is reflexive} \Leftrightarrow \underline{R}(X) \subseteq X \Leftrightarrow X \subseteq \overline{R}(X).$$

Now, Proposition 5 and Definition 6 lead us to the following remarkable

TABLE 3.3. Correspondence between accessibility relations and generalized lower and upper approximations.

(no condition)	$\overline{R}(X) = \underline{R}(X^C)^C$	$\underline{R}(X) = \overline{R}(X^C)^C$
(no condition)	$\underline{R}(X \cap Y) \subseteq \underline{R}(X) \cap \underline{R}(Y)$	$\overline{R}(X) \cup \overline{R}(Y) \subseteq \overline{R}(X \cup Y)$
(no condition)	$\underline{R}(X) \cap \underline{R}(Y) \subseteq \underline{R}(X \cap Y)$	$\overline{R}(X \cup Y) \subseteq \overline{R}(X) \cup \overline{R}(Y)$
(no condition)	$\underline{R}(U) = U$	$\overline{R}(\emptyset) = \emptyset$
(no condition)	$\underline{R}(X^C \cup Y) \subseteq (\underline{R}(X)^C \cup \underline{R}(Y))$	$\overline{R}(X)^C \cap \overline{R}(Y) \subseteq \overline{R}(X^C \cap Y)$
Seriality		$\underline{R}(X) \subseteq \overline{R}(X)$
Seriality	$\underline{R}(\emptyset) = \emptyset$	$\overline{R}(U) = U$
Reflexivity	$\underline{R}(X) \subseteq X$	$X \subseteq \overline{R}(X)$
Symmetry	$X \subseteq \underline{R}(\overline{R}(X))$	$\overline{R}(\underline{R}(X)) \subseteq X$
Transitivity	$\underline{R}(X) \subseteq \underline{R}(\underline{R}(X))$	$\overline{R}(\overline{R}(X)) \subseteq \overline{R}(X)$
Euclidness	$\overline{R}(X) \subseteq \underline{R}(\overline{R}(X))$	$\overline{R}(\underline{R}(X)) \subseteq \underline{R}(X)$

properties:

Proposition 7 *Let $\mathcal{M}=<U,R,V>$ be a Kripke model, then*

$$\|\Box p\|^{\mathcal{M}} = \underline{R}(\|p\|^{\mathcal{M}}),$$
$$\|\Diamond p\|^{\mathcal{M}} = \overline{R}(\|p\|^{\mathcal{M}}).$$

Thus, we can understand that the necessity and possibility operators can be interpreted by generalized lower and upper approximations, respectively. Further we have the following correspondence:

Proposition 8 *Given a Kripke model $\mathcal{M}=<U,R,V>$,*

$$
\begin{aligned}
Necessity: \quad & \|\Box p\|^{\mathcal{M}} = POS_R(\|p\|^{\mathcal{M}}), \\
Impossibility: \quad & \|\neg\Diamond p\|^{\mathcal{M}} = NEG_R(\|p\|^{\mathcal{M}}), \\
Contingency: \quad & \|\Diamond p \wedge \neg\Box p\|^{\mathcal{M}} = BN_R(\|p\|^{\mathcal{M}}).
\end{aligned}
$$

3.5   Validity

Validity of a sentence can be defined at the following three levels

Definition 9   *1. A sentence p is* valid in a Kripke model $\mathcal{M} = <U,R,V>$ *(written $\mathcal{M} \models p$) if and only if p is true at any world in $\mathcal{M}$:*

$$\mathcal{M} \models p \overset{def}{\Longleftrightarrow} \forall x \, (x \in W \Rightarrow \mathcal{M}, x \models p).$$

2. A sentence p is valid in a Kripke frame $\mathcal{F}$ *(written $\mathcal{F} \models p$) if and only if p is valid in any Kripke model on $\mathcal{F}$:*

$$\mathcal{F} \models p \overset{def}{\Longleftrightarrow} \forall \mathcal{M} \, ((\mathcal{M} \text{ on } \mathcal{F}) \Rightarrow \mathcal{M} \models p).$$

3. A sentence p is valid in a class $\mathbf{C}_F$ of Kripke frames *(written $\mathbf{C}_F \models p$) if and only if p is valid in any Kripke frame in $\mathbf{C}_F$:*

$$\mathbf{C}_F \models p \overset{def}{\Longleftrightarrow} \forall \mathcal{F} \, ((\mathcal{F} \text{ in } \mathbf{C}_F) \Rightarrow \mathcal{F} \models p).$$

The definitions of validity can be extended for a set of sentences so that every sentence in the set of sentences is valid: Given a set of sentences Γ

$$
\begin{aligned}
\mathcal{M} \models \Gamma \quad & \overset{def}{\Longleftrightarrow} \quad \forall p \, (p \in \Gamma \Rightarrow (\forall w \, (w \in W \Rightarrow \mathcal{M}, w \models p))), \\
\mathcal{F} \models \Gamma \quad & \overset{def}{\Longleftrightarrow} \quad \forall \mathcal{M} \, ((\mathcal{M} \text{ on } \mathcal{F}) \Rightarrow \mathcal{M} \models \Gamma), \\
\mathbf{C}_F \models \Gamma \quad & \overset{def}{\Longleftrightarrow} \quad \forall \mathcal{F} \, ((\mathcal{F} \text{ in } \mathbf{C}_F) \Rightarrow \mathcal{F} \models \Gamma).
\end{aligned}
$$

These definitions serves to describe validity of schemas.

Note that we can easily have

$$\mathcal{M} \models p \Leftrightarrow \|p\|^{\mathcal{M}} = U,$$

thus, in order to demonstrate the validity of p in a class $\mathbf{C}_F$ of Kripke frames, it is sufficient to prove the right side of the above formula for any $\mathcal{M}$ on $\mathcal{F}$ in $\mathbf{C}_F$. So, for example, is p has the form $\neg p'$, $p' \rightarrow p''$, or $p' \leftrightarrow p''$, then what we should show is $\|p\|^{\mathcal{M}} = \emptyset$, $\|p'\|^{\mathcal{M}} \subseteq \|p''\|^{\mathcal{M}}$, or $\|p'\|^{\mathcal{M}} = \|p''\|^{\mathcal{M}}$, because

$$\mathcal{M} \models \neg p' \Leftrightarrow \|p'\|^{\mathcal{M}} = \emptyset,$$
$$\mathcal{M} \models p' \rightarrow p'' \Leftrightarrow \|p\|^{\mathcal{M}} \subseteq \|p'\|^{\mathcal{M}}, \text{ or}$$
$$\mathcal{M} \models p' \leftrightarrow p'' \Leftrightarrow \|p'\|^{\mathcal{M}} = \|p''\|^{\mathcal{M}},$$

respectively. Thus, we have correspondence between the properties of accessibility relations and schemas shown in Tables 3.4 and 3.5 using formulas in Table 3.3.

TABLE 3.4. Correspondence among schemas, properties of accessibility relations, and generalized lower and upper approximations (1).

Df $\diamond = \diamond p \leftrightarrow \neg \Box \neg p$	(no condition)	$\overline{R}(X) = \underline{R}(X^C)^C$
M $= \Box(p \wedge q) \rightarrow (\Box p \wedge \Box q)$	(no condition)	$\underline{R}(X \cap Y) \subseteq \underline{R}(X) \cap \underline{R}(Y)$
C $= (\Box p \wedge \Box q) \rightarrow \Box(p \wedge q)$	(no condition)	$\underline{R}(X) \cap \underline{R}(Y) \subseteq \underline{R}(X \cap Y)$
N $= \Box\top$	(no condition)	$\underline{R}(U) = U$
K $= \Box(p \rightarrow q) \rightarrow (\Box p \rightarrow \Box q)$	(no condition)	$\underline{R}(X^C \cup Y) \subseteq (\underline{R}(X)^C \cup \underline{R}(Y))$
D $= \Box p \rightarrow \diamond p$	Seriality	$\underline{R}(X) \subseteq \overline{R}(X)$
P$\Box = \neg \Box \bot$	Seriality	$\underline{R}(\emptyset) = \emptyset$
T $= \Box p \rightarrow p$	Reflexivity	$\underline{R}(X) \subseteq X$
B $= p \rightarrow \Box \diamond p$	Symmetry	$X \subseteq \underline{R}(\overline{R}(X))$
4 $= \Box p \rightarrow \Box \Box p$	Transitivity	$\underline{R}(X) \subseteq \underline{R}(\underline{R}(X))$
5 $= \diamond p \rightarrow \Box \diamond p$	Euclidness	$\overline{R}(X) \subseteq \underline{R}(\overline{R}(X))$

3.6 What is a System of Modal Logic ?

Although modal logic is an extension of classical propositional logic, we must make clear what is meant by *propositional logic*, which is the basis for formulating systems of modal logic. According to, for example, Hughes and Cresswell ([5], p.111), there are two ways of defining a system of logic: (1) one by its axiomatic basis; and (2) the other in terms of its theorems. In this paper, we adopt the latter way, so (the system of) propositional logic can be identified with the set of 'all tautologies.'

Here, we assume that we can freely use all of the results of classical propositional logic; thus, the most convenient way of defining the set of tautologies is by using valuation functions in classical propositional logic. Note, however, that now the language is extended to $\mathcal{L}_{\text{ML}}(\mathcal{P})$. For example,

TABLE 3.5. Correspondence among schemas, properties of accessibility relations, and generalized lower and upper approximations (2).

Df$\Box = \Box p \leftrightarrow \neg \Diamond \neg p$	(no condition)	$\underline{R}(X) = \overline{R}(X^C)^C$
M$\Diamond = (\Diamond p \vee \Diamond q) \rightarrow \Diamond(p \vee q)$	(no condition)	$\overline{R}(X) \cup \overline{R}(Y) \subseteq \overline{R}(X \cup Y)$
C$\Diamond = \Diamond(p \vee q) \rightarrow (\Diamond p \vee \Diamond q)$	(no condition)	$\overline{R}(X \cup Y) \subseteq \overline{R}(X) \cup \overline{R}(Y)$
N$\Diamond = \neg \Diamond \bot$	(no condition)	$\overline{R}(\emptyset) = \emptyset$
K$\Diamond = (\neg \Diamond p \wedge \Diamond q) \rightarrow \Diamond(\neg p \wedge q)$	(no condition)	$\overline{R}(X)^C \cap \overline{R}(Y) \subseteq \overline{R}(X^C \cap Y)$
D $= \Box p \rightarrow \Diamond p$	Seriality	$\underline{R}(X) \subseteq \overline{R}(X)$
P $= \Diamond \top$	Seriality	$\overline{R}(U) = U$
T$\Diamond = p \rightarrow \Diamond p$	Reflexivity	$X \subseteq \overline{R}(X)$
B$\Diamond = \Diamond \Box p \rightarrow p$	Symmetry	$\overline{R}(\underline{R}(X)) \subseteq X$
4$\Diamond = \Diamond \Diamond p \rightarrow \Diamond p$	Transitivity	$\overline{R}(\overline{R}(X)) \subseteq \overline{R}(X)$
5$\Diamond = \Diamond \Box p \rightarrow \Box p$	Euclidness	$\overline{R}(\underline{R}(X)) \subseteq \underline{R}(X)$

a sentence $(p_1 \wedge \neg p_1)$ is of course counted as a tautology, but how about $(\Box p_1 \wedge \neg \Box p_1)$? It can also be counted as a tautology because it has the same form as the former one. So, the notion of tautologies should also be extended for the language $\mathcal{L}_{ML}(\mathcal{P})$.

Let us define the set of so-called *quasi-atomic* sentences by

$$\tilde{\mathcal{P}} \overset{\text{def}}{=} \mathcal{P} \cup \{\Box p \mid p \in \mathcal{L}_{ML}(\mathcal{P})\} \cup \{\Diamond p \mid p \in \mathcal{L}_{ML}(\mathcal{P})\} .$$

Then, a sentence $p \in \mathcal{L}_{ML}(\mathcal{P})$ is called an $\mathcal{L}_{ML}(\mathcal{P})$-*tautology* when $\tilde{v}(p) = T$ for any valuation

$$\tilde{v} : \tilde{\mathcal{P}} \rightarrow \{T, F\} ,$$

where T(true) and F(false) are classical truth values.[5] The set of all $\mathcal{L}_{ML}(\mathcal{P})$-tautologies is denoted by PL :

$$PL \overset{\text{def}}{=} \{p \in \mathcal{L}_{ML}(\mathcal{P}) \mid p \text{ is an } \mathcal{L}_{ML}(\mathcal{P})\text{-tautology}\} .$$

Definition 10 *A set of sentences Σ is said to be a* system of modal logic *just in case it contains all $\mathcal{L}_{ML}(\mathcal{P})$-tautologies:*

$$PL \subseteq \Sigma$$

and it is closed under the rule of modus ponens*:*

$$\mathbf{MP}.\ p, p \rightarrow q \in \Sigma \Rightarrow q \in \Sigma.$$

---

[5]Note that $\Box p$ and $\Diamond p$ can also take only one of either T or F.

$\mathcal{L}_{\text{ML}}(\mathcal{P})$ and PL are easily shown to be the largest and smallest systems of modal logic, respectively, in the sense of set inclusion:

Proposition 11 $PL \subseteq \Sigma \subseteq \mathcal{L}_{\text{ML}}(\mathcal{P})$.

Further, we have the following two propositions:

Proposition 12 *If, for any $i \in I$, Σ_i is a system of modal logic, then so is $\cap\, \{\Sigma_i \mid i \in I\}$.*

Proposition 13 *For a set Γ of sentences in $\mathcal{L}_{\text{ML}}(\mathcal{P})$,*

$$\cap\, \{\Sigma \mid \Gamma \subseteq \Sigma\}$$

is the smallest *system of modal logic that includes Γ.*

Definition 14 *A sentence p is said to be a* theorem *of a system Σ of modal logic (denoted $\vdash_\Sigma p$) just in case p is an element of Σ:*

$$\vdash_\Sigma p \overset{\text{def}}{\Longleftrightarrow} p \in \Sigma.$$

The notation is also applied to a schema **S** when any instance in **S** is a theorem of Σ:

$$\vdash_\Sigma \mathbf{S} \overset{\text{def}}{\Longleftrightarrow} \mathbf{S} \subseteq \Sigma.$$

3.7   Normal Systems of Modal Logic

Definition 15 *A system Σ of modal logic is said to be* normal *just in case it contains the schemas* **Df**$\diamond$ *and* **K** *:*

$$\mathbf{Df}\diamond \cup \mathbf{K} \subseteq \Sigma,$$

and it is closed under the rule of necessitation *:*

$$\mathbf{RN}.\ p \in \Sigma \Rightarrow \Box p \in \Sigma.$$

Thus, obviously, we have

$$\begin{aligned}
&\vdash_\Sigma \mathbf{Df}\diamond && (= \diamond p \leftrightarrow \neg\Box\neg p), \\
&\vdash_\Sigma \mathbf{K} && (= \Box(p \rightarrow q) \rightarrow (\Box p \rightarrow \Box q))
\end{aligned}$$

for any normal system Σ, and further we have

$$\begin{aligned}
\mathbf{M} \subseteq \Sigma, && \text{i.e.,} && \vdash_\Sigma \mathbf{M} && (= (\Box p \wedge \Box q) \rightarrow \Box(p \wedge q)), \\
\mathbf{C} \subseteq \Sigma, && \text{i.e.,} && \vdash_\Sigma \mathbf{C} && (= \Box(p \wedge q) \rightarrow (\Box p \wedge \Box q)), \\
\mathbf{N} \subseteq \Sigma, && \text{i.e.,} && \vdash_\Sigma \mathbf{N} && (= \Box\top).
\end{aligned}$$

Let us define the following set of sentences:

Definition 16 $K \stackrel{\text{def}}{=} \cap \{\Sigma \mid \Sigma : a\ normal\ system\}$.

Then, by Proposition 12, K is the smallest normal system. It has the following property :

Proposition 17 (Goldblatt [4])

$$\vdash_K p \quad \Leftrightarrow \quad \exists p_1, \cdots, p_m(= p) \in \mathcal{L}_{\text{ML}}(\mathcal{P})$$
$$such\ that\ \forall i\ (\leq m)$$
$$\begin{array}{ll}
(1) & p_i \in PL \\
or\ (2) & p_i \in \mathbf{Df}\Diamond \cup \mathbf{K} \\
or\ (3) & \exists j, k\ (< i)\ (p_k = (p_j \to p_i)) \\
or\ (4) & \exists j\ (< i)\ (p_i = \Box p_j)\ ).
\end{array}$$

So, from the point of view of an axiomatic basis, the system K is obtained by adding the axiom schemas **K** and **Df**$\Diamond$ and the rule **RN** to classical propositional logic. Further, we have

$$\begin{array}{llll}
\mathbf{M} \subseteq K, & \text{i.e.,} & \vdash_K \mathbf{M} & (= (\Box p \wedge \Box q) \to \Box(p \wedge q)), \\
\mathbf{C} \subseteq K, & \text{i.e.,} & \vdash_K \mathbf{C} & (= \Box(p \wedge q) \to (\Box p \wedge \Box q)), \\
\mathbf{N} \subseteq K, & \text{i.e.,} & \vdash_K \mathbf{N} & (= \Box\top), \\
\mathbf{D} \not\subseteq K, & \text{i.e.,} & \not\vdash_K \mathbf{D} & (= \Box p \to \Diamond p), \\
\mathbf{P} \not\subseteq K, & \text{i.e.,} & \not\vdash_K \mathbf{P} & (= \Diamond\top), \\
\mathbf{T} \not\subseteq K, & \text{i.e.,} & \not\vdash_K \mathbf{T} & (= \Box p \to p), \\
\mathbf{B} \not\subseteq K, & \text{i.e.,} & \not\vdash_K \mathbf{B} & (= p \to \Box\Diamond p), \\
\mathbf{4} \not\subseteq K, & \text{i.e.,} & \not\vdash_K \mathbf{4} & (= \Box p \to \Box\Box p), \\
\mathbf{5} \not\subseteq K, & \text{i.e.,} & \not\vdash_K \mathbf{5} & (= \Diamond p \to \Box\Diamond p).
\end{array}$$

Let us define the following set of sentences using schemas $\mathbf{S}_1, \cdots, \mathbf{S}_n$:

Definition 18 $KS_1 \cdots S_n \stackrel{\text{def}}{=} \cap \{\Sigma \mid (\Sigma : normal)\ \text{and}\ \mathbf{S}_1 \cup \cdots \cup \mathbf{S}_n \subseteq \Sigma\}$.

Also, by Proposition 13, $KS_1 \cdots S_n$ is the smallest normal system that contains schemas $\mathbf{S}_1, \cdots, \mathbf{S}_n$. This notation is called the *Lemmon code*, which is said to have been created by E. J. Lemmon. For $KS_1 \cdots S_n$, the following proposition holds:

Proposition 19 (Goldblatt [4])

$$\vdash_{KS_1\ldots S_n} p \quad \Leftrightarrow \quad \exists p_1, \cdots, p_m(= p) \in \mathcal{L}_{\text{ML}}(\mathcal{P})$$
$$such\ that\ \forall i\ (\leq m)$$
$$\begin{array}{ll}
(1) & p_i \in PL \\
or\ (2) & p_i \in \mathbf{Df}\Diamond \cup \mathbf{K} \\
or\ (3) & p_i \in \mathbf{S}_1 \cup \cdots \cup \mathbf{S}_n \\
or\ (4) & \exists j, k\ (< i)\ (p_k = (p_j \to p_i)) \\
or\ (5) & \exists j\ (< i)\ (p_i = \Box p_j)\ ).
\end{array}$$

According to this proposition, $KS_1 \ldots S_n$ is obtained by adding $\mathbf{S}_1, \cdots, \mathbf{S}_n$ as axiom schemas to K. By adding all combinations of schemas **D**, **T**, **B**, **4**, and **5** to K, we have the fifteen distinct normal systems[6] shown in Table 3.6.

TABLE 3.6. Normal systems and axiom schemas, where • and ∘ mean that a schema is an axiom schema and a theorem of a system, respectively.

	Traditional name	D	T	B	4	5	Lemmon code
1	K						K
2	D	•					KD
3	T	∘	•				KT
4				•			KB
5					•		$K4$
6						•	$K5$
7		•		•			KDB
8		•			•		$KD4$
9		•				•	$KD5$
10					•	•	$K45$
11		•			•	•	$KD45$
12				•	•	∘	$KB4$
				•	∘	•	$KB5\ (=KB4)$
13	B	∘	•	•			KTB
14	S4	∘	•		•		$KT4$
15	S5	∘	•	∘	∘	•	$KT5$
		∘	•	•	•	∘	$KTB4\ (=KT5)$
		•	∘	•	•	∘	$KDB4\ (=KT5)$
		•	∘	•	∘	•	$KDB5\ (=KT5)$

3.8   Soundness

Definition 20 *A system Σ of modal logic is said to be* sound *with respect to a class C_F of Kripke frames just in case, for any sentence p,*

$$\vdash_\Sigma p \Rightarrow C_F \models p,$$

that is,

$$\Sigma \subseteq \{p \in \mathcal{L}_{\mathrm{ML}}(\mathcal{P}) \mid C_F \models p\}\ .$$

---

[6] Note that there may be two systems with different axiom schemas that are just the same as a set of sentences. For example, $KDT=KT$, because **D** can be proved in KT. So the Lemmon code does not give a unique name for each system. Actually, the total number of distinct systems is less than the number of all combinations, that is, $2^5 = 32$.

For any class $\mathbf{C}_F$ of Kripke frames, the following can be shown:

$$p \in PL \Rightarrow \mathbf{C}_F \models p,$$
$$\mathbf{C}_F \models p \text{ and } \mathbf{C}_F \models p{\rightarrow}q \Rightarrow \mathbf{C}_F \models q,$$
$$\mathbf{C}_F \models p \Rightarrow \mathbf{C}_F \models \Box p,$$
$$\mathbf{C}_F \models \mathbf{Df}\Diamond (= \Diamond p{\leftrightarrow}\neg\Box\neg p),$$
$$\mathbf{C}_F \models \mathbf{K} (= \Box(p{\rightarrow}q){\rightarrow}(\Box p{\rightarrow}\Box q)).$$

So, for any class $\mathbf{C}_F$ of Kripke frames, the following set of sentences

$$\Sigma_{\mathbf{C}_F} = \{p \in \mathcal{L}_{\mathrm{ML}}(\mathcal{P}) \mid \mathbf{C}_F \models p\}$$

is a normal system, that is,

$$K \subseteq \Sigma_{\mathbf{C}_F},$$

which means the system K is sound with respect to any class $\mathbf{C}_F$ of Kripke frames.

For other schemas, their corresponding conditions should be imposed on accessibility relations. For example, let us say a frame $\mathcal{F}{=}{<}W, R{>}$ be *reflexive* if so is R. Then we can show the following correspondence: for any class $\mathbf{C}_{Fref}$ of reflexive frames,

$$\mathbf{C}_{Fref} \models \mathbf{T} (= \Box p{\rightarrow}p).$$

In general, if $\mathbf{S}_1, \cdots, \mathbf{S}_n$ are schemas that are valid respectively in classes $\mathbf{C}_{F1}, \cdots, \mathbf{C}_{Fn}$ of Kripke frames, then

$$\vdash_{KS_1 \cdots S_n} p \Rightarrow \mathbf{C}_{F1} \cap \cdots \cap \mathbf{C}_{Fn} \models p.$$

So, for example, the system $\mathrm{T}(=KT)$ is sound with respect to a class of reflexive frames, and the system $\mathrm{S5}(=KT5)$ is sound with respect to a class of reflexive euclidean frames. Since a reflexive euclidean relation is easily shown to be reflexive, symmetric, and transitive, i.e., an equivalence relation, so $\mathrm{S5}(=KT5)$ is sound with respect to any class of Kripke frames with an equivalence relation, that is, Pawlak approximation spaces. Soundness results for the fifteen systems are shown in Table 3.7.

3.9 Completeness

Definition 21 *A system* Σ *of modal logic is said to be* complete *with respect to a class* $\mathbf{C}_F$ *of Kripke frames* just in case

$$\mathbf{C}_F \models p \Rightarrow \vdash_\Sigma p,$$

that is,

$$\{p \in \mathcal{L}_{\mathrm{ML}}(\mathcal{P}) \mid \mathbf{C}_F \models p\} \subseteq \Sigma.$$

TABLE 3.7. Soundness and completeness for normal systems of modal logic (adapted from Chellas[2]).

	Serial	Reflexive	Symmetric	Transitive	Euclidean
K					
KD	•				
KT		•			
KB			•		
$K4$				•	
$K5$					•
KDB	•		•		
$KD4$	•			•	
$KD5$	•				•
$K45$				•	•
$KD45$	•			•	•
$KB4$			•	•	
$(=KB5)$			•		•
KTB	•	•	•		
$KT4$	•	•	•	•	
$KT5$		•			•
$(=KTB4)$		•	•	•	
$(=KDB4)$	•		•	•	
$(=KDB5)$	•		•		•

In general, it is more difficult to prove completeness than to show soundness. Although we do not explain it here in detail, the key point is construction of the so-called *canonical model* $\mathcal{M}_\Sigma$ for a system Σ using maximal consistent sets of sentences.

Let $\mathbf{C}_F$ be *the* class of Kripke frames. When proving the completeness of the system K with respect to $\mathbf{C}_F$:

$$\mathbf{C}_F \models p \Rightarrow \vdash_K p,$$

we can construct a canonical model $\mathcal{M}_K$ for K with the following remarkable property :

$$(*) \quad \mathcal{M}_K \models p \Leftrightarrow \vdash_K p.$$

Now the proof of completeness is easy: if a sentence p is valid in the class $\mathbf{C}_F$ of Kripke frames, i.e, $\mathbf{C}_F \models p$, then, since $\mathcal{M}_K$ is a model on a Kripke frame in $\mathbf{C}_F$ we have $\mathcal{M}_K \models p$, so, by $(*)$, $\vdash_K p$.

For other normal systems, let

$$\mathbf{C}_{FSER} = \text{the class of serial Kripke frames,}$$
$$\mathbf{C}_{FREF} = \text{the class of reflexive Kripke frames,}$$
$$\mathbf{C}_{FSYM} = \text{the class of symmetric Kripke frames,}$$
$$\mathbf{C}_{FTRN} = \text{the class of transitive Kripke frames,}$$
$$\mathbf{C}_{FEUC} = \text{the class of euclidean Kripke frames.}$$

Then, surprisingly enough, for KD (KT, KB, $K4$, $K5$), we can also construct canonical models $\mathcal{M}_{KD}$ ($\mathcal{M}_{KT}$, $\mathcal{M}_{KB}$, $\mathcal{M}_{K4}$, $\mathcal{M}_{K5}$) on a Kripke frame in C_{FSER} (C_{FREF}, C_{FSYM}, C_{FTRN}, C_{FEUC}, respectively). Thus, in a similar way, we have the following completeness results:

$$\mathsf{C}_{FSER} \models p \Rightarrow \vdash_{KD} p,$$
$$\mathsf{C}_{FREF} \models p \Rightarrow \vdash_{KT} p,$$
$$\mathsf{C}_{FSYM} \models p \Rightarrow \vdash_{KB} p,$$
$$\mathsf{C}_{FTRN} \models p \Rightarrow \vdash_{K4} p,$$
$$\mathsf{C}_{FEUC} \models p \Rightarrow \vdash_{K5} p,$$

by which we obtain the following correspondence:

$$\tau(\mathbf{D}) = \text{FSER},$$
$$\tau(\mathbf{T}) = \text{FREF},$$
$$\tau(\mathbf{B}) = \text{FSYM},$$
$$\tau(\mathbf{4}) = \text{FTRN},$$
$$\tau(\mathbf{5}) = \text{FEUC}.$$

Then, let $S_1, \cdots, S_n$ ($n \leq 5$) be any selection from schemas $\mathbf{D}$, $\mathbf{T}$, $\mathbf{B}$, $\mathbf{4}$, and $\mathbf{5}$, then we have the further completeness results:

$$\mathsf{C}_{\tau(\mathbf{S}_1)} \cap \cdots \cap \mathsf{C}_{\tau(\mathbf{S}_n)} \models p \Rightarrow \vdash_{KS_1 \cdots S_n} p.$$

Thus, Table 3.7 also shows completeness results for the fifteen systems. In particular, the system $S5 (=KT5)$ is found to be sound and complete with respect to the class of Kripke frames with an equivalence relation :

$$\mathsf{C}_{FREF} \cap \mathsf{C}_{FEUC} \models p \Rightarrow \vdash_{KT5} p,$$

where we note that

$$\mathsf{C}_{FREF} \cap \mathsf{C}_{FEUC} = \mathsf{C}_{FREF} \cap \mathsf{C}_{FSYM} \cap \mathsf{C}_{FTRN}.$$

Note that, in this sense, the system $KT5$ is sound and complete with respect to the class of reflexive, symmetric and transitive Kripke frames, that is, Pawlak approximation spaces.

3.10   Fuzzy Sets and Rough Sets

Crisp Subsets and Pawlak Approximation Spaces

Before discussing fuzzy subsets under rough set theory, first we will examine the relationship between crisp subsets and Pawlak approximation spaces.

(1) *From a crisp subset to a Pawlak approximation space*: A crisp subset A in the universe U itself gives us knowledge because it partitions U

into A and A^C. That is, a crisp subset itself gives us one of the simplest kinds of knowledge except for trivial knowledge. Such kind of knowledge is represented by an equivalence relation R_A defined by

$$aR_Ab \overset{\text{def}}{\Longleftrightarrow} a, b \in A.$$

The relation R_A has two non-empty equivalence classes. So, each crisp subset A has its own Pawlak approximation space $<U, R_A>$.

(2) *From a Pawlak approximation space to a crisp subset*: Conversely, given a Pawlak approximation space $<U, R>$ with an equivalence relation R on U that has just two non-empty equivalence classes, we cannot say which of the two classes the space describes. More information is needed to specify that which subset such Pawlak approximation space describes. For example, if we have at least one sample or seed element a^+ that belongs to the crisp subset we want to describe, we can specify one of the two subsets. So, we have the following structure by adding the element a^+ to the space:

$$<U, R_A, a^+>.$$

This means that we can generate one subset from available knowledge R with a sample element.

Fuzzy Subsets and Tolerance Relations

Can fuzzy subsets [16] be discussed in the same way as that of crisp subsets ? In general, elements in a fuzzy subset are related to each other with a tolerance (reflexive and symmetric) relation, because some elements in fuzzy sets cannot be distinguished, but the indistinguishability does not necessarily satisfy transitivity. Thus, one of the characteristics of fuzzy subsets is that *local indistinguishability* cannot induce *global transitivity* as in the crisp cases that we have seen in the previous subsection. Here, for simplicity, we further confine ourselves to typical tolerance relations that satisfy the following condition: for any x, y in U,

$\exists z_0 (= x), z_1, \cdots, z_{n-1}, z_n (= y) \in U$
such that $(|j - k| \leq 1 \Rightarrow z_j R z_k)$ and $(|j - k| \geq 2 \Rightarrow z_j \not\!R z_k)$.

We call the relation satisfying the condition a *non-transitive chain*. The condition has exactly the same nature in fuzzy theory. For example, Goguen wrote

We may include possibly imaginary men so that there shall be sufficiently many men of intermediate stature ([3], p.328)

when he considered a solution to the well-known *bald man paradox* using fuzzy theory.

(1) *From a fuzzy subset to a non-transitive chain*: Can a non-transitive

chain be generated from a given fuzzy subset ? First, we consider cases of rational-number-valued fuzzy subsets $\tilde{X}$ on the finite universe U:

$$\tilde{X} : U \to [0,1] \cap \mathbf{Q},$$

where $\mathbf{Q}$ denotes the set of rational numbers. For simplicity, we assume $x \neq y \Rightarrow \tilde{X}(x) \neq \tilde{X}(y)$.[7] The above-mentioned idea of imaginary elements described by Goguen [3] enables us to generate it by making an augmentation U^* of the original universe U.

1. Let $U^* = U$.
2. If there is no element in U^* that takes 1, then consider an imaginary element $x^\top$ such that $\tilde{X}(x^\top) = 1$ and let $U^* \leftarrow U^* \cup \{x^\top\}$.
3. Similarly, if there is no element in U^* that takes 0, then consider an imaginary element $a^\perp$ such that $\tilde{X}(x^\perp) = 0$ and let $U^* \leftarrow U^* \cup \{x^\perp\}$.
4. Let $U^* = \{x_1, \cdots, x_m\}$ with $0 = \tilde{X}(x_1) \leq \cdots \leq \tilde{X}(x_m) = 1$.
5. Notice that, if $x_i \neq x_j$, then

$$\exists \delta \in \mathbf{Q}^+ \, \forall x_i, x_j \in U^* \, \exists n \in \mathbf{N} \, (\tfrac{|\tilde{X}(x_i) - \tilde{X}(x_j)|}{\delta} = n),$$

where $\mathbf{Q}^+$ is the set of positive rational numbers and $\mathbf{N}$ is the set of natural numbers. Then generate a set of imaginary elements V so that

$$\forall x_i, x_j \in U^* \cup V \, (|\tilde{X}(x_i) - \tilde{X}(x_j)| = \delta)$$

and let $U^* \leftarrow U^* \cup V$.

Then if we define a relation on the augmented set U^* by

$$x R_{\tilde{X}} y \overset{\text{def}}{\Longleftrightarrow} |x - y| \leq \delta,$$

then the $R_{\tilde{X}}$ is obviously a non-transitive chain. A non-transitive chain on the finite universe with n elements is graphically represented by

$$x_1 \leftrightarrow x_2 \leftrightarrow x_3 \leftrightarrow \cdots \leftrightarrow x_{n-1} \leftrightarrow x_n.$$

In cases of real-number-valued fuzzy sets, we can present a similar discussion using approximation (not in the sense of Pawlak) because

$$\forall \epsilon \in \mathbf{R}^+ \, \exists \delta \in \mathbf{Q}^+ \, \forall x_i, x_j \in U^* \cup V \, (\exists n \in \mathbf{N} \, (|\tfrac{|\tilde{X}(x_i) - \tilde{X}(x_j)|}{\delta} - n| < \epsilon)),$$

where $\mathbf{R}^+$ is the set of positive real numbers.
(2) *From a non-transitive chain to a fuzzy subset:* First of all, we must note

---

[7]In the cases where it does not hold, take the quotient set of U by identifying elements with the same value.

that the well-known method of making *transitive closure* $\tilde{R}^*$ from a non-transitive chain $\tilde{R}$ is not promising at all because $\tilde{R}^*$ makes all elements related to each other:

$$\tilde{R}^* = R_U,$$

where R_U is the universal relation on U, that is, xR_Uy, for any x, y in U. Therefore, the method produces paradoxes such as *bold man*, which mean contradiction.

To avoid getting the universal relation on U, we must choose a consistent part of information conveyed by a tolerance relation as maximally as possible. Moreover, we assume that we have at least two instances, i.e., positive and negative ones, because the main characteristic of fuzziness is the existence of *intermediary* elements between positive and negative ones.

First, we introduce the concept of tolerance classes (cf. Klir and Yuan [6]). Given a tolerance relation R on U, a subset X in U is said to be a *tolerance class* just in case $\forall a, b \in X (aRb)$. A tolerance class is said to be *maximal* when it is not contained in any other tolerance class. Since we assume U is finite, there exists a finite number of maximal tolerance classes $X_1, \cdots, X_m$ for a torelance relation R. Thus, we can define m equivalence relations with two non-empty equivalence classes by

$$xR_iy \overset{\text{def}}{\Longleftrightarrow} [\,(x = y) \text{ or } (x, y \in X_i)].$$

and thus

$$R = R_1 \cup \cdots \cup R_m.$$

Let $\tilde{\mathcal{R}} = \{R_1, \cdots, R_m\}$. Then we can select maximal consistent subsets $\tilde{\mathfrak{R}}_i$ from $\tilde{\mathcal{R}}$ in the sense that

$$(\cup\tilde{\mathfrak{R}}_i)^* \neq R_U.$$

Thus, from one tolerance relation R, we have N equivalence relations ($N \geq 1$) that are not the universal relation:

$$(\cup\tilde{\mathfrak{R}}_1)^*, \cdots, (\cup\tilde{\mathfrak{R}}_m)^*.$$

Note that these generated relations have exactly two equivalence classes. Thus, we have k possible candidates for the proper subset that includes the positive instance x^+. Then we can define a fuzzy subset $\tilde{U}_1$ according to Resconi et al. [12] by calculating its membership grades by

$$\tilde{X}(x) \overset{\text{def}}{=} \frac{|\{i \mid x \in [x^+]_{(\cup\tilde{\mathfrak{R}}_i)^*}\}|}{m}.$$

3.11   Concluding Remarks

We conclude this chapter by evoking two possible directions of further generalization of approximation spaces.

(1) *Scott-Montague semantics:* The first direction is to introduce *Scott-Montague semantics* [2] for modal logic, whose classes characterize the so-called *classical systems* of modal logic [13], weaker ones than normal systems of modal logic. In this framework, we can find modal-logical places of Lin's neighborhood systems [7, 8], a further extension of rough set models, which is originated from the idea of general topology.

(2) *Multi-modal logics:* The second direction is to introduce Kripke frames for multi-modal logics [1], which may enable us to make systematic consideration of Pawlak's notion of a knowledge base $<U, \Re>$, where $\Re$ is a family of equivalence relations. Further details on multi-modal logics can be found in Sally Popkorn's book [11].

3.12   REFERENCES

[1] L. Catach, "Normal multimodal logics," *Proceedings of the National Conference on Artificial Intelligence*, pp.491-495, 1988.

[2] B. F. Chellas, *Modal Logic: An Introduction*, Cambridge University Press, 1980.

[3] J. A. Goguen, "The logic of inexact concepts," Synthese, Vol.19, pp.325-373, 1969.

[4] R. Goldbratt, *Logics of Time and Computation*, CSLI, 1987.

[5] G. E. Hughes and M.J.Cresswell, *A New Introduction to Modal Logic*, Routledge, London, 1996.

[6] G. J. Klir and B. Yuan, *Fuzzy Sets and Fuzzy Logic: Theory and Applications*, Prentice Hall, 1995.

[7] T. Y. Lin, "Granular computing on binary relation I: data mining and neighborhood systems," *Rough Sets in Knowledge Discovery 1: Methodology and Applications*, L. Polkowski and A. Skowron (eds.), Physica-Verlag, pp.107-121, 1998.

[8] T. Y. Lin, "Granular computing on binary relation II: rough set representations and belief functions," *Rough Sets in Knowledge Discovery 1: Methodology and Applications*, L. Polkowski and A. Skowron (eds.), Physica-Verlag, pp.122-140, 1998.

[9] Z. Pawlak, "Rough sets," Int. J. of Computer and Information Sciences, Vol.11, pp.341-356, 1982.

[10] Z. Pawlak, *Rough Sets: Theoretical Aspects of Reasoning about Data*, Kluwer, 1991.

[11] S. Popkorn, *First Steps in Modal Logic*, Cambridge University Press, 1994.

[12] G. Resconi, G. J. Klir, and U. St. Clair, "Hierarchical uncertainty metatheory based upon modal logic," *Int. J. of General Systems*, Vol.21, pp.23-50, 1992.

[13] K. Segerberg, *An Essay in Classical Modal Logic, Vols.1, 2, and 3*, University of Uppsala, 1971.

[14] Y. Y. Yao, "Two views of the theory of rough sets in finite universes," *Int. J. of Approximate Reasoning*, Vol.15, pp.291-317, 1996.

[15] Y. Y. Yao, S. K. M. Wang, and T. Y. Lin, "A review of rough set models," *Rough Sets and Data Mining : Analysis for Imprecise Data*, T. Y. Lin and N. Cercone (eds.), Kluwer, pp.47-75, 1996.

[16] L. A. Zadeh, "Fuzzy sets," *Information and Control*, Vol.8, pp.338-353, 1965.

4

Fuzzy Cognitive Maps: Analysis and Extensions

Zhi-Qiang Liu

4.1 Introduction

Fuzzy cognitive map (FCM) [5] was a modification of the cognitive map of Axelrod [1]. FCMs can be used in knowledge representation and inference which are essential to any intelligent system. FCM encodes rules in its networked structure in which all concepts are *causally* connected. Rules are fired based on a given set of initial conditions and the structure of the FCM. The resulting *map pattern* represents the causal inference of the FCM. In FCMs, we are able to represent all concepts and arcs (edges) connecting the concepts by symbols or numerical values. Moreover, in such a framework it is possible to handle different types of uncertainties effectively and to combine readily several FCMs into a single FCM that takes the knowledge from different experts into consideration [6]. FCM provides a mechanism for handling causality between events/objects in a more natural fashion. Indeed, FCM is a flexible and realistic representation scheme for dealing with knowledge. This scheme is potentially useful in the development of human-centered systems that require *soft-knowledge* in the sense that system concepts, their relationships, and the meta-system knowledge can be represented only to a certain degree. In addition, subtle (spatial and temporal) variations in the knowledge base can often result in completely different outcomes or decisions [25]. Many recently developed systems and successful applications have shown that fuzzy cognitive maps represent a promising paradigm for the development of *functional* intelligent systems [10, 15, 19, 20].

Recently, fuzzy cognitive maps have found many applications in different areas, including geographic information systems [10, 18, 19, 20], fault detection [15, 16], policy analysis [17], etc.. Furthermore, a few modifications of the conventional FCM have been proposed: Silva [22] proposed new forms of combined matrices for FCM; Hagiwara [3] extended FCM by permitting nonlinear and time delay on the arcs; Schneider [21] presented a method for automatically construct FCM.

Liu *et al.* have carried extensive research on FCM, in particular, the inference properties of FCM, contextual FCMs (CFCMs) based on the object-oriented paradigm for decision support, and application of CFCMs to geographic information systems (GIS) [10, 11, 18, 19, 20]. Despite all these developments, progress in the detailed investigation of the basic behavior of inference patterns and the analysis has been little. As a consequence, it is in general very difficult to develop large-scale intelligent systems using fuzzy cognitive maps [10, 19, 20].

In this chapter, we present the basic properties of FCM which is followed by discussions on recent developments and applications.

4.2   Fuzzy Cognitive Maps

In traditional expert systems, a tree-type structure is usually used for knowledge representation, which requires graph search. This form of knowledge representation has the following major problems:

- it is very difficulty in modeling dynamical behavior of the system;

- it is almost impossible to achieve real-time performance that is a major requirement in most human-centered systems;

- it is in most cases inadequate and inaccurate for representing knowledge, especially, when we are considering causal effects and uncertainties;

- it is computationally too complex to be of any practical use in real applications.

In 1976, Axelrod [1] proposed *cognitive maps* for representing social knowledge. These maps are able to capture the descriptive aspect of the human knowledge and assist decision making in some simple problems. However, the original cognitive maps were unable to handle variations and partial effects between concepts, because they are based on rigid structure with fixed values ("+" for positive causality or "−" for negative causality) for causality between concepts. Because causality is fuzzy and admits degrees [5], we can improve the cognitive map by including fuzzy notions in it.

The FCM is a digraph in which vertexes represent concepts and arcs (edges) between the vertexes indicate causal relationships between the concepts. For instance, the town planner is to make a decision on building a regional shopping center based on the following information: location, public transport, roads, parking facilities, other regional shopping centers, and so on. For this task, the town planner relies on expert knowledge to construct the FCM in which concepts must be defined first and necessary arcs established between the concepts. For example, if it is a prominent location it has an *increased* possibility that public transport is available; on the other hand, if there is a primary road, it will *lack* On-street parking; and so on. Based on these conditions, we can come out with an FCM shown in Figure 4.1.

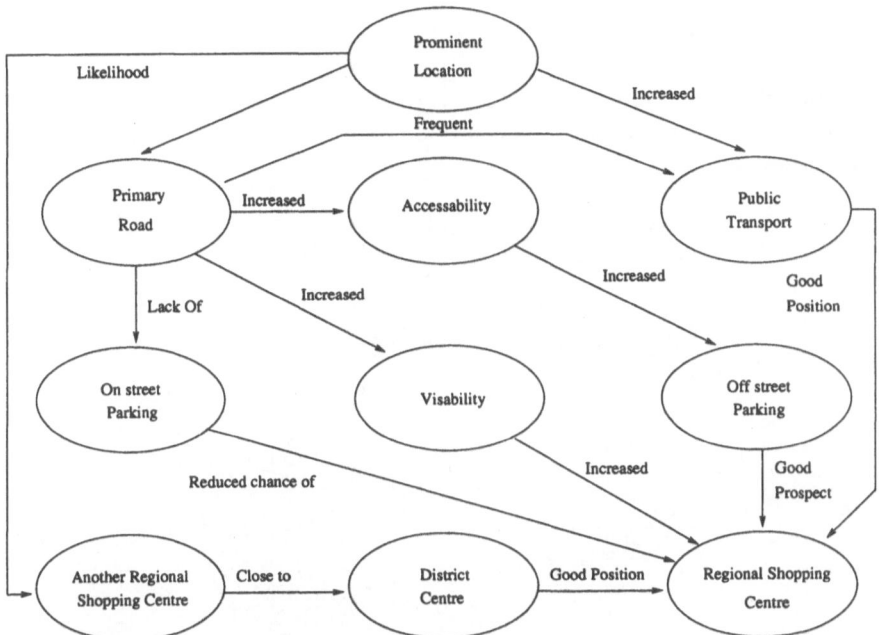

FIGURE 4.1. An FCM sub-fragment of a larger FCM based on the goal: "is a good position for a regional shopping a center?"

4.2.1 Causality and Logical Implication

In general causality is different than the classic logical implication. For instance, if "*X causes Y*" is interpreted as logical implication, "*X implies Y*", then in the classic sense, we would be able to replace "*X causes Y*" with "*not-Y causes not-X*"; which however, in reality does not usually hold. For example, it is a fact that "Drink-and-Drive *causes* Road Accident", but the inverse is not usually true, "No Road Accident *causes* Non Drink-and-Drive". This shows that causality is not implication and should not be confused with.

- Concept C_i is defined as the union of quantity set Q_i and its *disquantity* set $\sim Q_i$:

$$C_i = Q_i \cup \sim Q_i.$$

- C_i causes C_j iff $Q_i \Rightarrow Q_j$ and $\sim Q_i \Rightarrow \sim Q_j$

- C_i causally decrease C_j iff $Q_i \Rightarrow \sim Q_j$ and $\sim Q_i \Rightarrow Q_j$

Modifiers: A subtle point is that modifiers of causal quantities need not be negated, for example

"`Lower employment rate causes social in-stability.`"

but not

"`Lower employment rate causes dis-social stability.`"

In this example, *social* is the modifier of the concept *stability*, which should not be negated.

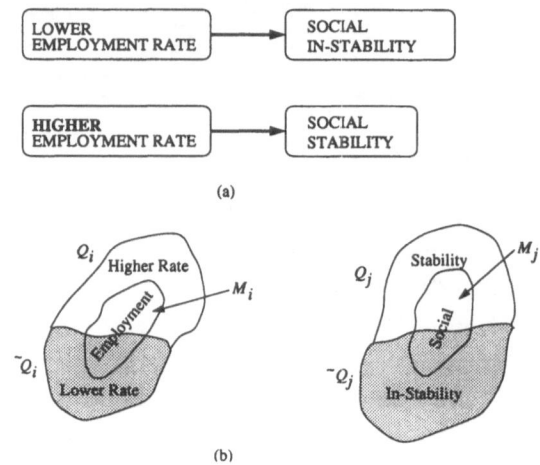

(a)

(b)

FIGURE 4.2. (a) Modifier sets (`Employment` and `Social`) in a simple FCM, (b) Their relationships with the quantifier sets.

Taking modifiers into account, we may define M_i as the modifier set for the ith concept. Therefore

$$C_i = (Q_i \cup \sim Q_i) \cap M_i.$$

This leads naturally to the following definition:

- C_i causes C_j iff

$$(Q_i \cap M_i) \Rightarrow (Q_j \cap M_j)$$

and

$$(\sim Q_i \cap M_i) \Rightarrow (\sim Q_j \cap M_j);$$

- C_i causally decreases C_j iff

$$(Q_i \cap M_i) \Rightarrow (\sim Q_j \cap M_j)$$

and

$$(\sim Q_i \cap M_i) \Rightarrow (Q_j \cap M_j).$$

4.2.2 Building Fuzzy Cognitive Maps

The FCM models the world as a collection of *concepts* and causal relationships between the concepts. A simple FCM has causal edge weights in {-1, 0, 1}. In general a causal weight is a function

$$\mu : X \to \mathbb{R}.$$

For instance, "*Improving* living standard causes *longer* life expectancy" represents a positive cause: $\mu = 1$ or in general $\mu > 0$.

The causal relationship between, say, C_i and C_j is represented by an arc (or edge) e_{ij} that shows how much C_i causes C_j. Usually, $e_{ij} \in [-1, 1]$: if $e_{ij} > 0$, C_i causes C_j to *increase*; if $e_{ij} < 0$, C_i causes C_j to *decrease*; and if $e_{ij} = 0$, C_i and C_j have no causal relationship. However, in most FCM applications, the values $e_{ij} \in \{-1, 0, 1\}$ or $e_{ij} \in \{0, 1\}$. That is, if causality occurs, it will have just positive, negative, or zero effect. This simple implementation gives a quick first-order approximation to an expert's knowledge and is useful at the early stage of analysis and design.

Let's consider the three concepts shown in Figure 4.3: drink and drive, accident rate, and police check points. The partial causal relationships between the concepts are drawn from some observations or experiences; for instance, drink-and-drive can cause road accident rate to *increase* with a degree of 0.7, random police check points can *curb* drink-and-drive to some extent (-0.4) and *reduce* (-0.2) the accident rate.

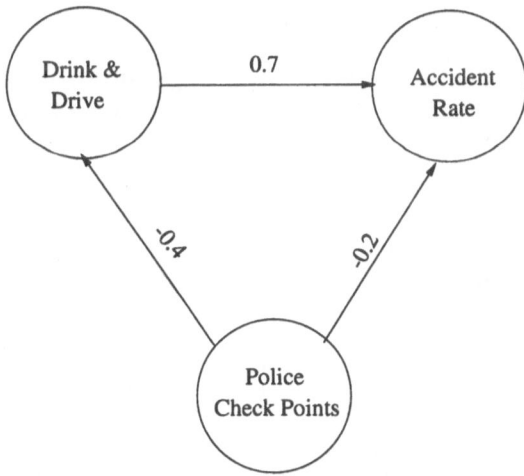

FIGURE 4.3. Drink-&-drive can cause accident rate to increase whereas police check points can reduce drink-&-drive and road accidents.

We can construct FCMs from reports or expert knowledge in some specific problem domains. From the FCM constructed, we obtain a *causal connection* matrix. The entry at i, j is the signed degree to the which the ith concept (the source concept) influences the jth concept (the affected concept) in the absence of other influences. The effect $(i \rightarrow j)$ is not necessarily symmetric, that is in general $e_{ij} \neq e_{ji}$. As a result, the connection matrix is a square and usually asymmetric matrix. For the trivial FCM in Figure 4.3, we have

$$
\begin{array}{c c c c}
 & d & a & p \\
d & 0 & 0.7 & 0 \\
a & 0 & 0 & 0 \\
p & -0.4 & -0.2 & 0
\end{array}
$$

where d stands for the concept drink-and-drive in Figure 4.3, a is for accident rate, and p for police check points.

4.2.3   Causal Inference in FCM

The FCM encodes rules in its networked structure. The rules are fired based on a given set of initial conditions and on the underlying dynamics. The result of firing the concepts represents the causal pattern of FCM. The inference process in FCMs is characterized by causal propagation and causal combination on the FCM. In virtually all applications, the resulting inference is represented as a limit-cycle bit pattern for binary or trivalent concept states.

INFERENCE PATTERN WITH LIMIT CYCLE

An initial condition is provided and propagated through the map until a static pattern is reached. We use the following recursive formula to calculate the inference pattern,

$$
C_{t+1} = f_{T_i}(C_t \times E), \tag{4.1}
$$

where C_t is the state vector, E is the causal connection matrix, and $f_{T_i}(.)$ is the vertex function,

$$
x_i = f_{T_i}(\mu) = \begin{cases} 1 & \mu > T_i, \\ 0 & \mu \leq T_i, \end{cases} \tag{4.2}
$$

where x_i is called the state of the concept C_i.

Let $C_1, C_2, \cdots, C_n$ be the causal concepts that are associated with a causal edge function,

$$
e_{ij} = e(C_i, C_j), i, j \in \{1, n\}
$$

that defines the effect C_i imparts to C_j; $e_{ij} = 0$ represents an absence of a causal relationship or effect.

After each iteration, the appropriate states x_i of the concepts C_i are updated. This process is repeated until the state vector C settles to a stationary state vector that does not undergo any further change. This is what we called *limit cycle* and represents the final inference pattern for the FCM given the initial state vector. This set of concepts forms a resonant state of "hidden pattern" that corresponds to the expert's response to what-if questions about the system's behavior. Different scenarios can be examined by varying the initial state vector. Since the vertex function operation is deterministic and there are 2^n possible binary states for an FCM with n concepts. Therefore, the recursive formula (4.1) will converge in at most 2^n iterations. In most applications, FCMs converge to a limit cycle in only a few iterations. To illustrate this process, let's have a look at the following example shown in Figure 4.4, which is a simple trivalent FCM describing politics in South Africa during the apartheid era [8].

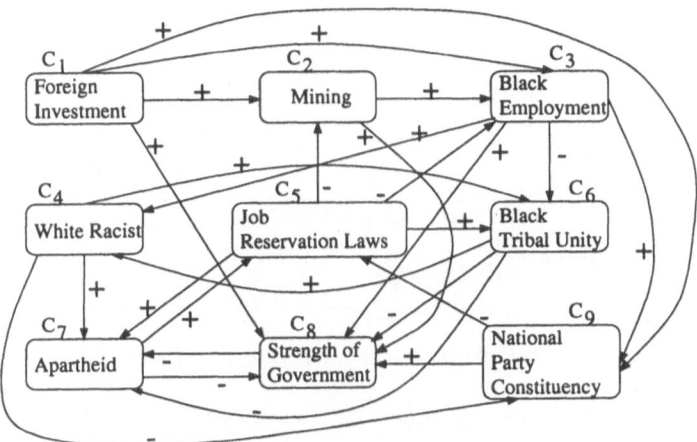

FIGURE 4.4. A small fuzzy cognitive map describing politics in South Africa during the apartheid era.

In this example, there are nine concepts. From the FCM in Figure 4.4, we obtain the causal connection matrix E as follows

$$
E = \begin{array}{c|ccccccccc}
 & C_1 & C_2 & C_3 & C_4 & C_5 & C_6 & C_7 & C_8 & C_9 \\
C_1 & 0 & 1 & 1 & 0 & 0 & 0 & 0 & 1 & 1 \\
C_2 & 0 & 0 & 1 & 0 & 0 & 0 & 0 & 1 & 0 \\
C_3 & 0 & 0 & 0 & 1 & 0 & -1 & 0 & 1 & 1 \\
C_4 & 0 & 0 & 0 & 0 & 0 & 1 & 1 & 0 & -1 \\
C_5 & 0 & -1 & -1 & 0 & 0 & 1 & 1 & 0 & 0 \\
C_6 & 0 & 0 & 0 & 1 & 0 & 0 & -1 & -1 & 0 \\
C_7 & 0 & 0 & 0 & 0 & 1 & 0 & 0 & -1 & 0 \\
C_8 & 0 & 0 & 0 & 0 & 0 & 0 & -1 & 0 & 0 \\
C_9 & 0 & 0 & 0 & 0 & -1 & 0 & 0 & 1 & 0 \\
\end{array}
$$

Let's ask "*What if foreign countries invest in South Africa to a high degree?*" We start with the foreign investment policy:

$$D_1 = [x_1, x_2, x_3, x_4, x_5, x_6, x_7, x_8, x_9],$$
$$= [\boxed{1}\,0\,0\,0\,0\,0\,0\,0\,0].$$

In this case, C_1 is the persistent concept, with $x_1 = 1$, which will stay in the inference patterns and is indicated as $\boxed{1}$. In the following, we set the threshold to 1.

$$D_1 E = [0\,1\,1\,0\,0\,0\,0\,1\,1],$$
$$\rightarrow [\boxed{1}\,1\,1\,0\,0\,0\,0\,1\,1] = D_2;$$

$$D_2 E = [0\,1\,2\,1\,-1\,-1\,-1\,4\,1],$$
$$f_{T_i}(D_2 E) = [0\,1\,1\,1\,0\,0\,0\,1\,1],$$
$$\rightarrow [\boxed{1}\,1\,1\,1\,0\,0\,0\,1\,1] = D_3;$$

$$D_3 E = [0\,1\,2\,1\,-1\,0\,0\,4\,1],$$
$$f_{T_i}(D_3 E) = [0\,1\,1\,1\,0\,0\,0\,1\,1],$$
$$\rightarrow [\boxed{1}\,1\,1\,1\,0\,0\,0\,1\,1] = D_3.$$

We have reached the stable state. D_3 represents the answer

$$D_3 = [x_1, x_2, x_3, x_8, x_9]$$

corresponding to the following set of concepts being activated:

$$\{C_1, C_2, C_3, C_8, C_9\},$$

which concludes that with the increased foreign investment South Africa will experience increased mining, improved employment in black population, strengthen the apartheid government and the National Party's constituency. This is a reasonable analysis and perhaps many political analysts would agree.

4.2.4   Combining Fuzzy Cognitive Maps

Expert systems usually employ one domain expert. This is due largely to the lack of effective search strategies in the rule base and difficulties in maintaining a large tree structure that represents the knowledge base. The result of merging two trees is not a tree but a cyclic graph. However, we can combine several FCMs generated by different experts relatively easily by carrying out some simple matrix addition. The combination is made through the augmented connection matrices for individual FCMs: $F_1, \cdots, F_m$. Each augmented matrix F_i has n rows and n columns, where n is the total number of distinct concepts used by the experts. There are no restrictions on the number of FCMs involved in the combination nor on the size of each FCM as long as its connection matrix is augmented for combination.

$$F = \sum_{i=1}^{m} F_i. \tag{4.3}$$

Let's consider a simple sewage system [22].

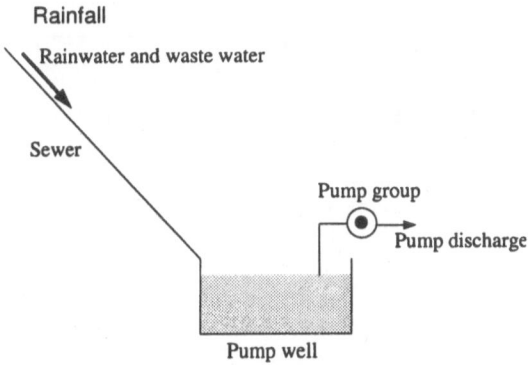

FIGURE 4.5. A simple sewage system.

The following 10 concepts are important to the problem in this example:

- C1: present rainfall intensity,
- C2: water storage quantity,
- C3: water inflow,
- C4: rate of change of water level,
- C5: rate of change of water level (initial),
- C6: present water level,
- C7: water level,
- C8: present pump discharge,
- C9: pump discharge,
- C10: change pump discharge.

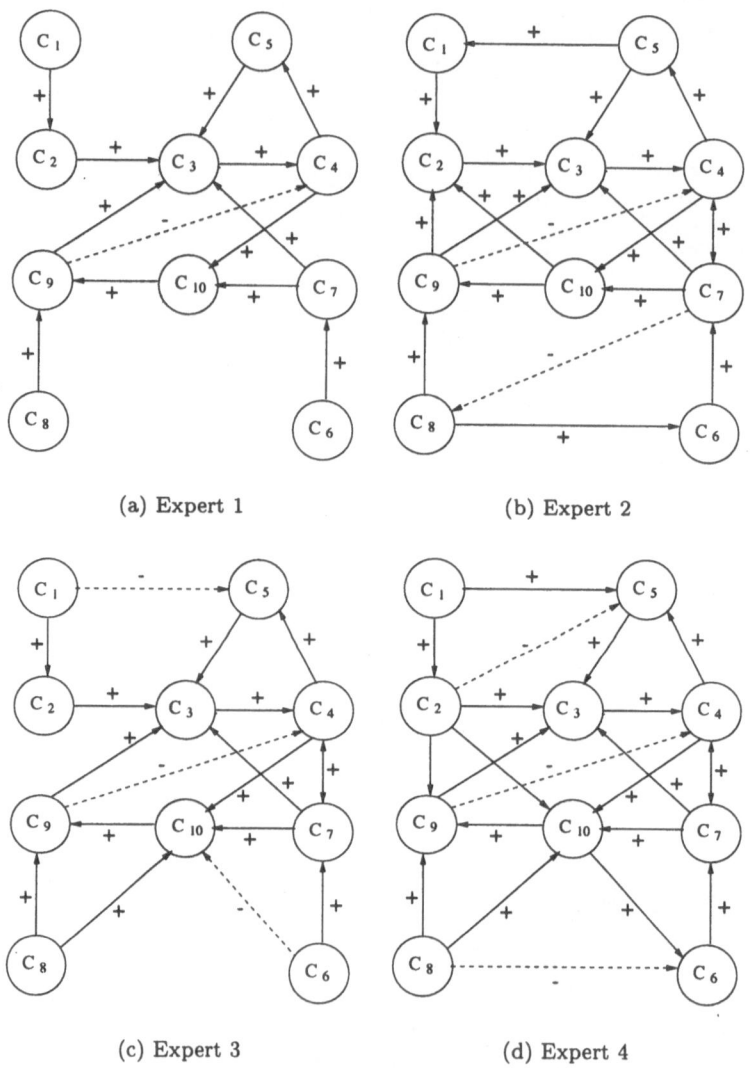

(a) Expert 1

(b) Expert 2

(c) Expert 3

(d) Expert 4

FIGURE 4.6. Fuzzy cognitive maps generated by four experts for the sewage problems. [22]

Given the four experts' FCMs, it is possible to combine them into one FCM and a single connection matrix:

$$
\begin{bmatrix}
0 & 4 & 0 & 0 & 0 & 0 & 0 & 0 & 0 & 0 \\
0 & 4 & 0 & -1 & 0 & 0 & 0 & 0 & 0 & 1 \\
0 & 0 & 0 & 4 & 0 & 0 & 0 & 0 & 0 & 0 \\
0 & 0 & 0 & 0 & 4 & 0 & 4 & 0 & 0 & 4 \\
1 & 0 & 4 & 0 & 0 & 0 & 4 & 0 & 0 & -1 \\
0 & 0 & 0 & 0 & 0 & 0 & 0 & 0 & 0 & 0 \\
0 & 0 & 4 & 4 & 0 & 0 & 0 & -1 & 0 & 4 \\
0 & 4 & 0 & 0 & 0 & 0 & 0 & 0 & 4 & 0 \\
0 & 1 & 4 & -4 & 0 & 0 & 0 & 0 & 0 & 0 \\
0 & 1 & 0 & 0 & 0 & 1 & 0 & 0 & 4 & 0
\end{bmatrix}
$$

To perform inference analysis, we simply follow the procedures discussed in Section 4.2.3.

4.3   Extensions to FCM

There are four major problems with the original fuzzy cognitive maps:

1. Pairwise interaction, whereas, in reality two or more concepts may act together and inseparable. Let's consider three concepts: C_1, C_2, and C_3 for which we may have six pairwise causal relationships: $C_i \to C_j$ where $i \neq j$. However, in real applications, some concepts will have to form a single AND-type causal action to implement antecedent and consequent.

2. Lack of time concept (i.e., delays), in reality a causal effect can only take place *after* some time, for instance, in economy, oversupply will result in inflation; however, this will not happen immediately; a new tax policy will affect the pricing of electronic goods, but it may take a few months for the consumer to see the effect.

3. Static and linear causal relationship between causal concepts. In particular, in most applications, the edge value is binary or trivalent which will not effectively model the non-linear and dynamic nature of the problem. For instance, increased income will result increased consumption, in turn it will stimulate production, and so on; however, these increases (positive causes) cannot go on linearly with a constant positive multiplier, say, 0.3.

4. Thresholding causality will result in inadequate inference and in some cases lead to inaccurate conclusion.

To solve these problems, recently many researchers have proposed new approaches to these problems. In this section, we will look at three important extensions to the original FCM.

4.3.1   FCM with Non-linear Edge Functions

There are four fundamental elements in a causal system: the cause, the causal relationship, the dynamics, and the effect. The fuzzy cognitive map improves cognitive map (CM) by describing the strength of the causal relationship. However, it is unable to model the non-linearity and time delays in the causes. Recently, Hagiwara proposed the extended fuzzy cognitive map (E-FCM) [3] that includes time delays and some nonlinear properties as the edge functions. For instance, migration, say, to Silicon Valley ("gold rush" over the Internet) over *time* is not a linear process; rather it is non-linear: initially opportunities were abundant, many people rushed in. As time went by, more and more people moved in. Now the residents are facing reduced resources and opportunities and some are moving out, which will in turn slow down the migration; eventually the valley's population will become stable, as shown in Figure 4.7(a). Similarly, the infrastructure will have to cope with the growth of the population for awhile, but given the limited resources and budget, the infrastructure can only keep pace with the growth of the population to a certain point after which the infrastructure will no longer be able to cope with a large population and start to deteriorate as illustrated by Figure 4.7(b).

In the E-FCM, the total impact to node C_j at time t is represented as follows,

$$f_j = \sum_{i=1}^{n} w_{ij}(C_j(t - d_{ij}))C_i(t - d_{ij})$$

where $C_i(t)$ is a concept which is a function of t; d_{ij} is the time it takes for C_i to have any effect on C_j; and $w_{ij}(.)$ is a weight function from C_i to C_j and is in general non-linear.

Using E-FCMs, Hagiwara constructed a simple simulation of municipal management with only seven concepts. The results have shown that indeed non-linear modeling can greatly influence the behavior of the FCM and that with good modeling it is possible to develop an E-FCM that outperforms the conventional FCM. However, many questions still remain when time delay and non-linear weight functions are introduced in FCMs; for instance, systematic modeling process, stability, goal state reachability, performance, criteria for optimality, and so on. Recently, we have proposed dynamic cognitive networks (DCNs) in an attempt to systematically address these problems and have obtained some encouraging initial results [9, 13, 14].

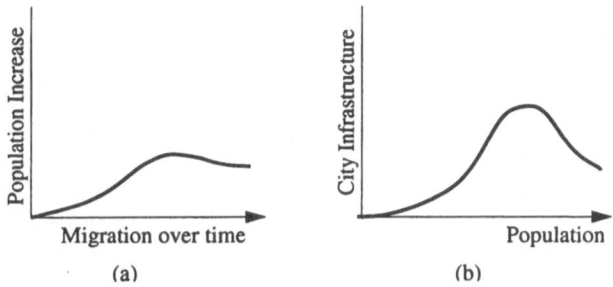

FIGURE 4.7. (a)Population growth versus migration, (b) city infrastructure as a function of population growth.

4.3.2   FCM with Constant Time-Delays

In some problems, time delays can be modeled as discrete values, for instance, in a stable society and economy, it is relatively easy to predict the time required to see the effect of implementing a tax policy, say, goods and services tax (GST). In such cases, we may set the time delays in the causal effects as some constants. Each edge can be modeled as $e_{ij} \in [-1, 1]$ and time delay $t_{ij} \in [a, b]$ where a and b are constants and $0 < a \leq b$. For example, if it takes six months for C_i to have an effect on C_j, we may express $t_{ij} = t(C_i \to C_j) = 0.5$ year or $t_{ij} = 0.5$ if it takes two years between C_j and C_k we simple write $t_{jk} = t(C_j \to C_k) = 2$ years or $t_{ij} = 2$. Figure 4.8 shows an FCM with constant time delays in the causal effects [4]. In the figure, we used $e_{ij}(t_{ij})$ to express the weight and the associated time delay, e.g., -0.7(2) means $e_{BC} = -0.7$ and $t_{BC} = 2$.

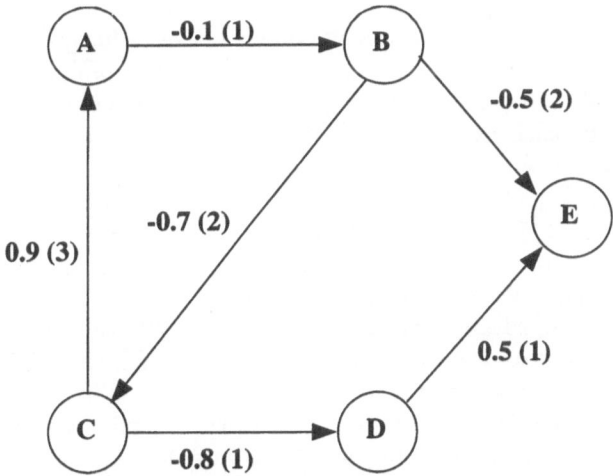

FIGURE 4.8. A hypothetical FCM with constant time delays: $e_{ij}(t_{ij})$.

In order to compute causal inference using Equation (4.1), we must convert the FCM into a new FCM with unit-time delays. This can be done by inserting dummy concepts (nodes) in the FCM; for example, if C_i influences C_j with two delay units, a dummy concept node C_{i0} must be inserted: $C_i \to C_{i0} \to C_j$, so that each edge will have a unit-time delay. In general, if the time delay between C_i and C_j is $m \geq 1$, then $m-1$ dummy concepts (nodes) must be inserted between C_i and C_j. When $e_{ij} < 0$, care must be taken to make sure that the overall causal effect from C_i to C_j remains unchanged. Figure 4.9 shows the new FCM converted from that in Figure 4.8. Kim and Park have demonstrated that the two FCMs are equivalent in terms of their inference behaviors [4].

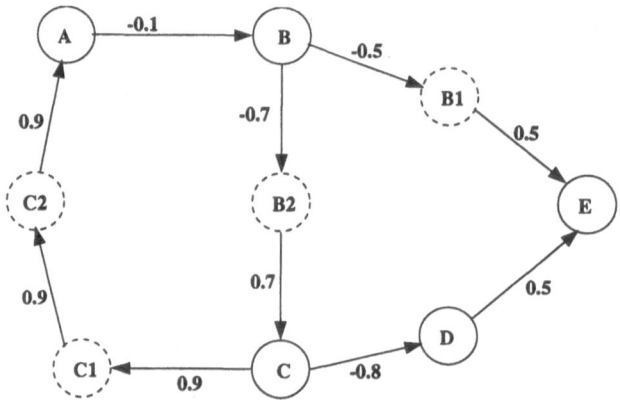

FIGURE 4.9. Converted FCM with unit-time delay.

4.3.3 Weighted Combination of FCMs

In Section 4.2.4 we discussed how to combine several fuzzy cognitive maps constructed by different experts and we treated the opinions of all expert equally. In reality, however, some experts may be more credible than others. We can weight each expert with a non-negative credibility weight w_i as follows,

$$F = \sum_i w_i F_i.$$

To determine w_i is a very difficult problem.

One possible approach: *map* experience directly into the credibility weight. However, this approach may counter-productive when there is little correlation between experience and accuracy of the map: What is the mapping scheme? How accurate and reliable is it? In their study of expert credibility [23, 24], instead of assigning merits to individual experts, Taber and Siegel proposed to judge the FCMs constructed by the experts on their own merits. They made two assumptions:

1. Concurrence of an expert with others implies a high level of experience.

2. The maps contain a sizable measure of expertise.

The second assumption provides a background for comparison. If it is false, the knowledge base is largely random; if true, then an expert agreeing with the panel of experts (assumption 1) tends to reinforce the universal belief.

The individual FCMs are excited with state vectors. These vectors are antecedents. The FCMs predict the consequents. The FCMs with the same or similar responses for all the state vectors receive high credibility. A Hamming matrix $H_q(i, i)$ is generated by computing the distance [1] between experts i and j for all stimulus vectors. The weight is estimated as follows:

$$w_i = 1 - \frac{h_i}{\sum_{i=1}^{i=N_e} h_i},$$

where N_e is the number of experts and

$$h_i = \frac{\sum_{j=1}^{j=N_e} mean(i, j)}{N_e},$$

where the mean is

$$mean(i, j) = \frac{\sum_{q=1}^{q=N_q} H_q(i, j)}{N_q},$$

where $H_q(i, j)$ is a Hamming matrix between experts i and j, N_q is the number of what-if questions (state vectors).

4.4   Analysis of Fuzzy Cognitive Maps

FCM represents a very promising paradigm for the development of *functional* intelligent systems [10, 15, 19, 20]. In real-world applications, fuzzy cognitive maps are usually very large and complex, containing a large number of concepts and arcs. However, the current method for constructing and analyzing fuzzy cognitive maps are inadequate and infeasible in practice. Furthermore, as the FCM is a typical, nonlinear system, the combination of several inputs or initial states may result in new patterns with unexpected behaviors. In this section,[2] we present some recent results on the analysis of the causal inferences in FCMs. An FCM can be divided into several *basic*

---

[1] The Hamming distance between two vectors is the number of bits that are different, e.g., the Hamming distance between 0100 and 0111 is 2.

[2] For conciseness, we have omitted the proofs for the theorems in this section. Please refer to [11, 12] for the proofs.

FCMs. The dynamic of a basic FCM is determined by its *key* vertexes. A group of recurrence formulas is given to describe the dynamics of the key vertexes.

4.4.1 FCM and Its State Space

The FCM is a digraph in which nodes represent concepts and arcs between the nodes indicate causal relationships between the concepts. The connectivity of the FCM can be conveniently represented by a connection matrix

$$\mathbf{E} = \begin{bmatrix} \cdots & \cdots & \cdots \\ \cdots & e_{ij} & \cdots \\ \cdots & \cdots & \cdots \end{bmatrix},$$

where e_{ij} is the value of the arc from node i to node j, i.e., the value of arc a_{ij}. Based on causal inputs, the concepts in the FCM can determine their states. This gives FCMs the ability to infer causally. As suggested by Kosko [7, 8], this can be determined simply by a threshold. In general, we define a vertex function as follows.

Definition 22 *A vertex function* f_{T,v_i} *is defined as:*

$$x_i = f_{T,v_i}(\mu_i) = \begin{cases} 1 & \mu_i > T, \\ 0 & \mu_i \leq T, \end{cases}$$

where μ_i *is the total input of* v_i, *i.e.,* $\mu_i = \sum_k e_{ik} \cdot x_k$.

Usually, if $T = 0$, f_{T,v_i} is denoted as f_{v_i}, or simply, f_i. For the sake of simplicity and without loss of generality, throughout this section we assume $T = 0$ unless specified otherwise.

Given this definition, an FCM can be defined as a weighted digraph with vertex functions. $\mho$ denotes an FCM, $v(\mho)$, or simply, v stands for vertex (or concept) of $\mho$; $V(\mho)$ represents the set containing all the vertices of $\mho$; $x(\mho)$ or x is the state of $v(\mho)$; $\phi = (x_1, \cdots, x_n)^T$ denotes the state of $\mho$, where x_i is the state of vertex v_i and n is the number of vertices of $\mho$; $a(\mho)$ or a stands for an arc in $\mho$; $A(\mho)$ represents the set containing all the arcs of $\mho$; $v^O(a(\mho))$ is the start vertex of $a(\mho)$ and $v^I(a(\mho))$ is the end vertex of $a(\mho)$.

As every vertex may take a value in $\{0, 1\}$, the state space of the FCM is $\{0, 1\}^n$, denoted by $X^0(\mho)$ or X^0. If the state of $\mho$ is ϕ after k inference steps from an initial state ϕ_0, we say that ϕ can be *reached* from ϕ_0, or $\mho$ can reach ϕ after k inference steps. Although the initial state ϕ_0 of the FCM may be any state, i.e., $\phi_0 \in X^0 = \{0, 1\}^n$, some states can never be reached no matter which initial state it is set to. For example, state $(* \ * \ 1 \ 1)^T$ cannot be reached in the FCM shown in Figure 4.10, where $*$ means that it can be any value in $\{0, 1\}$.

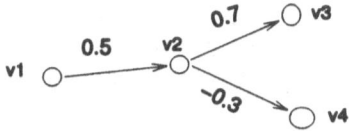

FIGURE 4.10. State space of fuzzy cognitive map.

We define $X(\mho)$ or X as the *reachable* state set of $\mho$ which contains all states that $\mho$ can reach. $X^{\infty}(\mho)$ or X^{∞} is defined as the state set of the states which can be reached by $\mho$ after 2^n inference steps. Obviously

$$X^{\infty}(\mho) \subseteq X(\mho) \subseteq X^0(\mho).$$

It is easy to see which state can be reached by the trivial FCM in Figure 4.10. However, it will be difficult if the FCM contains a large number of concepts and complex connections. Whether a state can be reached in a given FCM is a fundamental and important problem in the analysis and design of FCMs.

Clearly, if a state $\phi^* \in X(\mho)$, there exists a state ϕ_0 such that ϕ^* can be reached in one step inference if ϕ_0 is set as the initial state. Since $\phi^* \in X(\mho)$, there exists a state sequence in $X(\mho)$,

$$\phi^*(0), \phi^*(1), \cdots, \phi^*(k) = \phi^*.$$

Renumbering the vertices of $\mho$ we can write ϕ^* slightly differently,

$$\overline{\phi} = (\phi_+^T, \phi_-^T)^T,$$

where $\phi_+ = (1, 1, \cdots, 1)^T, \phi_- = (0, 0, \cdots, 0)^T$. Denote $\overline{E}$ as the adjacency matrix of the renumbered FCM, $n_+ = dim(\phi_+), n_- = dim(\phi_-), \overline{E_+} = (\overline{E}_1^T, \cdots, \overline{E}_{n_+}^T)^T, \overline{E_-} = (-\overline{E}_{n_++1}^T, \cdots, -\overline{E}_{n_++n_-}^T)^T. \phi^* \in X(\mho)$ if and only if

$$\begin{cases} \overline{E_+} \cdot \psi > 0, \\ \overline{E_-} \cdot \psi \geq 0, \end{cases}$$

has a solution ψ in $\{0, 1\}^n$.

From the above discussion, we may develop an algorithm to determine whether a state can be reached from a particular initial state [11, 12].

4.4.2   Causal Module of FCM

In general, an FCM may contain a large number of vertexes with very complicated connections. It is difficult to be handled directly, if at all possible. However, an FCM can be divided into several *basic* modules. Every causal module is a smaller FCM. Vertexes (or concepts) of a causal module infer each other and are closely connected. Basic FCM modules are the minimum FCM entities that cannot be divided further.

An FCM $\mho$ is "divided" as FCM $\mho_1$ and FCM $\mho_2$ if

$$1) V(\mho) = V(\mho_1) \cup V(\mho_2),$$

$$2) A(\mho) = A(\mho_1) \cup A(\mho_2) \cup B(\mho_1, \mho_2),$$

where

$$B(\mho_1, \mho_2) = \{ a(\mho) \quad | \; V^I(a(\mho)) \in V(\mho_1), V^O(a(\mho)) \in V(\mho_2), \\ or \; V^I(a(\mho)) \in V(\mho_2), V^O(a(\mho)) \in V(\mho_1) \},$$

$$V(\mho_1) \cap V(\mho_2) = \varnothing,$$

$$A(\mho_1) \cap A(\mho_2) = \varnothing,$$

$$A(\mho_1) \cap B(\mho_1, \mho_2) = \varnothing,$$

$$A(\mho_2) \cap B(\mho_1, \mho_2) = \varnothing.$$

This operation is denoted as

$$\mho = \mho_1 \leftarrow \frac{B(\mho_1, \mho_2)}{\oplus} \rightarrow \mho_2.$$

Particularly, we consider the causal relationships in subgraphs:

$$B(\mho_1, \mho_2) = \{a(\mho) | V^I(a(\mho)) \in V(\mho_1), V^O(a(\mho)) \in V(\mho_2)\},$$

this case is described as "$\mho_2$ is caused by $\mho_1$", or "$\mho_1$ causes $\mho_2$". Such a division is called a *regular division*, and is denoted as

$$\mho = \mho_1 \frac{B(\mho_1, \mho_2)}{\oplus} \rightarrow \mho_2.$$

Definition 23 *An FCM containing no circle is called a simple FCM .*

A simple FCM has no feedback mechanism and its inference pattern is usually trivial.

Definition 24 *A basic FCM is an FCM containing at least one circle, but cannot be regularly divided into two FCMs, both of which contain at least one circle.*

From the Definitions 23 and 24, we can see that an FCM containing circles can always be regularly divided into basic FCMs. In general, the basic FCMs of $\mho$ can be ordered concatenately as follows.

$$\mho = \mho_1 \frac{B(\mho_1, \mho_2)}{\oplus} \to \mho_2 \ \frac{B(\mho_2, \mho_3)}{\oplus} \to \mho_3 \ \cdots \ \frac{B(\mho_{m-1}, \mho_m)}{\oplus} \to \mho_m.$$

We establish a formal result in the following theorem.

Theorem 25 *Suppose FCM $\mho$ can be regularly divided into m basic FCMs, then*

$$\mho = (\cup_{i=1}^m \mho_i) \ \cup \ (\cup_{i=2}^m \cup_{j=1}^i \ B(\mho_j, \mho_i)), \tag{4.4}$$

where

$$V(\mho_i) \cap V(\mho_j) \ = \ \oslash \quad i \neq j,$$

$$A(\mho_i) \cap A(\mho_j) \ = \ \oslash \quad i \neq j.$$

$$B(\mho_i, \mho_j) = \{a(\mho)|V^I(a(\mho)) \in V(\mho_i), V^O(a(\mho)) \in V(\mho_j)\}.$$

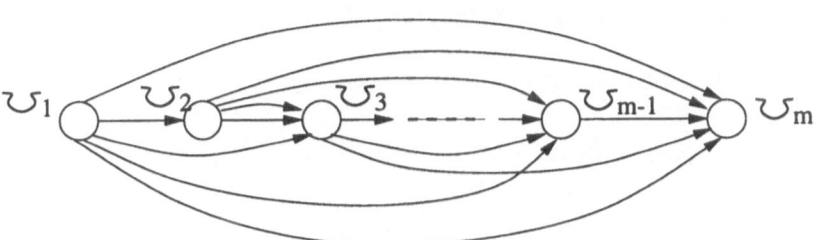

FIGURE 4.11. Causal modules of FCM.

The inference pattern of a basic FCM $\mho_i$ is determined by its input (external) and initial state. The inputs of $\mho_i$ can be determined once the inference pattern of $\mho_k, k < i$ are known. Subsequently, the inference pattern of $\mho_i$ can be analyzed. If we know the inference patterns of the basic FCMs individually, we will be able to obtain the inference pattern of the entire FCM, because the basic FCMs collectively contain all the concepts of the FCM:

$$V(\mho) = \cup_{i=1}^m V(\mho_i).$$

The following theorem determines if an FCM is not a basic FCM.

Theorem 26 *Suppose that FCM $\mho$ is input-path standardized and trimmed of affected branches. If a vertex v_0 of $\mho$ has at least two input arcs and v_0 does not belong to any circle, then $\mho$ is not a basic FCM.*

Theorem 26 provides not only a rule for determining whether an FCM is a basic FCM, but also presents an approach to regularly dividing an FCM if it is not a basic FCM: $In(v_0)$ and $Out(v_0)$. This is illustrated in Figure 4.12. The FCM in Figure 4.12(a) can be regularly divided into two FCMs as shown in Figures 4.12(b) and 4.12(c).

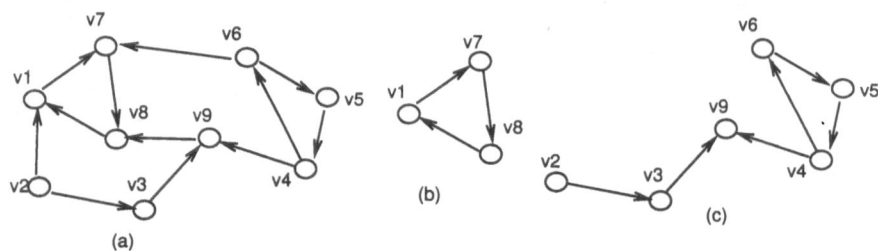

FIGURE 4.12. Regularly divided and basic FCMs: According to vertex v_9, FCM in (a) is divided as $Out(v_9)$ in (b) and $In(v_9)$ in (c).

More specifically such a division can be done by the following algorithm.

Algorithm 1

Step 1: If $\mho$ is a simple FCM stop.

Step 2: Select a vertex v of the $\mho$, mark v.

Step 3: Form $In(v)$.

Step 4: If there is no circle in $In(v)$, go to step 7.

Step 5: Form $Out(v)$.

Step 6: If there is no circle in $Out(v)$, go to step 7.

Step 7: $\mho$ is regularly divided into $In(v_9)$ and $Out(v_9)$, stop.

Step 8: If there is no unmarked vertices, $\mho$ is a basic FCM, stop.

Step 9: Select an unmarked vertex v, mark v, go to step 2.

In general, an FCM can be regularly divided into basic FCMs by repeatedly implementing Algorithm 1.

4.4.3   Inference Patterns of Basic FCMs

The inference pattern of an FCM can be obtained by recursively calculating

$$\phi(k+1) = f[E\phi(k)] = (f_1[E_1\phi(k)], \cdots, f_n[E_n\phi(k)])^T.$$

However, since in most real applications the FCM contains a large number of vertexes and complicated connections, the state sequence can be very long and difficult to analyze. It will be most useful to draw out properties or recursive formula for the state sequence.

Proposition 27 *If an FCM is a simple FCM, it will become static after L inference iterations unless it has an external input sequence, where L is the length of the longest path of the FCM.*

The Proof of Proposition 27 is obvious. Consequently, the following is true.

Corollary 1 *Vertexes except the end vertex of an input path will become static after L inference iterations unless it has an external input sequence, where L is the length of the path.*

In this section, all FCMs are assumed as basic FCMs, with input paths being standardized and affected branches trimmed. For the sake of simplicity, we assume that all FCMs do not have external input sequences unless they are specifically indicated. FCMs with external input sequences can be analyzed in the similar way.

In our study of the inference pattern of the FCM, we found that some vertexes may play more important roles than others. We define these vertexes as *key* vertexes. The state of every vertex in the FCM can be determined by the state of key vertexes. In the following part of this section, the definition of key vertex is followed by some discussions of the properties of key vertex.

Definition 28 *A vertex is called as a key vertex if*

1. *it is a common vertex of an input path and a circle, or*

2. *it is a common vertex of two circles with at least two arcs pointing to it which belongs to the two circles, or*

3. *it is any vertex on a circle if the circle contains no other key vertexes.*

Proposition 29 *Every circle contains at least one key vertex.*

Given Property 3 in Definition 28, the correctness of Proposition 29 is obvious.

Lemma 30 *If any vertex, $v_0 \in V(\mho)$, is not on an input path, then it is on a circle.*

Lemma 31 *Suppose that v_0 is not a key vertex and not on an input path, then there is one and only one key vertex (denoted as v^*) can affect v_0 via a normal path, and there is only one normal path from v^* to v_0.*

Again suppose the key-vertex set is $\{v_1, \cdots, v_r\}$. If $v_j (1 \leq j \leq r)$ satisfies (1) of Definition 28, it is the end vertex of an input branch, denote $v'(j)$ as the input vertex. Denote the normal paths from v_i to v_j as $\overline{P_{i,j}^0}, \overline{P_{i,j}^1}, \cdots, \overline{P_{i,j}^{r(i,j)}}$, where $r(i,j)$ is the number of normal paths from v_i to v_j, $\overline{P_{i,j}^0} = P_{i,j}^{<>}$ means that there is no normal path from v_i to v_j, and $f_{\overline{P_{i,j}^0}} = 0$.

Theorem 32 *If v_j $(1 \leq j \leq r)$ is not the end of an input branch,*

$$x_j(l) = f_{T_j, v_j} (\sum_{l=1}^{r} \sum_{s=0}^{r(l,j)} f_{\overline{P_{l,j}^s}} \{x_l(l - L(\overline{P_{l,j}^s})) \}). \qquad (4.5)$$

If v_j is the end of an input branch,

$$\begin{aligned} x_j(l) &= f_{T_j^1, T_j^2, v_j} (\sum_{l=1}^{r} \sum_{s=0}^{r(l,j)} f_{\overline{P_{l,j}^s}} \{x_l(k - L(\overline{P_{l,j}^s})) \} \\ &+ f_{\overline{P_{v(j),j}}} (x_{v(j)}(l - L(\overline{P_{v(j),j}})))). \end{aligned} \qquad (4.6)$$

Therefore the inference pattern of the FCM is can be determined by the recursive formula in terms of key vertexes. After the states of key vertexes are determined, the states of the remaining vertexes can be determined by states of key vertexes as follows. In turn, the state of the entire FCM is also determined.

Theorem 33 *Suppose that $K_v(\mho)$ is the key vertex set of $\mho$, $I_v(\mho)$ is the input vertex set of $\mho$, $v_0 \in V(\mho)$, $v_0 \notin K_v(\mho)$ and $v_0 \notin I_v(\mho)$. There exists one and only one vertex x^*, $x^* \in K_v(\mho)$ or $x^* \in I_v(\mho)$ such that there exists a path, $P^*(v^*, v_0)$ from v^* to v_0 via no vertexes in $K_v(\mho)$.*

Theorem 34 *$P^*(v^*, v_0)$ is a normal path.*

As $P^*(v^*, v_0)$ is a normal path,

$$x_0(l) = f_{P^*(v^*, v_0)} (x^*(l - L(P^*(v^*, v_0)))).$$

Thus the state of the entire FCM is determined. With the help of Theorems 32, 33, and 34, we discuss some important causal inference properties of typical FCMs in the following sections.

4.4.4   Inference Pattern of General FCMs

When considering the inference pattern of a general FCM($\mho$), we should first regularly divide it into basic FCMs ($\mho_i, 1 \leq i \leq m$) and then determine the inference pattern one by one according to the causal relationships between of them. For the basic FCM($\mho_i$), the external input should be formed according to the inference pattern of $\mho_j$, $1 \leq j < i$. Then the input paths should be standardized. After this process, we delete all the affected branches to simplify the FCM($\mho_i$) for further analysis.

From the simplified FCM, we need to construct the key-vertex set. If the basic FCM contains only one circle, and every vertex on the circle has only one input arc, then the key vertex set contains only one vertex. It can be any vertex on the circle according to Definition 11. If the basic FCM contains more than one circle, the key vertexes are the circle vertexes that have at least two input arcs. This can be judged from E. If the ith column of E has at least two non-zero elements, v_i has at least two input arcs. If $T_{ii} \neq 0$, as the following proposition indicates, v_i is on a circle.

Proposition 35 *Vertex v_i is on a circle if and only if $T_{ii} \neq 0$, where T_{ii} is the ith row and ith column element of matrix $T = \sum_{i=k}^{n} E^k$.*

Following Equations (4.5) and (4.6) in Theorem 32, we can obtain the inference pattern of key vertexes. Subsequently the states of other vertexes including those on the affected branches can be determined accordingly.

The steps for analyzing the inference pattern of an FCM are given below.

Algorithm 2

Step 1:   Divide the $\mho$ regularly into basic FCMs: $\mho_1, \cdots, \mho_m$:

$$\mho = (\cup_{i=1}^{m} \mho_i) \cup (\cup_{i=2}^{m} \cup_{j=1}^{i} B(\mho_j, \mho_i)).$$

Step 2: $i = 1$.

Step 3:   Delete the attached branches of $\mho_i$.

Step 4:   Construct new inputs from basic FCMs $\mho_j$ by $B(\mho_j, \mho_i)$ ($j < i$).

Step 5:   Standardize the input paths as normal paths.

Step 6:   Construct the key-vertex set.

Step 7:   Determine the state-sequence formula of the key vertexes according to Theorem 4.5.

Step 8:   Determine the state-sequence formula of the remaining vertexes.

Step 9:   Determine the state-sequence formula of the affected branch.

Step 10: $i = i + 1$, if $i < m$, go to step 2, else stop.

This algorithm is illustrated by the following example. In the example, all the arcs are assumed to be positive, i.e. $e_{ij} > 0$.

Example
The FCM $\mho_a$ shown in Figure 4.13 can be simplified as $\mho_b$ by trimming off the affected branches.

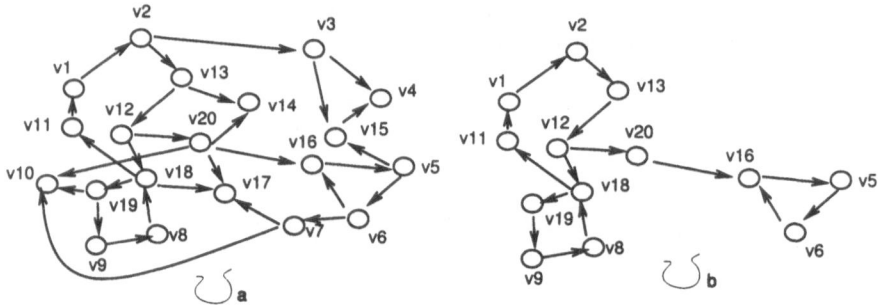

FIGURE 4.13. Example of analyzing inference pattern of FCMs: $\mho_a$ is simplified as $\mho_b$ by trimming the affected branches of $\mho_a$.

FCM $\mho_c$ and $\mho_d$ in Figure 4.14 are the two basic FCMs in $\mho_b$ shown in Figure 4.13. Figure 4.15 shows $\mho_e$ which is $\mho_c$ minus the only affected branch, $AB(v_{20})$. v_{18} is the only key vertex of $\mho_e$:

$$x_{18}(l) = f_{18}(x_{18}(l-4) + x_{18}(l-6)).$$

As 2 is the common factor of 4 and 6, the final state sequence of v_{18} is: $0, 1, 0, 1, \cdots, 0, 1, 0, 1, \cdots$. The remaining vertex states of $\mho_e$ can be completely determined by x_{18}. For example,

$$x_1(l) = x_{18}(l-2).$$

The final state sequence of v_1 is also $0, 1, 0, 1, \cdots, 0, 1, 0, 1, \cdots$.
The affected branch $AB(v_{20})$ of $\mho_c$ contains only one vertex: v_{20}. Its state is determined by $\mho_e$, or more specifically, by v_{12}.

$$x_{20}(l) = x_{12}(l-1).$$

The final state sequence of v_{20} is also: $0, 1, 0, 1, \cdots, 0, 1, 0, 1, \cdots$.

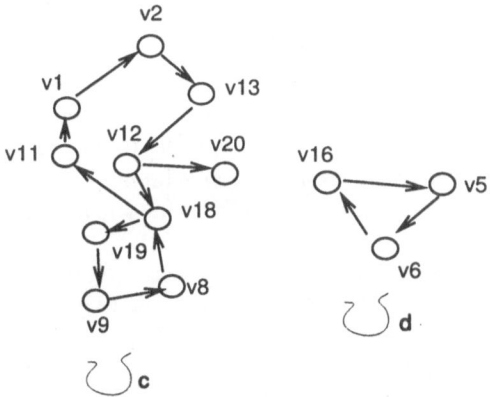

FIGURE 4.14. Example of analyzing inference pattern of FCMs: $\mho_b$ (see Figure 4.13) can be regularly divided into two basic FCMs: $\mho_c$ and $\mho_d$.

After all the state patterns of $\mho_c$ are determined, we can reconstruct inputs for basic FCM $\mho_d$. It is shown in Figure 4.15 as $\mho_f$. The key vertex of $\mho_f$ is v_{16}. From Theorem 32, we have

$$x_{16}(l) = f_{16}(x_{16}(l-3) + x_{20}(l-1)).$$

As the common factor of 2 and 3 is 1, the final state of $\mho_f$ is

$$x_6 \equiv x_{16} \equiv x_5 \equiv 1.$$

With all the state patterns of $\mho_b$ being determined, it is easy to obtain the state pattern for the vertexes $(v_3, v_4, v_{15}, v_{14}, v_7, v_{17}, v_{10})$ in the affected branches of $\mho_a$ by the state of $\mho_b$ and it is omitted here.

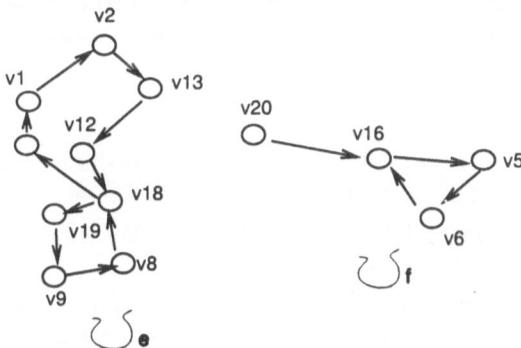

FIGURE 4.15. Example of analyzing inference pattern of FCMs: $\mho_e$ is obtained by deleting the affected branch of $\mho_c$ (see Figure 4.14). $\mho_f$ is derived from $\mho_d$ (see Figure 4.14) by reconstructing the input according to the inference pattern of $\mho_c$.

4.5   Conclusions

In this chapter we briefly presented the structure of FCMs and inference process. We also introduced some interesting modifications to the conventional FCM, in particular, non-linear weight functions, constant time delays, and the weighed FCM combination scheme. These modifications represent the initial and *ad hoc* attempts to tackle the major problems outlined in Section 4.3.

Although in real-world applications, fuzzy cognitive maps can be extremely complex, in Section 4.4 we have shown that we can *regularly* divide a given FCM into basic FCM modules. The inference pattern of a basic FCM is characterized by the state sequence of a few *key* vertexes whose behaviors are described by a general recursive formula.

Indeed, in many decision making processes, we often face the problem of reaching a goal state under some constraints such as contexts, structures, and resources. In such cases, we will have to find whether there exist some initial states (such as certain initiatives) that lead us to the goal state. This turns out to be an NP-hard problem in FCMs. On one hand it represents a major problem in terms of reachability, on the other hand it shows that fuzzy cognitive maps are certainly a rich framework that is applicable in many areas. We are currently investigating *approximate* conditions that enable us to overcome the NP problem. We will further study the dynamic inferential behaviors of fuzzy cognitive maps under the influence of time delays and variable causal relationships between concepts.

Fuzzy cognitive maps represent a very promising inference structure based on causality between concepts, which is particularly important in human-centered *servant modules* that must have the cognitive capability in decision making. However, as discussed in this chapter, many interesting and significant problems still remain open for further investigation, especially, goal-state design and reachability, learning, and dynamic modeling.

4.6   REFERENCES

[1] R. Axelrod, *Structure of Decision: the Cognitive Maps of Political Elites*, Princeton, NJ:Princeton University Press, 1976.

[2] J. A. Dickerson and B. Kosko, "Virtual Worlds as Fuzzy Cognitive Maps," pp.471-477, 1993.

[3] M. Hagiwara, "Extended Fuzzy Cognitive Maps," *Proceedings of International Conference on Fuzzy Systems*, pp.795-801, 1992.

[4] S.H. Kim and K.S. Park, "Fuzzy Cognitive Maps Considering Time Relationships," *International Journal of Human Computer Studies*, Vol.42, Issue 2, pp.157-168, 1995.

[5] B. Kosko, "Fuzzy Cognitive Maps," *International Journal Man-machine Studies*, Vol.24, pp.65-75, 1986.

[6] B. Kosko, "Adaptive Inference in Fuzzy Knowledge Networks," *Proceedings of the First Int. Conf. on Neural Networks*, Vol.2, pp261-268, 1987.

[7] B. Kosko, "Fuzzy System as Universal Approximators," *Proceedings of the 1st IEEE International Conference on Fuzzy Systems*, pp.1153-1162, 1992.

[8] B. Kosko, *Neural Networks and Fuzzy Systems: A Dynamical Systems Approach to Machine Intelligence*, Prentice-Hall, Englewood Cliffs, 1992.

[9] Z.Q. Liu, *Fuzzy Cognitive Networks: Theory and Applications*, Spriner-Verlag, Berling, (to appear) Dec. 2000.

[10] Z.Q. Liu and R. Satur, "Contextual Fuzzy Cognitive Map for Decision Support in Geographic Information Systems," (to appear) *IEEE Trans. Fuzzy Systems*, Sept. 1999.

[11] Z.Q. Liu and Y. Miao, "Fuzzy Cognitive Map and Its Causal Inferences," *Proceedings of IEEE International Conference on Fuzzy Systems*, Vol.3, pp.1540-1545, FUZZ-IEEE'1999, Seoul, S. Korea, Aug.22-25, 1999.

[12] Y. Miao and Z.Q. Liu, "On Causal Inference in Fuzzy Cognitive Maps," (to appear) *IEEE Tran. Fuzzy Systems*, Nov. 1999.

[13] Y. Miao and Z.Q. Liu, "Dynamical Cognitive Network as an Extension of Fuzzy Cognitive Map," (in press) *Proceedings of International Conference on Tools in Artificial Intelligence*, Chicago, IL, USA, Nov. 9-11, 1999.

[14] Y. Miao and Z.Q. Liu, "Dynamical Cognitive Networks," submitted to *IEEE Tran. Fuzzy Systems*, 1999.

[15] T. D. Ndousse and T. Okuda, "Computational Intelligence for Distributed Fault Management in Networks Using Fuzzy Cognitive Maps," *Proceedings of IEEE International Conference on Communications Converging Technologies for Tomorrow's Application*, IEEE New York, Vol.3, pp.1558-1562, 1996.

[16] C. E. Pelaez and J. B. Bowles, "Applying Fuzzy Cognitive Maps Knowledge-Representation to Failure Modes Effects Analysis," *Proceedings of Annual Reliability and Maintainability Symposium*, pp.450-456, 1995.

[17] K. Perusich, "'Fuzzy Cognitive Maps for Policy Analysis," *Proceedings of International Symposium on Technology and Society Technical Expertise and Public Decisions*, IEEE New York, pp.369-373, 1996.

[18] R. Satur, Z.Q. Liu, and M. Gahegan, "Multi-layered Fuzzy Cognitive Map Applied to Context Dependent Learning," *Proceedings of International Joint Conference of 4th IEEE International Conference on*

Fuzzy Systems and 2nd International Fuzzy Engineering Symposium IEEE FUZZ-IEEE/IFES'95, Yokohama, Japan, pp.561-568, March 20-24, 1995.

[19] R. Satur and Z.Q. Liu, "A Context-driven Intelligent Database Processing System Using Object Oriented Fuzzy Cognitive Maps," *International Journal of Intelligent Systems*, Vol.11, No.9, pp.671-689, 1996.

[20] R. Satur and Z.Q. Liu, "A Contextual Fuzzy Cognitive Map Framework for Geographic Information Systems," (to appear) *IEEE Trans. Fuzzy Systems*, Sept. 1999.

[21] M. Schneider, E. Shnaider, A. Kandel, and G. Chew, "Constructing Fuzzy Cognitive Maps," *Proceedings of IEEE International Conference on Fuzzy Systems*, IEEE New York, pp.2281-2288, 1995.

[22] P.C. Silva, "New Forms of Combinated Matrices in Fuzzy Cognitive Maps," *Proceedings of International Conference on Neural Networks*, Vol.2, pp.771-776, 1995.

[23] R.W. Taber and M.A. Siegel, "Estimation of Expert Weights Using Fuzzy Cognitive Maps," *Proceedings of 1st International Conference on Neural Networks*, Vol.2, pp.319-325, 1987.

[24] R.W. Taber, "Knowledge Processing with Fuzzy Cognitive Maps," *Expert Systems with Applications*, Vol.2, Issue 1, pp.83-87, 1991.

[25] W.A. Woods, "Important Issues in Knowledge Representation," *Knowledge-Based Systems: Fundamentals and Tools*, O. N. Garcia and Y. Chien (Eds), IEEE Computer Society Press, 1991.

5

Methods in Hard and Fuzzy Clustering

Sadaaki Miyamoto
Kazutaka Umayahara

5.1 Introduction

Clustering, also referred to as cluster analysis, is a class of unsupervised classification methods for data analysis. There have been numerous studies of clustering, which are both theoretical and applicational. Applications to scientific classifications, engineering problems, behavioral sciences, etc., have been investigated and usefulness of this technique has been appreciated.

From the end of 1960's to the early 1970's, fuzzy set theory [47] was applied to clustering and a number of fuzzy clustering techniques have been developed [2, 3, 8, 9, 39, 48]. We can divide the methods into two main categories: (1) nonhierarchical fuzzy clustering and (2) hierarchical fuzzy clustering in accordance with the corresponding classes of methods in hard clustering [1]. As we will see later, the roles of fuzziness in these two categories are different.

The best-known method of fuzzy nonhierarchical clustering is the fuzzy c-means [3, 8, 9]. Fuzzy c-means refers to a class of algorithms. For example, recently new methods of fuzzy c-means have been proposed using regularization, which we will discuss below.

In contrast, the transitive closure [42, 48] induces a hierarchical classification. It has been shown [10] that this method is equivalent to a well-known method of the single link [1, 11] in ordinary agglomerative clustering. Thus the fuzzy hierarchical method is not a generalization of a crisp technique, but providing a new viewpoint for agglomerative clustering.

This chapter is organized as follows. First, we give a brief review of basic methods in clustering, namely, the hard c-means and the single link in agglomerative methods. The single link is described as an option of a general algorithm for agglomerative clustering.

Second, we introduce the fuzzy c-means and its variants. We emphasize fuzzy classification functions, which are important in the development of fuzzy clustering algorithms and often inadequately discussed in the literature.

In addtion, we review some major techniques in hard clustering such as the vector quantization [17] and the EM algorithm for a mixture density model [37].

It is also shown that the transitive closure of a fuzzy relation is the algebraic representation of the single link, whereas the connected component of a fuzzy graph shows the geometric aspect of the same method.

5.2   Basic Methods in Clustering

Clustering in general implies that objects to be classified should be put into a number of classes on the basis of their mutual distance. For a given distance, objects with smaller distances should be given the same class label, whereas those with larger distances should be put into different classes. This means that computation of the distance between objects is a integral part of the solution to a given clustering problem. There have been many studies of this problem of measuring and defining distances between objects in the literature. Readers may refer to standard textbooks [1, 3, 11, 14].

Clustering techniques can be divided into a number of classes, among which a most convenient way herein is to mention hierarchical and non-hierarchical clustering.

Notice that our purpose is not to show various techniques, but to focus on fuzzy and related methods. In accordance with this purpose, we begin by discussing two basic methods in hard clustering, one is the hard c-means in nonhierarchical clustering, and another is the single link in hierarchical clustering. As the single link has been considered to be an option in a general agglomerative procedure which is a subclass of hierarchical clustering, we refer to agglomerative clustering instead of hierarchical clustering.

Objects denoted by $X = \{x_1, \dots, x_n\}$ are classified into clusters which are subsets $G_1, \dots, G_c$ of X. In classical methods such as the hard c-means, clusters form a *partition* of X, namely,

$$\bigcup_{i=1}^{c} G_i = X, \qquad G_i \cup G_j = \emptyset \quad (i \neq j). \tag{5.1}$$

In contrast, fuzzy clustering does not necessarily assume such a partition. Indeed, the method of fuzzy c-means uses a fuzzy partition, as we will see below.

It should also be noticed that an object x_i is not necessarily in the Euclidean space in general. This should in particular be remarked when an agglomerative method such as the single link, the complete link, *etc.*, is used which do not require the assumption of the Euclidean space.

In contrast, the method of c-means assumes the Euclidean space. In the latter case, x_i is assumed to be a point in p-dimensional Euclidean space:

$$x_i = (x_i^1, \dots, x_i^p) \in \mathbf{R}^p, \quad i = 1, \dots, n.$$

Terms and symbols

Let us review symbols and terms used in clustering. As noted earlier, the objects or individuals for clustering is $X = \{x_1, \dots, x_n\}$. The number of objects is thus n. Generic elements in X are sometimes denoted by $x, y \in X$. The collection of clusters are $\mathcal{G} = \{G_1, \dots, G_c\}$, i.e., the number of clusters is c. Generic elements of $\mathcal{G}$ are denoted by $G, G' \in \mathcal{G}$.

We have mentioned in the beginning that mutual distance is used for clustering. A distance between two objects $x, y \in X$ is a real-valued measure $d \colon X \times X \to \mathbf{R}$. Most distance satisfies

(i) $d(x, y) \geq 0$ and $d(x, x) = 0$.

(ii) $d(x, y) = d(y, x)$.

The triangular inequality

$$d(x, z) \leq d(x, y) + d(y, z)$$

is not necessarily satisfied. The word of *dissimilarity* is also used for distance interchangeably.

Sometimes a similarity measure denoted by $s(x, y)$ is used instead of a distance. A similarity measure means that a large value of $s(x, y)$ implies x and y are similar, while a small value of $d(x, y)$ implies x and y are not similar. There is no theoretical difference between $d(x, y)$ and $s(x, y)$, since a strictly monotone decreasing function $h \colon \mathbf{R} \to \mathbf{R}$ will transform a distance into a similarity and vice versa: $s(x, y) = h(d(x, y))$ or $d(x, y) = h(s(x, y))$. Nevertheless, we will use the both measures in accordance with the context for the sake of simplicity.

In particular, a similarity is assumed to be normalized hereafter:

(i) $0 \leq s(x, y) \leq 1$ and $s(x, x) = 1$.

(ii) $s(x, y) = s(y, x)$.

Such general measures are used in hierarchical clustering, while the basic method of c-means in the nonhierarchical clustering assumes the square of the Euclidean norm for measuring the distance between two objects $x = (x^1, \dots, x^p)$ and $y = (y^1, \dots, y^p)$:

$$d(x, y) = \|x - y\|^2 = \sum_{j=1}^{p} (x^j - y^j)^2.$$

Some studies assume the L_1 metric which is also called the Manhattan distance or city block distance:

$$d(x, y) = \|x - y\|_1 = \sum_{j=1}^{p} |x^j - y^j|.$$

A class of clustering algorithms is based on the idea of optimizing an objective function. The methods are accordingly called *optimal clustering*. Optimization problems considered in the sequel are written like

$$\min_{s \in S} \mathcal{F}(s)$$

in which $\mathcal{F}$ is the objective function and S is the admissible region of the feasible solutions. Sometimes the admissible region is described by constraints such as $S = \{x : g_k(x) \leq 0, k = 1, \ldots, K\}$.

When $\bar{s}$ is the optimal solution for this problem:

$$\mathcal{F}(\bar{s}) = \min_{s \in S} \mathcal{F}(s),$$

we also write

$$\bar{s} = \arg\min_{s \in S} \mathcal{F}(s).$$

The term of cluster centers will frequently be used below. A coordinate of a cluster center is given by a weighted average of the coordinates of the objects in the cluster. A typical case is the mean value, i.e., the center v for a cluster G is given by

$$v = \frac{1}{|G|} \sum_{x \in G} x \tag{5.2}$$

where $|G|$ is the number of elements in G. In this case v is called the center of gravity or *centroid* for the cluster. The centroid for G is also denoted by $M(G)$.

Hard c-means

Hard c-means should not be understood as a single algorithm. Rather, there are different versions of hard c-means. Indeed, this term implies a family of clustering methods that use c points of cluster centers which are more or less *means* or centroids of the clusters. As MacQueen [21] described, objects are allocated to the nearest centers of the clusters and centers are updated by this method.

Remark. Hard c-means is referred to as k-means in most literature of nonfuzzy clustering. In contrast, the symbol c is used instead of k in discussing fuzzy clustering. Both symbols imply the number of clusters. As the principal aim here is to discuss fuzzy clustering, we use the term of c-means.

A simple algorithm of the hard c-means is as follows.

Algorithm F

F1 c cluster centers are, or equivalently, an initial partition is randomly generated.

F2 Each object is allocated to the cluster of the nearest center.

F3 If no change of cluster memberships for all objects, stop. Otherwise calculate the centroids for the c new clusters, and go to **F2**.

This simple algorithm is the starting point for our discussion of fuzzy c-means. Notice that the number c of clusters should be specified, as is the case for most nonhierarchical clustering methods.

Remark that the centroid $M(G)$ minimizes the sum of squared distance for all objects in G:

$$M(G) = \arg\min_{v} \sum_{x \in G} \|x - v\|^2. \tag{5.3}$$

Moreover the nearest center allocation in **F2** is

$$x_k \in G_\ell \iff v_\ell = \arg\min_{q} \|x_k - v_q\|^2. \tag{5.4}$$

These allocation and calculation thus solve the above optimization problems.

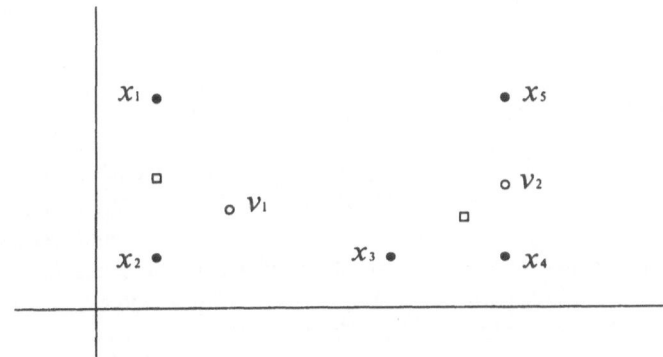

FIGURE 5.1. A simple example for hard c-means

Example 1. Let us consider a simple example of dividing five objects on a plane into two clusters. Figure5.1 shows the objects by $x_1, \ldots, x_5$. Suppose $G_1 = \{x_1, x_2, x_3\}$ and $G_2 = \{x_4, x_5\}$ are initial clusters in **F1**. Then the two centroids v_1 and v_2 are shown by the small circles.

In **F2**, we find that x_3 is nearer to v_2, and cluster membership is changed. The new centers should thus be calculated in **F3**, which are shown by small squares.

Repeating **F2**, we observe the cluster membership is not changed. Hence we have $G_1' = \{x_1, x_2\}$ and $G_2' = \{x_3, x_4, x_5\}$.

Notice that centroids are updated after all objects are reallocated in **F2**. Another algorithm updates a centroid immediately after the membership of an object is changed.

Algorithm HCM

HCM0 An initial partition is randomly generated and centroids $v_1, \ldots, v_c$ are calculated.

HCM2 For all x_k, $k = 1, \ldots, n$, repeat **HCM2-1**.

HCM2-1 Assume $x_k \in G_q$. Calculate

$$v_r = \arg \min_{1 \le j \le c} \|x_k - v_j\|$$

If $r \neq q$, update v_r, v_q, G_r and G_q as follows:

$$v_r := \frac{|G_r|}{|G_r| + 1} v_r + \frac{x_k}{|G_r| + 1}$$

$$v_q := \frac{|G_r|}{|G_r| - 1} v_q - \frac{x_k}{|G_r| - 1}$$

$$G_r := G_r \cup \{x_k\}$$

$$G_q := G_q - \{x_k\}$$

HCM3 If no change of membership for all x_k, $k = 1, \ldots, n$, stop, else go to **HCM2**.

Let us consider Example 1 again. We apply **HCM** with the same initial partition. For x_1 and x_2, the centroids do not change. When the algorithm processes x_3, this object moves from G_1 to G_2, and the centers and cluster memberships in **HCM2-1** are updated. The centroids are moved to the positions of the small squares. After that, x_4 and x_5 do not change memberships. By the condition in **HCM3**, **HCM2** is repeated again, and this time no change of the memberships occurs for any object, and thus the algorithm terminates.

These two algorithms are related to more recent methods based on different ideas. Namely, the former (algorithm **F**) leads us to fuzzy c-means, while the latter (algorithm **HCM**) is compared with a vector quantization method [17].

Moreover there are other versions of the hard c-means algorithms. We omit the detail of them; they are described in standard literature [1, 7, 14].

Agglomerative clustering

Hierarchical clustering techniques are another class of methods most frequently used in various applications. Hierarchical techniques are further divided into agglomerative methods and divisive methods [11]. Agglomerative methods have more frequently been used and moreover in relation to fuzzy set theory, we require this class of techniques alone.

In agglomerative hierarchical clustering, no strict optimization of an objective function is attempted, Furthermore, the number of clusters is not fixed. Rather, agglomerative methods show us the process by which clusters are merged one by one into larger clusters.

Well-known agglomerative methods such as the single link, the complete link, the average link, the centroid method, and the Ward method [1, 11] can be described as options in a general procedure.

Remark again that some methods accept general distance measures (and also similarity measures) regardless whether they are the Euclidean or not. In contrast, the other class of methods are based on the Euclidean space. In agglomerative techniques, the single link, the complete link, and the average link can use general measures, while the centroid method and the Ward method are in the latter class. We discuss the former three method and omit the other two. Notice also that a method which accepts general measures of similarity or distance will be called distance/similarity based.

We hereafter consider similarity $s(x_i, x_j)$ instead of dissimilarity, since similarity is more convenient for discussing the relation between the single link and the transitive closure of a fuzzy relation. Remark that a similarity is obtained from dissimilarity by an appropriate transformation $s(x_i, x_j) = h(d(x_i, x_j))$ For example, $s(x_i, x_j) = 1/d(x_i, x_j)$ or $s(x_i, x_j) = \max_{p,q} d(x_p, x_q) - d(x_i, x_j)$.

The characteristic of an agglomerative method is that a pair of clusters $G, G' \in \mathcal{G}$ is selected and merged at a time, whereby the number of clusters is reduced by one: $K = K - 1$. (The number of clusters is not specified and the variable parameter K is used.) The merging is repeated: the initial clusters are the objects themselves (i.e., $\mathcal{G} = \{\{x_1\}, \dots, \{x_n\}\}$, $K = n$), and the final cluster is the set X ($\mathcal{G} = X$, $K = 1$). The process of pairwise merging of clusters in the agglomerative procedure is exhibited by a tree called a dendrogram [1, 11, 14]. The way of forming a dendrogram is too complicated to be shown here (see, e.g., [24]), but how the merging proceeds is described. The key is to define a similarity between clusters G and G', which is denoted by $s(G, G')$. There are different methods of defining similarity between clusters, which is the option of updating the similarity in **AC3** in the following.

Agglomerative Clustering Procedure AC

AC1 (initial partition)

Let $K = n$, $G_i = \{x_i\}$, and

$$s(G_i, G_j) = s(x_i, x_j)$$

for all i, j.

AC2 (merge)
Find

$$s(G_p, G_q) = \max_{i;j} s(G_i, G_j) \tag{5.5}$$

and merge:

$$G_r = G_p \cup G_q.$$

AC3 (update similarity)
Calculate $s(G_r, G_i)$ for all clusters $i \neq r$; $K = K - 1$; if $K = 1$, stop, else go to **AC2**.

In the single link method, $s(G_r, G_i)$ in **AC3** is calculated by

$$s(G_r, G_i) = \max\{s(G_p, G_i), s(G_q, G_i)\}. \tag{5.6}$$

If we use the complete link method, the similarity in **AC3** should be updated by

$$s(G_r, G_i) = \min\{s(G_p, G_i), s(G_q, G_i)\}. \tag{5.7}$$

The average link method uses

$$s(G_r, G_i) = \frac{|G_p|}{|G_r|} s(G_p, G_i) + \frac{|G_q|}{|G_r|} s(G_q, G_i). \tag{5.8}$$

Remark. When we use a distance $d(x, y)$, the symbol $s(\cdot, \cdot)$ should be $d(\cdot, \cdot)$, and (5.5) is replaced by

$$d(G_p, G_q) = \min_{i,j} d(G_i, G_j). \tag{5.9}$$

Moreover the updating formulas are as follows.

(the single link)

$$d(G_r, G_i) = \min\{d(G_p, G_i), d(G_q, G_i)\}.$$

(the complete link)

$$d(G_r, G_i) = \max\{d(G_p, G_i), d(G_q, G_i)\}.$$

(the average link)

$$d(G_r, G_i) = \frac{|G_p|}{|G_r|} d(G_p, G_i) + \frac{|G_q|}{|G_r|} d(G_q, G_i).$$

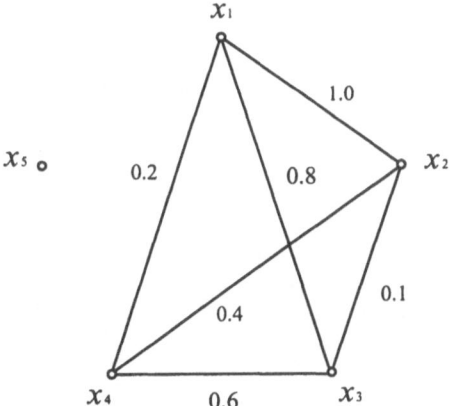

FIGURE 5.2. A simple example for the single link

Example 2. Let us consider the five objects $\{x_1, \ldots, x_5\}$ shown by dots in Fig.5.2. These are not the points on the Euclidean plane, but vertices of a weighted graph. A number on a edge between two vertices x, y implies the value of the similarity measure $s(x, y)$. Thus, $s(x_1, x_2) = 1.0$, $s(x_1, x_3) = 0.8$, and so on. The vertex x_5 is not connected to any other vertices, which means $s(x_5, x_k) = 0$, $k = 1, 2, 3, 4$.

Applying the procedure **AC**, we find the initial partition is $G_i = \{x_i\}$, $k = 1, \ldots, 5$. In **AC2**, we find the pair $G_1 = \{x_1\}$ and $G_2 = \{x_2\}$ is most similar, and hence merged $G'_1 := G_1 \cup G_2$. Then the similarity between clusters are updated by (5.6). We have $s(G'_1, G_3) = 0.8$, $s(G'_1, G_4) = 0.4$, and $s(G'_1, G_5) = 0$. Next pair to be merged is G'_1 and G_3, and we have $G''_1 = G'_1 \cup G_3$, $s(G''_1, G_4) = 0.6$, $s(G''_1, G_5) = 0$. These merging process is expressed by a tree (dendrogram) shown as Fig.5.3. In this tree the leaves are objects and a branch implies that the corresponding clusters are merged. The vertical coordinate of a branch is the level of the similarity measure at which the merging occurs. Thus, $\{x_1\}$ and $\{x_2\}$ are merged at the level 1.0, $\{x_1, x_2\}$ and $\{x_3\}$ are merged at the level 0.8, and so on.

It is not surprising that a weighted graph is considered in this example. As we will see later, the fundamental structure in the single link is a fuzzy graph.

5.3   Fuzzy c-Means

Although there have been many proposals for methods of fuzzy clustering [2, 3, 14, 39], many studies are concentrated on fuzzy c-means.

To begin with, we note that the algorithm **F** can be formulated as an

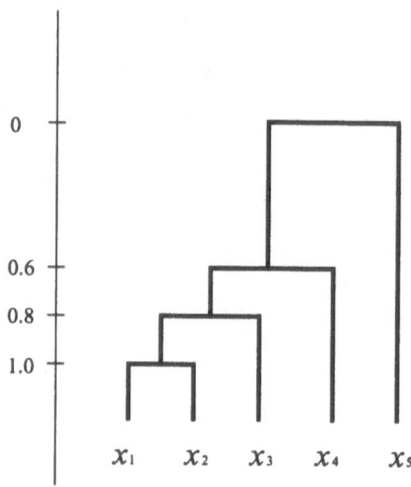

FIGURE 5.3. The dendrogram for the example in Fig.5.2.

optimization problem, as Bezdek [3] discussed.

Recall that n objects to be clustered are represented by points $x_k = (x_k^1, \ldots, x_k^p)$, $k = 1, \ldots, n$, in a p-dimensional Euclidean space. Let $U = (u_{ij})$ be $c \times n$ binary valued matrix, $v_i = (v_i^1, \ldots, v_i^p)$ is the center for cluster i and $V = (v_1, \ldots, v_c)$. (v_i is not necessarily the centroid of a cluster hereafter.)

Consider the objective function

$$J_1(U, V) = \sum_{k=1}^{n} \sum_{i=1}^{c} u_{ik} \|x_k - v_i\|^2. \tag{5.10}$$

($\|x_k - v_i\|$ is the Euclidean norm as noted before).

Also let

$$M_c = \{(u_{ik}) | u_{ik} \in \{0, 1\}, \sum_{i=1}^{c} u_{ik} = 1 \text{ for all } k\}.$$

Let us put $J(U, V) = J_1(U, V)$ and $M = M_c$. Consider the following procedure.

Procedure FC

FC1 Generate an initial $\bar{U}$ and $\bar{V}$.

FC2 Solve

$$\min_{U \in M} J(U, \bar{V})$$

and let the optimal solution be $\bar{U}$.

FC3 Solve

$$\min_{V} J(\bar{U}, V)$$

and let the optimal solution be $\bar{V}$.

FC4 If the solution $(\bar{U}, \bar{V})$ is convergent, stop. Otherwise, go to **FC2**.

It is not difficult to see that the algorithm **F** is equivalent to the procedure **FC**. Indeed, given $\bar{V} = (\bar{v}_1, \dots, \bar{v}_c)$, the optimal solution in **FC2** is given by

$$\bar{u}_{ik} = 1 \iff \bar{v}_i = \arg\min_{\ell} \|x_k - \bar{v}_\ell\| \tag{5.11}$$

$$\bar{u}_{jk} = 0 \quad j \neq i. \tag{5.12}$$

The optimal solution $\bar{V}$ given $\bar{U}$ is

$$\bar{v}_i = \frac{\displaystyle\sum_{k=1}^{n} \bar{u}_{ik} x_k}{\displaystyle\sum_{k=1}^{n} \bar{u}_{ik}}$$

If we define

$$G_i = \{x_k \in X : \bar{u}_{ik} = 1\}$$

Then the equation for $\bar{U}$ means the allocation to the nearest center and $\bar{v}_i$ is the centroid of G_i.

It should be remarked that the above alternative optimization procedure **FC** is used throughout the description of different methods of fuzzy c-means. Indeed, the fundamental idea of fuzzy c-means is to alter the objective function and the constraint. Different options for $J(U, V)$ and M produce various methods of fuzzy c-means.

Standard fuzzy c-means

In the standard method, the binary valued matrix U is generalized into real-valued matrix: Thus the constraint is the next set:

$$M_f = \{(u_{ik}) | u_{ik} \in [0,1], \sum_{i=1}^{c} u_{ik} = 1 \text{ for all } k\}.$$

If we use J_1 and M_f in **FC**, we still have crisp optimal solutions, since the objective function J_1 is linear with respect to u_{ik}, and the optimization in **FC2** is a linear programming. The optimal solution is attained at an extremal point, and hence a crisp solution is given by (5.11) and (5.12).

For finding an interesting *fuzzification*, to make the objective function nonlinear is thus necessary, Bezdek [3] and Dunn [8, 9] have thus introduced a parameter $m(> 1)$ into the objective function:

$$J_m(U, V) = \sum_{k=1}^{n} \sum_{i=1}^{c} (u_{ik})^m \|x_k - v_i\|^2. \tag{5.13}$$

The standard fuzzy c-means algorithm [3, 8, 9] is **FC** using $J(U, V) = J_m(U, V)$ and $M = M_f$.

It is well-known that the solution in **FC2** and that in **FC3** are

$$u_{ik} = \left[\sum_{j=1}^{c} \left(\frac{\|x_k - \bar{v}_i\|^2}{\|x_k - \bar{v}_j\|^2} \right)^{\frac{1}{m-1}} \right]^{-1}, \tag{5.14}$$

$$v_i = \frac{\sum_{k=1}^{n} (\bar{u}_{ik})^m x_k}{\sum_{k=1}^{n} (\bar{u}_{ik})^m}, \tag{5.15}$$

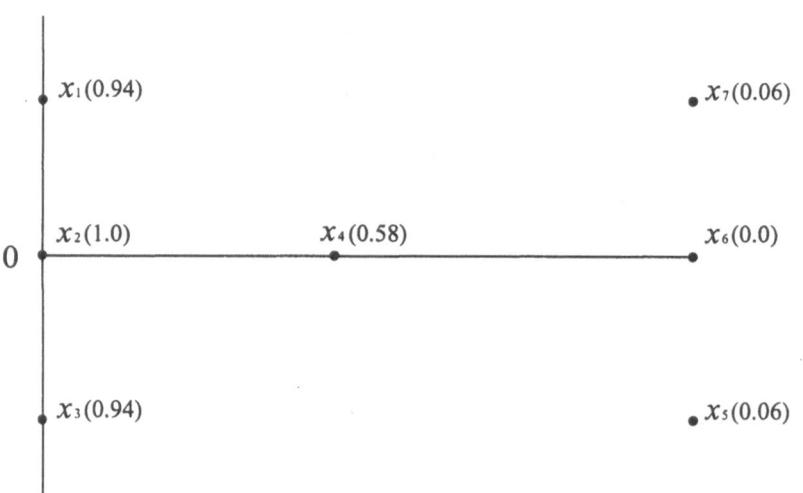

FIGURE 5.4. Seven objects on a plane and two fuzzy clusters

Example 3. Figure 5.4 show seven points on a plane for clustering. We divide them into two fuzzy clusters, that is, $n = 7$ and $c = 2$. The numbers in the parentheses are membership values for a cluster, say G_1, which have been derived from the standard method of fuzzy c-means. The membership for the other cluster is $u_{2k} = 1 - u_{1k}$, $k = 1, \ldots, 7$.

Methods of regularization

Recently, Li and Mukaidono [20] have proposed a new method of fuzzy c-means using the concept of maximum entropy [46]. Similar discussions are found in Masulli *et. al.* [22] and Rose *et. al.* [38].

We discuss the entropy method from the viewpoint of regularization of an ill-posed problem and show the usefulness of this concept.

Regularization is an old technique to solve ill-posed problems of functional equations [43, 44] and has been applied to many real problems. Regularization in general implies modification of a given problem that is singular in some sense into a regular problem. The singular problem is difficult to solve but the latter problem is easier to handle. The latter problem is called the regularization of the original problem when the solution of the regularized problem approximates the original solution.

We recognize fuzzy c - means to be a regularization for the crisp c - means. It is strange to say that the well-known method of the hard c - means is singular, since the solution of the crisp c - means is by no means ill-posed. Although the crisp c - means is a well-posed problem, the crisp solution is characterized by the extremal points when the optimal clustering is generalized into the fuzzy membership case [3]. We regard extremal points as singular solutions, whereas intermediate points are considered to be regular solutions. The fuzzification by using the parameter m is thus a regularization.

Let us consider if there are other methods of regularizing the function J_1. A typical regularization in the optimization problem is to use an additional term. An original objective function, say J', is regularized by using an additional function K and a positive parameter λ:

$$J^\lambda = J' + \lambda^{-1} K.$$

We find this type of regularization is also useful in fuzzy clustering. In applying this idea to fuzzy c-means, we should specify J' and K. The 'hard' objective function J_1 should obviously be J'; We consider two options for the regularizing function K.

Regularization by entropy

Miyamoto and Mukaidono [28] have reformulated the maximum entropy method [20] by using the regularization. Namely, they put $J' = J_1$ and

$$K = \sum_{k=1}^{n} \sum_{i=1}^{c} u_{ik} \log u_{ik}.$$

In other words,

$$J^\lambda = \sum_{k=1}^{n} \sum_{i=1}^{c} u_{ik} \|x_k - v_i\|^2 + \lambda^{-1} \sum_{k=1}^{n} \sum_{i=1}^{c} u_{ik} \log u_{ik}.$$

Accordingly they use the procedure **FC** with $J = J^\lambda$ and $M = M_f$. The solutions in **FC2** and that in **FC3** are

$$u_{ik} = \frac{e^{-\lambda\|x_k - \bar{v}_i\|^2}}{\sum\limits_{j=1}^{c} e^{-\lambda\|x_k - \bar{v}_j\|^2}}, \tag{5.16}$$

$$v_i = \frac{\sum\limits_{k=1}^{n} \bar{u}_{ik} x_k}{\sum\limits_{k=1}^{n} \bar{u}_{ik}}. \tag{5.17}$$

Proof for the solution in FC2

Let us relax the constraint to $\sum_i u_{ik} = 1$, i.e., by deleting $0 \le u_{ik} \le 1$. Moreover put $d_{ik} = \|x_k - v_i\|^2$ for simplicity.

Define the Lagrangean

$$L = \sum_{k=1}^{n}\sum_{i=1}^{c} u_{ik} d_{ik} + \lambda^{-1} \sum_{k=1}^{n}\sum_{i=1}^{c} u_{ik} \log u_{ik} + \sum_{k=1}^{n} \nu_k \left(\sum_{i=1}^{c} u_{ik} - 1\right).$$

where ν_k is the Lagrange multipliers.

From

$$\frac{1}{2}\frac{\partial L}{\partial u_{ik}} = d_{ik} + \lambda^{-1}(1 + \log u_{ik}) + \nu_k = 0,$$

we have

$$u_{ik} = \exp(-\lambda d_{ik} - \lambda \nu_k - 1). \tag{5.18}$$

Remark that the above solution satisfies $0 \le u_{ik} \le 1$. Hence (5.18) provides the solution for the stationary point under the constraint M_f.

Determining ν_k from $\sum_i u_{ik} = 1$, we obtain the optimal solution (5.16). Notice that the objective function is convex with respect to each u_{ik}.

Regularization by quadratic term

Miyamoto and Umayahara [31] propose

$$K = \frac{1}{2} \sum_{k=1}^{n}\sum_{i=1}^{c} u_{ik}^2.$$

Thus, the optimization of

$$J^\lambda = \sum_{k=1}^{n}\sum_{i=1}^{c} u_{ik}\|x_k - v_i\|^2 + \frac{1}{2}\lambda^{-1}\sum_{k=1}^{n}\sum_{i=1}^{c} u_{ik}^2.$$

is considered in **FC** with $M = M_f$. The solution in **FC3** is given by (5.17), but the optimal U is not expressed by a simple formula. Rather, it is calculated by an algorithm.

Derivation of the optimal solution in FC2

First we derive the necessary condition for the optimal solution. Put $w_{ik}^2 = u_{ik}$. Then the constraint is represented by the single equation $\sum_i w_{ik}^2 = 1$.

Define the Lagrangean:

$$L = \sum_{k=1}^n \sum_{i=1}^c w_{ik}^2 d_{ik} + \frac{1}{2}\lambda^{-1} \sum_{k=1}^n \sum_{i=1}^c w_{ik}^4 - \sum_{k=1}^n \mu_k \left(\sum_{i=1}^c w_{ik}^2 - 1\right)$$

$(d_{ik} = \|x_k - v_i\|^2)$.

From

$$\frac{1}{2}\frac{\partial L}{\partial w_{ik}} = w_{ik}(d_{ik} + \lambda^{-1}w_{ik}^2 - \mu_k) = 0,$$

we have $w_{ki} = 0$ or $w_{ki}^2 = \lambda(\mu_k - d_{ki})$. In terms of u_{ki},

$$u_{ki} = 0 \quad \text{or} \quad u_{ki} = \lambda(\mu_k - d_{ki}). \tag{5.19}$$

Notice that the right hand side of the last equation should be nonnegative.

The above solution has been derived from the necessary condition for optimality. Hence we should find the optimal solution from the set of u_{ki} that satisfies (5.19).

Simplification of the problem is convenient in order to find the optimal solution. Let

$$J^{(k)}(U, v) = \sum_{i=1}^c u_{ki}d_{ki} + \frac{1}{2}\lambda^{-1} \sum_{i=1}^c u_{ki}^2 \tag{5.20}$$

Then, $J^\lambda = \sum_{k=1}^n J^{(k)}$ and each $J^{(k)}$ can independently be minimized from other $J^{(k')}$. Hence we need only consider optimization of $J^{(k)}$ for a fixed k. We assume

$$d_{k,k_1} \leq \cdots \leq d_{k,k_c} \tag{5.21}$$

and moreover we write

$$d_{k1} \leq d_{k2} \leq \cdots \leq d_{kc} \tag{5.22}$$

instead of (5.21) for simplicity.

Suppose that we have found u_{ki}, $i = 1, \ldots, c$, that satisfies the condition (5.19). Let I be a set of indices such that u_{ki} is positive, i.e.,

$$I = \{i : u_{ki} > 0\}, \tag{5.23}$$

and $|I|$ be the number of elements in I. Then, from (5.19) and $\sum_i u_{ki} = 1$, we have

$$\lambda |I| \mu_k - \lambda \sum_{\ell \in I} d_{k\ell} = 1. \tag{5.24}$$

This implies that, for $i \in I$,

$$u_{ki} = |I|^{-1}(\lambda \sum_{\ell \in I} d_{k\ell} + 1) - \lambda d_{ki} > 0 \tag{5.25}$$

while $u_{ki} = 0$ for $i \notin I$.

In this way, the problem is to find the optimal choice of the index set I, since when I is found, it is easy to calculate the membership values by (5.25). It can be proved that the optimal choice of the index set has the form of $I = \{1, \dots, K\}$ for some $K \in \{1, 2, \dots, c\}$. Moreover let

$$f_\ell^L = \lambda(\sum_{j=1}^{L}(d_{kj} - d_{k\ell})) + 1, \quad \ell = 1, \dots, L. \tag{5.26}$$

Then whether or not f_L^L is positive determines the optimal index set. The details of the discussion is found in [31], which is omitted here, and the final algorithm is given as follows.

Algorithm for the optimal solution in FC2

Under the assumption (5.22), the solution that minimizes $J^{(k)}$ is given by the following procedure.

1. Calculate f_L^L for $L = 1, \dots, c$ by (5.26). Let $\bar{L}$ be the smallest number such that $f_{L+1}^{L+1} \leq 0$. (Namely, $I = \{1, 2, \dots, \bar{L}\}$ is the optimal index set.)

2. For $i = 1, \dots, \bar{L}$, put

$$u_{ki} = \bar{L}^{-1}(\lambda \sum_{\ell=1}^{L} d_{k\ell} + 1) - \lambda d_{ki}$$

and for $i = \bar{L} + 1, \dots, c$, put $u_{ki} = 0$.

Classification functions

It is necessary to remark classification rules or classification functions induced from hard and fuzzy c-means.

Classification in this section means that a new object is observed and should be classified after clusters of former objects have been established. It should be noticed that the hard c-means are based on the nearest centroid classification, as shown in algorithm **F**. Indeed, this algorithm can briefly rewritten as follows.

F1 Generate initial clusters.

F2 *Nearest allocation rule* is applied.

F3 If not convergent, *update the cluster center using the centroids*, and go to **F2**.

Thus, when a new object x is observed, it should be allocated to the cluster of the nearest center. (For simplicity we assume that the center nearest to x is unique.)

This implies that fuzzy c-means induce fuzzy classification rules which are determined by the membership calculation rules. Namely, in the standard fuzzy c-means, the rule by which x is allocated to cluster i is given by the function:

$$U_i^s(x) = \left[\sum_{j=1}^{c} \left(\frac{\|x - \bar{v}_i\|^2}{\|x - \bar{v}_j\|^2} \right)^{\frac{1}{m-1}} \right]^{-1} \qquad x \neq \bar{v}_i, \qquad (5.27)$$

$$U_i^s(x) = 1 \qquad x = \bar{v}_i. \qquad (5.28)$$

This function has already been used by Keller *et. al.* [16] as a fuzzy classification rule. It is obvious to see that $U_i^s(x)$ interpolates the membership values for $x_1, \ldots, x_n$:

$$U_i^s(x_k) = u_{ik}, \quad k = 1, \ldots, n.$$

If we rewrite the algorithm **FC** using the classification function, we have

FC'1 Generate initial clusters.

FC'2 Fuzzy classification function $U_i^s(x_k)$ is used to determine the membership u_{ik}.

FC'3 Calculate cluster centers by (5.15).

FC'4 If not convergent, go to **FC'2**.

For the entropy method, we have the classification function in the same way:

$$U_i^e(x) = \frac{e^{-\lambda \|x - \bar{v}_i\|^2}}{\sum_{j=1}^{c} e^{-\lambda \|x - \bar{v}_j\|^2}}. \qquad (5.29)$$

This function also has the property of interpolating memberships.

We can write the procedure **FC'** for the entropy method: use $U_i^e(x_k)$ in **FC'2** and (5.17) in **FC'3**.

For the quadratic regularization method, the classification function is calculated by an algorithm. For a given point $x \in R^p$, assume

$$d(x, v_{j_1}) \leq d(x, v_{j_2}) \leq \cdots \leq d(x, v_{j_c}).$$

where $d(x, v_{j_\ell}) = \|x - v_{j_\ell}\|^2$.

Calculate

$$f_i^L(x) = \lambda \sum_{\ell=1}^{L} (d(x, v_{j_\ell}) - d(x, v_i)) + 1$$

for $L = 1, \ldots, c$. Let $\bar{L}$ be the first L such that $f_i^{L+1}(x) \le 0$. Then, let

$$U_i^q(x) = \bar{L}^{-1}(\lambda \sum_{\ell=1}^{L} d(x, v_{j_\ell}) + 1) - \lambda d(x, v_i)$$

for $i = 1, \ldots, \bar{L}$ and $U_i^q(x) = 0$ for $i = \bar{L} + 1, \ldots, c$,

We thus have obtained two views of the membership u_{ik}: one is the optimal solution and the other is the classification function.

We can now investigate theoretical properties of the classification functions, by which characteristics of the above methods of fuzzy c-means are disclosed [32]. Namely, the locations of the maximum of the classification functions and their values as $\|x\| \to \infty$ are considered. Moreover properties of the classification functions are related to the Voronoi diagram in computational geometry [35]. (Kohonen [17] mentions the Voronoi diagram in regard to classification using the vector quantization and the learning vector quantization.)

Notice first that the maximum value of $U_i^s(x)$ is at the center v_i by the definition, whereas the maximum value of $\hat{U}_i^e(x)$ is not necessarily at v_i.

Define regions in R^p in which the value of the classification function for class i is greatest among $j = 1, \ldots, c$:

$$\begin{aligned} S_i &= \{y \in R^p : U_i^s(y) > U_j^s(y), \ j \ne i\}, \\ E_i &= \{y \in R^p : U_i^e(y) > U_j^e(y), \ j \ne i\}. \end{aligned}$$

The nearest prototype allocation used in the hard c-means implies that the region in which an object y is allocated to the cluster i is given by

$$K_i = \{y \in R^p : \|y - v_i\| > \|y - v_j\|, \ j \ne i\}.$$

It should be remarked that not only K_i but also S_i and E_i are the Voronoi sets [17, 35] in R^p with the respective centers v_i.

We note a difference between the behaviors of the two functions as $\|x\| \to \infty$.

Proposition 1. Assume that S_i and E_i are unbounded. Then we can move x toward infinity ($\|x\| \to \infty$) in S_i and E_i. Then,

$$\lim_{\|x\| \to \infty} U_i^s(x) = \frac{1}{c}, \tag{5.30}$$

$$\lim_{\|x\| \to \infty} U_i^e(x) = 1. \tag{5.31}$$

Another notable property in the entropy regularization is that the classification function has convex level sets.

Proposition 2. $U_i^e(x)$ is a convex fuzzy set, in other words, an arbitrary α-cut for $U_i^e(x)$

$$\{U_i^e(\cdot)\}_\alpha = \{x \in \mathbf{R}^p \mid \hat{U}_i^e(x) \geq \alpha\}$$

is a convex set.

(Proof for Propositions 1 and 2) The following equation (where $\langle \cdot, \cdot \rangle$ is the scalar product) makes it easier to see that these Propositions hold.

$$\{U_i^e(x)\}^{-1} = \sum_{j=1}^c e^{\lambda\{\|x-v_i\|^2 - \|x-v_j\|^2\}} = 1 + Const \cdot \sum_{j \neq i} e^{2\lambda\langle x, v_j - v_i \rangle}$$

$$(5.32)$$

Assume S_i and E_i are unbounded. The proof for equation (5.30) is straightforward and is omitted. In order to see (5.31) is valid, when the vector x goes to infinity in E_i, $\langle x, v_j - v_i \rangle$ is negative for all $j \neq i$. Hence we have from (5.32)

$$\{U_i^e(x)\}^{-1} \to 1, \quad \text{as } \|x\| \to \infty, \ x \in E_i,$$

from which the second equation in Proposition 1 follows.

As for Proposition 2, note that $e^{2\lambda\langle x, v_j - v_i \rangle}$ is a convex function and a finite sum of convex functions is also a convex function. Hence $\{U_i^e(x)\}^{-1}$ is a convex function from (5.32). A level set for $\{U_i^e(x)\}^{-1}$ is hence convex, from which it is immediate to see that an arbitrary α-cut for $U_i^e(x)$ is convex.

Theoretical properties for $U_i^q(x)$ is more difficult to see in general. A remarkable feature is that $U_i^q(x)$ is piecewise linear, since when $U_i^q(x) > 0$,

$$\lambda^{-1}(\bar{L}\, U_i^q(x) - 1) = \sum_{\ell=1}^L (\|x - v_{j_\ell}\|^2 - \|x - v_i\|^2)$$

$$= \sum_{\ell=1}^L (2\langle v_i - v_{j_\ell}, x \rangle + \|v_{j_\ell}\|^2 - \|v_i\|^2).$$

It should also be noted that, roughly speaking, when a center, say v_K, is sufficiently far from x and another center is nearer to x than v_K, $U_K^q(x) = 0$. Such a property cannot be observed in the foregoing two methods.

Variations of fuzzy c-means

Numerous variations of fuzzy c-means have been proposed. Most of them are based on the alternative optimization algorithm **FC** in which parameters and objective functions are changed in accordance with different purposes. A typical variation uses the objective function

$$J(U,V) = \sum_{k=1}^{n} \sum_{i=1}^{c} (u_{ik})^m D_{ik} \tag{5.33}$$

in which D_{ik} means not only a distance between x_k and v_i, but also other measures of relatedness between x_k and cluster i. It should be noted that the formula for U is similar to (5.15) and the derivation is straightforward:

$$u_{ik} = \left[\sum_{j=1}^{c} \left(\frac{D_{ik}}{D_{jk}} \right)^{\frac{1}{m-1}} \right]^{-1}.$$

Therefore the calculation of the optimal parameters in **FC3** is the problem to be solved.

In the fuzzy c-varieties by Bezdek [3], D_{ik} is the distance between x_k and the linear variety for the cluster i. The algorithm includes calculation of eigenvalues and eigenvectors by which the linear varieties are parameterized.

In fuzzy c-switching regressions [12], D_{ik} is the error between the observed value and the predicted value by the switching regression.

Recent studies (e.g.,[33, 45]) handle identification of dimensions of the clusters. Complicated forms of D_{ik} which include distances between x_k and varieties of different dimensions with parameters are used for this purpose; the details are omitted here.

In other studies D_{jk} is a distance between x_k and v_i. However, the distance is not Euclidean: L_1-space based fuzzy c-means have been considered and different algorithms for the cluster centers have been developed [4, 15, 26]. It should be noted that the center is the median in the case of crisp L_1-based c-means, and hence it is the weighted median for the fuzzy case. Other spaces have also been discussed: Bobrowski and Bezdek [4] consider L_∞ space; Miyamoto and Agusta [29] deal with L_p spaces.

It should furthermore be remarked that the regularization methods using

$$J = \sum_{k=1}^{n} \sum_{i=1}^{c} u_{ik} D_{ik} + \lambda^{-1} K$$

(K may either be the entropy or the quadratic term) express the corresponding variations of the new fuzzy c-means. We will thus obtain fuzzy c-varieties, fuzzy c-regressions, L_1-based c-means, etc., by the regularization methods.

There are still other methods related to fuzzy c-means, most of which are omitted here. Possibilistic approach proposed in [18] uses a different constraint in which the condition of fuzzy partition $\sum_i u_{ik} = 1$ is not assumed. Since the optimal solution U for J_m is trivial, they modified the objective function as $J_m + K$ by adding $K = \sum_{i=1}^{c} \eta_i \sum_{k=1}^{n} (u_{ik} - 1)^m$. This method may be viewed another regularization, but the idea is different from the beforementioned methods.

5.4   Other Nonhierarchical Methods

Vector quantization and crisp c-means

The Self-Organizing Map which is abbreviated as SOM [17] is now extensively studied. The SOM is the mapping of the signal patterns onto a low-dimensional space and is not the method of clustering itself. However, clustering in this framework has been attempted by many researchers. Moreover, Kohonen himself mentions clustering in his book many times [17].

Here we see a glimpse of a basic method, the vector quantization method (VQ), which is related to SOM and also can be interpreted as a variation of the hard c-means. We moreover refer to the learning vector quantization (LVQ) which is also used for clustering. It should be noted that the initial purposes of the VQ and LVQ methods are not clustering of a finite number of objects.

Before the description of these methods, we note that a learning rate factor $\alpha(t)$ is used throughout the discussion of the VQ, LVQ, and SOM. This factor was introduced in the *stochastic approximation* in which the factor should satisfy

$$\sum_{t=1}^{\infty} \alpha(t) = \infty, \qquad \sum_{t=1}^{\infty} \alpha^2(t) < \infty$$

(e.g., $\alpha = Const/t$) in order that the approximation is asymptotically optimal.

There are, however, a variety of other choices for $\alpha(t)$. We omit the details, as readers can see them in Kohonen [17].

Vector quantization

Let $t = 1, 2, \ldots$ be the discrete time variable. Assume that $x(t) \in \mathbf{R}^p$ $(t = 1, 2, \ldots)$ is an infinite sequence of stochastic signal obtained from a probabilistic distribution. The vector quantization is a signal approximation method for the probability density of the distribution using a finite number of *codebook* vectors $m_i \in \mathbf{R}^p$, $i = 1, \ldots, K$. (Remark that K need not be the number of clusters in this context.)

It has been proved that a certain function of the histogram of codebook vectors approximates the density function when the following algorithm is used and the number of the codebook vectors is large.

The algorithm VQ(Kohonen [17])

(i) Set initial (codebook) vectors $m_i(1)$, $i = 1, \ldots, K$. Repeat **(ii)** and
    (iii) for $t = 1, 2, \ldots$ until convergence.

(ii)

$$\ell = \arg \min_{1 \leq i \leq K} \|x(t) - m_i(t)\|.$$

(iii)

$$\begin{aligned} m_\ell(t+1) &:= m_\ell(t) + \alpha(t)[x(t) - m_\ell(t)], \\ m_i(t+1) &:= m_i(t), \quad i \neq \ell. \end{aligned}$$

Application of the VQ algorithm to clustering is immediate. We should
define the infinite sequence $x(t)$ from the finite number of objects $x_1, \ldots, x_n$
for clustering. A simple choice is

$$\begin{aligned} x(1) &= x_1, \ x(2) = x_2, \ldots, x(n) = x_n, \\ x(n+1) &= x_1, \ x(n+2) = x_2, \ldots, \\ x(2n+1) &= x_1, \ldots \end{aligned} \tag{5.34}$$

The number K should be taken to the number of clusters: $K = c$. Thus we
have a simple algorithm of clustering. When we compare this algorithm (of
clustering) with **HCM**, we note that the nearest center allocation is used
in the both (**(ii)** in VQ), while the updating formulae for the centers are
different.

Learning vector quantization

Although Kohonen describes the LVQ as a method of *supervised learning*,
the method is easily modified to that of clustering.

Here the LVQ1 [17] is modified to a clustering algorithm. LVQ1 updates
the reference vectors m_i using a formula similar to the VQ, the algorithm
derived from LVQ1 should be a combination of the nearest center allocation
and the updating scheme in LVQ1. Notice that the objects are interpreted
as the sequence $x(t)$ using (5.34).

A clustering algorithm derived from LVQ1

(i) Set initial values for m_i, $i = 1, \ldots, c$. Generate an initial clusters
    $G_1, \ldots, G_c$ using the nearest center allocation, i.e., allocate each x_k
    to the cluster of the nearest center. (Alternatively, we can generate
    randomly an initial partition and calculate the centers as the cen-
    troids.) Repeat (ii) and (iii) for $t = 1, 2, \ldots$ until convergence.

(ii)

$$\ell = \arg \min_{1 \leq i \leq K} \|x(t) - m_i(t)\|.$$

(iii)

If $x(t) \in G_\ell$, then

$$m_\ell(t+1) := m_\ell(t) + \alpha(t)[x(t) - m_\ell(t)].$$

If $x(t) \notin G_\ell$, then

$$m_\ell(t+1) := m_\ell(t) - \alpha(t)[x(t) - m_\ell(t)].$$

For $i = 1, \ldots, c, i \neq \ell$,

$$m_i(t+1) := m_i(t), \quad i \neq \ell.$$

Pal *et. al.* [34] consider a modified algorithm for clustering in the framework of LVQ.

Mixture densities and the EM algorithm

The mixture density model is very different from others in the sense that it is based on the traditional statistics. We should hence discuss the maximum likelihood estimates and the EM algorithm [6, 37] for solving the clustering problems in this model.

Probability density functions for most standard distributions are unimodal, while clustering problems require handling multimodal distributions.

Let us suppose that many data items of a variable are collected and the histogram is found to have two modes of maximal values.

Apparently, the histogram is approximated by adding two densities of the normal distributions. Hence the probability density for this distribution is

$$p(x) = \alpha_1 p_1(x) + \alpha_2 p_2(x),$$

where α_1 and α_2 are nonnegative numbers such that $\alpha_1 + \alpha_2 = 1$ and moreover

$$p_i(x) = \frac{1}{\sqrt{2\pi}\sigma_i} e^{-\frac{(x-\mu_i)^2}{2\sigma_i^2}}, \qquad i = 1, 2.$$

Suppose we have good estimates of the parameters α_i, μ_i, and σ_i, $i = 1, 2$. After having a good approximation of the mixture distribution we can solve the clustering problem using the Bayes formula for posterior probability.

Let $P(X|C_i)$ and $P(C_i)$, $i = 1, \ldots, m$, is the conditional probability of event X given that class C_i occurs, and the prior probability of the class C_i, respectively. We assume that exactly one of the class C_i, $i = 1, \ldots, m$, necessarily occurs. Then the Bayes formula is

$$P(C_i|X) = \frac{P(X|C_i)P(C_i)}{\sum_{j=1}^{m} P(X|C_j)P(C_j)}. \tag{5.35}$$

When we apply this formula to the above example of two normal distributions, we can put

$$P(X) = P(a < x < b), \quad P(C_i) = \alpha_i, \quad P(X|C_i) = \int_a^b p_i(x)dx \quad (i = 1, 2).$$

Then the Bayes formula tells us

$$P(C_i|X) = \frac{\alpha_i \int_a^b p_i(x)dx}{\sum_{j=1}^2 \alpha_j \int_a^b p_j(x)dx}.$$

Assume we have an observation y. Taking two numbers a, b such that $a < y < b$, we have the probability of the class C_i given X by the above formula. Going to the limit $a \to y$ and $b \to y$, we have the probability of the class C_i given the data y by the following:

$$P(C_i|y) = \frac{\alpha_i p_i(y)}{\sum_{j=1}^2 \alpha_j p_j(y)}. \tag{5.36}$$

As the last equation provides us the probability of allocating an observation to all classes, we can thus solve the clustering problem.

Since we consider mixture of normal distributions in this section, the above formula (5.36) is immediately generalized to the case of m classes. We should now solve the problem of how to obtain good estimates of the parameters. The EM algorithm should be introduced for this purpose.

The EM algorithm

In this section we consider a general class of mixture distribution given by

$$p(x|\Phi) = \sum_{j=1}^m \alpha_i p_i(x|\phi_i) \tag{5.37}$$

in which $p_i(x|\phi_i)$ is the probability density corresponding to class C_i, and ϕ_i is a vector parameter to be estimated. (Readers can suppose $p_i(x|\phi_i)$ is a normal distribution and $\phi_i = (\mu_i, \sigma_i)$.) Moreover Φ represents the whole sequence of the parameters to be estimated, i.e., $\Phi = (\alpha_1, \ldots, \alpha_m, \phi_1, \ldots, \phi_m)$. We assume that observed data $x_1, \ldots, x_n$ are mutually independent samples taken from the population having this mixture distribution. We use these symbols $x_1, \ldots, x_n$ for both given data and variables for the sample distribution. This abuse of terminology simplifies the description and no confusion arises.

A classical method of solving parameter estimation problems is to use the maximum likelihood. From the assumption of independence the sample distribution is given by

$$\prod_{k=1}^{n} p(x_k|\Phi)$$

Suppose x_k is the observed data, then the above is a function of the parameter Φ. The maximum likelihood is the method of using the estimate of the parameter that maximizes the above function. For convenience of calculations, the log-likelihood is used:

$$L(\Phi) = \log \prod_{k=1}^{n} p(x_k|\Phi) = \sum_{k=1}^{n} \log p(x_k|\Phi). \tag{5.38}$$

Notice again that $x_1, \ldots, x_n$ are given data. Thus, the maximum likelihood estimate is given by

$$\hat{\Phi} = \arg\max_{\Phi} L(\Phi). \tag{5.39}$$

For simple distributions, the maximum likelihood estimates are sufficient, but the calculation of the maximum likelihood estimates is difficult for the present mixture distribution.

We therefore should consider more advanced algorithm. For such purposes, it has been found that the EM algorithm is useful [6, 37]. The EM algorithm is an iteration process in which an Expectation step (E-step) and a Maximization step (M-step) are repeated until convergence. Let us introduce the idea of the EM algorithm with some definitions and assumptions.

1. In addition to the observed data $x_1, \ldots, x_n$, we assume another *complete data* $y_1, \ldots, y_n$. Accordingly $x_1, \ldots, x_n$ is called *incomplete data*. For simplicity, we write $\mathbf{x} = (x_1, \ldots, x_n)$ and $\mathbf{y} = (y_1, \ldots, y_n)$. $\mathbf{y}$ is not observed directly, but a partial observation is the incomplete data. We are thus assuming an implicit mapping from $\mathbf{y} \to \mathbf{x} = \mathcal{X}(\mathbf{y})$. Given $\mathbf{x}$ the set of all $\mathbf{y}$ such that $\mathbf{x} = \mathcal{X}(\mathbf{y})$ is denoted by $\mathcal{Y}(\mathbf{x})$.

2. We assume that $\mathbf{x}$ and $\mathbf{y}$ have the distributions $g(\mathbf{x}|\Phi)$ and $f(\mathbf{y}|\Phi)$, respectively.

3. Assume that an estimate Φ' for Φ is given. Define function $Q(\Phi|\Phi')$ by the following.

$$Q(\Phi|\Phi') = E(\log f(\mathbf{y}|\Phi)|\mathbf{x}, \Phi') \tag{5.40}$$

where $E(\log f|\mathbf{x}, \Phi')$ is the conditional expectation given $\mathbf{x}$ and Φ'.

Let us assume that $k(\mathbf{y}|\mathbf{x}, \Phi')$ is the conditional probability density of $\mathbf{y}$ given $\mathbf{x}$ and Φ'. It then follows that

$$Q(\Phi|\Phi') = \int_{\mathcal{Y}(\mathbf{x})} k(\mathbf{y}|\mathbf{x}, \Phi') \log f(\mathbf{y}|\Phi) dy. \qquad (5.41)$$

We are now ready to describe the EM algorithm.

The EM algorithm

(O) Set an initial value $\Phi^{(0)}$ for the estimate. Let $\ell = 0$. Repeat the following **E** and **M** until a convergence criterion is satisfied.

(E) (Expectation)
Calculate $Q(\Phi|\Phi^{(\ell)})$.

(M) (Maximization)
find the maximizing solution

$$\bar{\Phi} = \arg\max_{\Phi} Q(\Phi|\Phi^{(\ell)}).$$

Let $\ell := \ell + 1$ and $\Phi^{(\ell)} = \bar{\Phi}$.

It should be noted that the **E** and **M** steps can be replaced by a single step of repeating

$$\Phi^{(\ell+1)} = \arg\max_{\Phi} Q(\Phi|\Phi^{(\ell)})$$

for $\ell = 0, 1, \ldots$.

Application to the mixture densities

The above algorithm should be applied to the present class of the mixture distributions. We have a few questions to be answered for this purpose.

First question is what the complete data are in this case. Suppose we have the information, in addition to x_k, from which class C_i the observation has been obtained. Then the estimation problem seems to become simpler with this information. Hence we assume $y_k = (x_k, i_k)$ in which i_k means the number of class C_{i_k} from which x_k has been obtained.

Given this information, the density $f(\mathbf{y}|\Phi)$ is given by

$$f(\mathbf{y}|\Phi) = \prod_{k=1}^{n} \alpha_{i_k} p_{i_k}(x_k|\phi_{i_k}).$$

On the other hand, $k(\mathbf{y}|\mathbf{x}, \Phi')$ is calculated as follows.

$$k(\mathbf{y}|\mathbf{x}, \Phi') = \frac{f(\mathbf{y}|\Phi')}{g(\mathbf{x}|\Phi')} = \prod_{k=1}^{n} \frac{\alpha'_{i_k} p_{i_k}(x_k|\phi'_{i_k})}{p(x_k|\Phi')}.$$

Notice that $g(\mathbf{x}|\Phi') = \prod\limits_{k=1}^{n} p(x_k|\Phi')$.

We calculate the function Q using these. It should be remarked that $\mathcal{Y}(\mathbf{x})$ is reduced to the set

$$\{\,(i_1,\dots,i_n) : 1 \le i_\ell \le m,\ \ell = 1,\dots,m\,\},$$

whence the integral in (5.41) is replaced by summations. We thus have

$$Q(\Phi|\Phi') = \sum_{i_1=1}^{m} \cdots \sum_{i_n=1}^{m} \sum_{k=1}^{n} \log[\alpha_{i_k} p_{i_k}(x_k|\phi_{i_k})] \prod_{k=1}^{n} \frac{\alpha'_{i_k} p_{i_k}(x_k|\phi'_{i_k})}{p(x_k|\Phi')}.$$

After some straightforward calculation we have

$$Q(\Phi|\Phi') = \sum_{i=1}^{m} \sum_{k=1}^{n} \log[\alpha_i p_i(x_k|\phi_i)] \frac{\alpha'_i p_i(x_k|\phi'_i)}{p(x_k|\Phi')}.$$

For simplicity, put

$$\psi_{ik} = \frac{\alpha'_i p_i(x_k|\phi'_i)}{p(x_k|\Phi')}, \quad \Psi_i = \sum_{k=1}^{n} \psi_{ik}.$$

and note that $\sum\limits_{i=1}^{m} \Psi_i = n$. It then follows that

$$Q(\Phi|\Phi') = \sum_{i=1}^{m} \Psi_i \log \alpha_i + \sum_{i=1}^{m} \sum_{k=1}^{n} \psi_{ik} \log p_i(x_k|\phi_i).$$

To obtain the optimal α_i, we must take the constraint $\sum\limits_{i=1}^{m} \alpha_i = 1$ into account. Hence the Lagrangean with the Lagrange multiplier λ is introduced:

$$L = Q(\Phi|\Phi') - \lambda(\sum_{i=1}^{m} \alpha_i - 1)$$

Using

$$\frac{\partial L}{\partial \alpha_i} = \frac{\Psi_i}{\alpha_i} - \lambda = 0, \tag{5.42}$$

and taking the sum of $\lambda \alpha_i = \Psi_i$ with respect to $i = 1,\dots,m$, we have $\lambda = n$. Thus, the optimal solution is

$$\alpha_i = \frac{\Psi_i}{n} = \frac{1}{n} \sum_{i=1}^{n} \frac{\alpha'_i p_i(x_k|\phi'_i)}{p(x_k|\Phi')}, \qquad i = 1,\dots,m. \tag{5.43}$$

We have not specified the density functions until now. We now proceed to consider normal distributions and estimate the means and variances. For simplicity we first derive solutions for the univariate normal distributions. After that we show the solutions for multivariate normal distributions.

For the univariate normal distributions,

$$p_i(x|\phi_i) = \frac{1}{\sqrt{2\pi}\sigma_i} e^{-\frac{(x-\mu_i)^2}{2\sigma_i{}^2}}, \quad i = 1, \ldots, m$$

where $\phi_i = (\mu_i, \sigma_i)$. For the optimal solutions we should minimize

$$J = \sum_{i=1}^{m} \sum_{k=1}^{n} \psi_{ik} \log \frac{1}{\sqrt{2\pi}\sigma_i} e^{-\frac{(x_k-\mu_i)^2}{2\sigma_i{}^2}}.$$

From

$$\frac{\partial J}{\partial \mu_i} = -\sum_{k=1}^{n} \psi_{ik} \frac{x_k - \mu_i}{\sigma_i^2} = 0,$$

we have

$$\mu_i = \frac{1}{\Psi_i} \sum_{k=1}^{n} \psi_{ik} x_k, \quad i = 1, \ldots, m. \tag{5.44}$$

In the same manner, from

$$\frac{\partial J}{\partial \sigma_i} = \sum_{k=1}^{n} \psi_{ik} \frac{(x_k - \mu_i)^2}{\sigma_i^3} - \sum_{k=1}^{n} \psi_{ik} \frac{1}{\sigma_i} = 0,$$

We have

$$\sigma_i^2 = \frac{1}{\Psi_i} \sum_{k=1}^{n} \psi_{ik} (x_k - \mu_i)^2 = \frac{1}{\Psi_i} \sum_{k=1}^{n} \psi_{ik} x_k^2 - \mu_i^2, \quad i = 1, \ldots, m, \tag{5.45}$$

in which μ_i is given by (5.44).

Let us consider multivariate normal distributions for p_i:

$$p_i(x) = \frac{1}{2\pi^{\frac{p}{2}} |\Sigma_i|^{\frac{1}{2}}} e^{-\frac{1}{2}(x-\mu_i)^T \Sigma_i{}^{-1}(x-\mu_i)}$$

in which $x = (x^1, \ldots, x^p)^T$ and $\mu_i = (\mu_i^1, \ldots, \mu_i^p)^T$ are vectors, and $\Sigma_i = (\sigma_i^{j\ell})$ $(1 \le j, \ell \le p)$ is the covariance matrix; $|\Sigma_i|$ is the determinant of Σ_i.

By the same manner as above, the optimal α_i is given by (5.43); the optimal solutions for μ_i and Σ_i are as follows [37].

$$\mu_i = \frac{1}{\Psi_i} \sum_{k=1}^{n} \psi_{ik} x_k, \quad i = 1, \ldots, m, \tag{5.46}$$

$$\Sigma_i = \frac{1}{\Psi_i} \sum_{k=1}^{n} \psi_{ik} (x_k - \mu_i)(x_k - \mu_i)^T, \quad i = 1, \ldots, m. \tag{5.47}$$

Proof of (5.46) and (5.47)

For the purpose of convenience, let jth component of a vector μ_i be μ_i^j and $i\ell$ component of a matrix Σ_i be $\sigma_i^{j\ell}$, and furthermore, jth component of a vector a is also written as $(a)^j$; a matrix of which ij component is f^{ij} is denoted by $[f^{ij}]$. Thus, $\mu_i^j = (\mu_i)^j$ and $\Sigma_i = [\sigma_i^{j\ell}]$.

Let

$$J_i = \sum_{k=1}^{n} \psi_{ik} \log \frac{1}{2\pi^{\frac{p}{2}}|\Sigma_i|^{\frac{1}{2}}} e^{-\frac{1}{2}(x_k-\mu_i)^T \Sigma_i^{-1}(x_k-\mu_i)}$$

and remark that we should find the solutions of $\frac{\partial J_i}{\partial \mu_i^j} = 0$ and $\frac{\partial J_i}{\partial \sigma_i^{j\ell}} = 0$.

It is immediate to see that the solution of $\frac{\partial J_i}{\partial \mu_i^j} = 0$ is given by (5.46), and hence the detail is omitted.

Let us consider the solution of

$$2\frac{\partial J_i}{\partial \sigma_i^{j\ell}} = -\sum_{k=1}^{n} \psi_{ik} \frac{\partial}{\partial \sigma_i^{j\ell}} \left((x_k - \mu_i)^T \Sigma_i(x_k - \mu_i) \right) - \frac{\partial}{\partial \sigma_i^{j\ell}} (\log |\Sigma_i|) = 0.$$

For solving this, we note the following.

(a) Let the adjoint for $\sigma_i^{j\ell}$ in the matrix Σ_i be *Adj* $\sigma_i^{j\ell}$. We then have

$$\left[\frac{\partial}{\partial \sigma_i^{j\ell}} \log |\Sigma_i| \right] = \left[\frac{1}{|\Sigma_i|} \frac{\partial}{\partial \sigma_i^{j\ell}} |\Sigma_i| \right]$$

$$= \left[\frac{1}{|\Sigma_i|} Adj\, \sigma_i^{j\ell} \right]$$

$$= \Sigma^{-1}$$

(b) For calculating $\frac{\partial}{\partial \sigma_i^{j\ell}}\Sigma_i^{-1}$, let $E^{j\ell}$ is the matrix in which the $j\ell$-component alone is the unity and other elements are zero. Then,

$$\frac{\partial}{\partial \sigma_i^{j\ell}}(\Sigma_i^{-1}\Sigma_i) = \left(\frac{\partial}{\partial \sigma_i^{j\ell}}\Sigma_i^{-1} \right) \Sigma_i + \Sigma_i^{-1} \left(\frac{\partial}{\partial \sigma_i^{j\ell}}\Sigma_i \right)$$

$$= \left(\frac{\partial}{\partial \sigma_i^{j\ell}}\Sigma_i^{-1} \right) \Sigma_i + \Sigma_i E^{j\ell} = 0$$

whereby we have

$$\frac{\partial}{\partial \sigma_i^{j\ell}}\Sigma_i^{-1} = \Sigma_i^{-1} E^{j\ell}\Sigma_i^{-1}.$$

Suppose a does not contain an element in Σ_i. We then obtain

$$
\begin{aligned}
\left[\frac{\partial}{\partial \sigma_i^{j\ell}}(a^T \Sigma_i^{-1} a)\right] &= [a^T \Sigma_i^{-1} E^{j\ell} \Sigma_i^{-1} a] \\
&= [(\Sigma_i^{-1} a)^T E^{j\ell}(\Sigma_i^{-1} a)] \\
&= [(\Sigma_i^{-1} a)^j (\Sigma_i^{-1} a)^\ell] \\
&= (\Sigma_i^{-1} a)(\Sigma_i^{-1} a)^T = \Sigma_i^{-1} a a^T \Sigma_i^{-1}
\end{aligned}
$$

Using (a) and (b) in the last equation, we obtain

$$
2\frac{\partial J_i}{\partial \sigma_i^{j\ell}} = -\sum_{k=1}^n \psi_{ik} \Sigma_i^{-1}(x_k - \mu_i)(x_k - \mu_i)^T \Sigma_i^{-1} + \left(\sum_{k=1}^n \psi_{ik}\right) \Sigma_i^{-1} = 0
$$

Multiplication of Σ_i to the above equation from the right and the left lead us to

$$
-\sum_{k=1}^n \psi_{ik}(x_k - \mu_i)(x_k - \mu_i)^T + \Psi_i \Sigma_i = 0.
$$

We thus have (5.47).

Dissimilarity based methods in optimal clustering

Sometimes an object is not represented by a point in a finite-dimensional space, but a dissimilarity or similarity matrix $(d(x_i, x_j))$ is directly provided. In applications to psychological studies, such dissimilarity data are frequently used. Moreover application to information retrieval requests such a dissimilarity based technique [24]. (Remark: some studies refer to similarity and others dissimilarity. Generally, transformation between similarity and dissimilarity is trivial in the methodological sense.)

There is a family of methods of optimal hard clustering that uses the values of dissimilarity measures but not the intrinsic properties of the underlying space. In other words, the methods do not care about the spaces. A typical formulation is to minimize

$$
f(C) = \sum_{k=1}^K \frac{1}{|G_k|^\alpha} \sum_{x_i, x_j \in G_k} d(x_i, x_j). \tag{5.48}
$$

with respect to all partitions $G_1, \ldots, G_K$ of $\{x_1, \ldots, x_n\}$. Since this problem is NP-hard for $\alpha \neq 0$, heuristic algorithms should be used. Genetic algorithms have been applied to such problems [23, 36]. Miyamoto and Katoh [27] apply a wider class of metaheuristic techniques to this problem and assert that simpler algorithms of the local search and multistart local search are as effective as the genetic algorithm and less time-consuming than the latter. Moreover they propose a dissimilarity based fuzzy c-means algorithm [30].

The old method by Ruspini [39] also is a dissimilarity-based technique. Namely, a given set of dissimilarity or similarity is approximated by a fuzzy partition.

The method of additive clustering [41] which is well-known in psychometrics has been transformed into fuzzy additive clustering by Sato and Sato [40] with applications to psychometrics.

5.5   A Numerical Example

A set of data is prepared for the purpose of illustrating and comparing performances of different methods in nonhierarchical clustering.

Figure 5.5 shows an artificial example of about 200 points shown by x on a plane. We divide these points into four clusters using the hard c-means, the LVQ method, the three methods of fuzzy clustering, and the mixture density model.

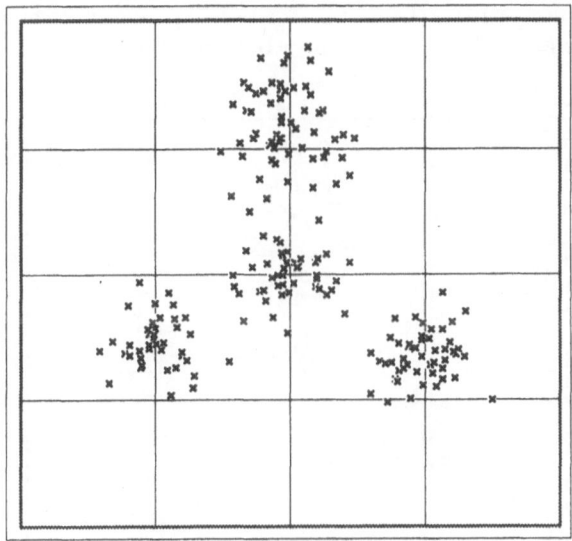

FIGURE 5.5. An example of about 200 points on a plane

Figures 5.6–5.10 depicts the four clusters obtained from the hard c-means, the standard fuzzy c-means, the entropy method, the quadratic regularization, and the LVQ method, respectively.

The four clusters are distinguished by the four symbols of x, •, *, and +. Readers can see that all methods provide similar results of acceptable clusters. In case of fuzzy clustering, each point is crisply allocated to the cluster of maximum membership among the four group.

In hard c-means, the algorithm **F** is used but the **HCM** algorithm was found to output the same clusters.

In the standard fuzzy c-means, the parameter $m = 1.8$ and the convergence criterion

$$\sum_{i=1}^{c}\sum_{k=1}^{n} |\bar{u}_{ik} - \hat{u}_{ik}| < \epsilon \qquad (5.49)$$

were used, where $(\hat{u}_{ik})$ is the optimal solution in the previous loop, and moreover we set $\epsilon = 10^{-4}$.

For the entropy method and the quadratic method $\lambda^{-1} = 0.2$ and (5.49) were used with $\epsilon = 10^{-4}$.

In the LVQ method $\alpha = 2.0/t$ was used.

Ten trials for each methods using different initial clusters that had been randomly generated resulted in the same clusters shown above, except the LVQ method: in the case of the LVQ, 4 out of 10 trials were in failure.

We observed no significant difference of number of loops and runtime until convergence, except the LVQ. In the LVQ, the number of loops until convergence is generally longer than other methods.

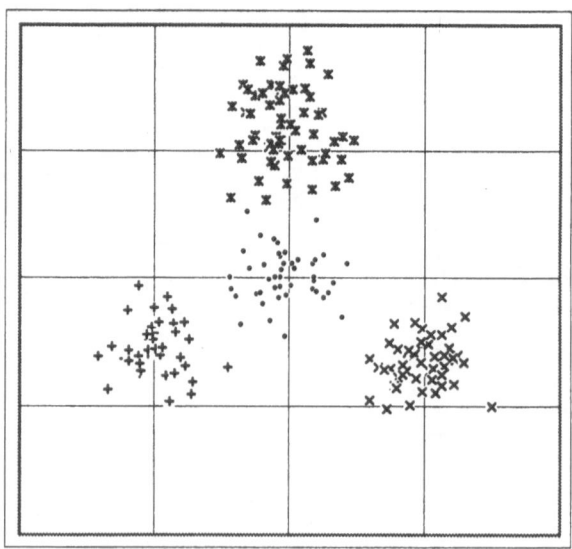

FIGURE 5.6. Four clusters obtained by the hard c-means

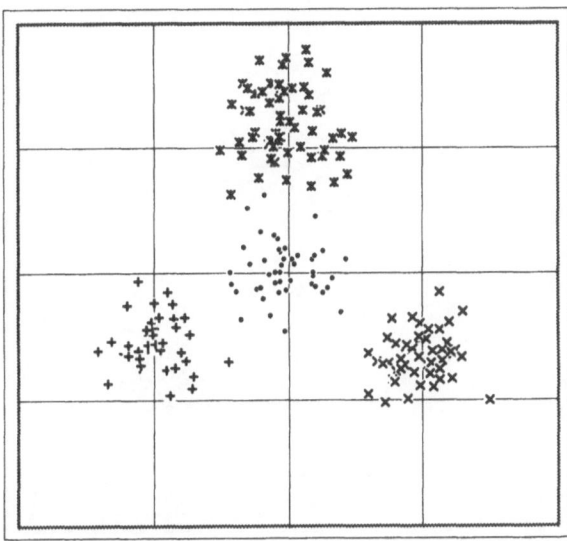

FIGURE 5.7. Four clusters obtained by the standard fuzzy c-means

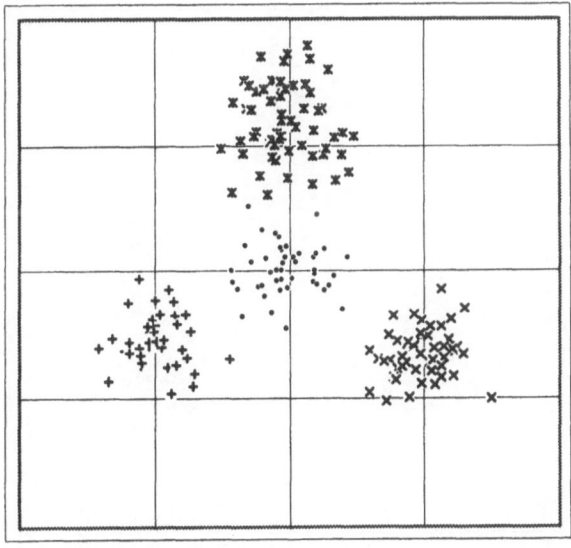

FIGURE 5.8. Four clusters obtained by the entropy method

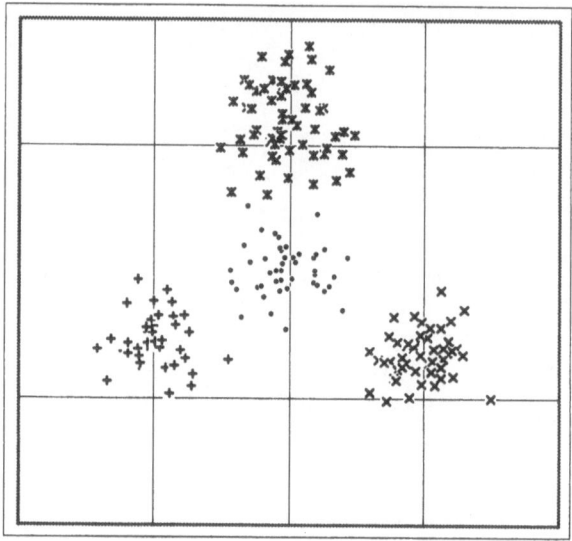

FIGURE 5.9. Four clusters obtained by the quadratic regularization

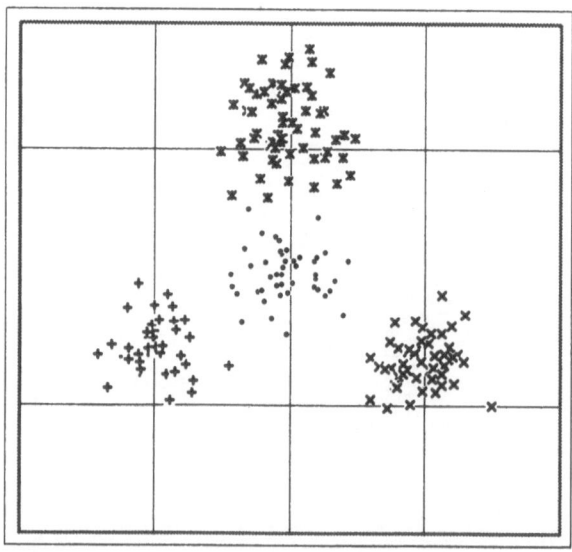

FIGURE 5.10. Four clusters obtained by the LVQ method

Figures 5.11 and 5.12 show contours of classification functions for two clusters obtained by the standard fuzzy c-means, whereas Figures 5.13 and 5.14 are classification functions by the entropy method. The classification functions by the quadratic method are similar to those by the entropy method and hence omitted.

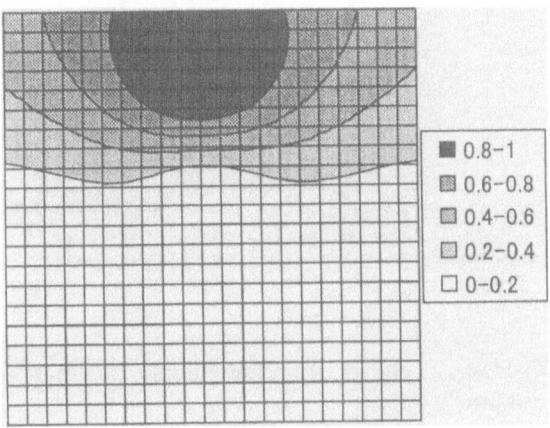

FIGURE 5.11. Contours of classification function for the 'north' cluster by the standard fuzzy c-means

5.6   Fuzzy Hierarchical Clustering

It has been discussed that the single link method has the most desirable properties in the theoretical sense among agglomerative techniques. It also is proved that the single link is closely related to the minimum spanning tree of a network [1, 24]. We briefly show that the transitive closure of a fuzzy relation and connected components of a fuzzy graph are in a sense equivalent to the single link.

Transitive closure and single link

Zadeh [48] has defined the transitive closure of a fuzzy relation and a similarity relation; the latter means a fuzzy relation T on X having the following three properties:

(I) (reflexivity)

$$T(x, x) = 1, \quad \text{for all } x \in X.$$

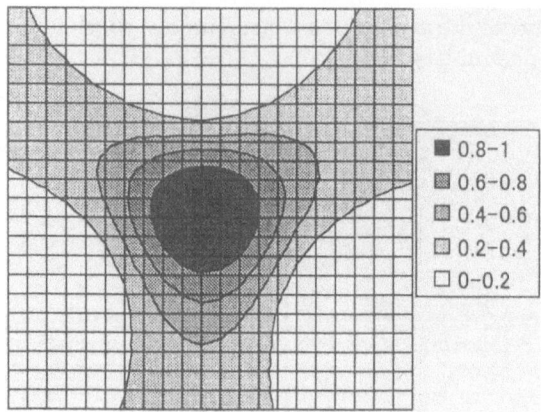

FIGURE 5.12. Contours of classification function for the 'central' cluster by the standard fuzzy c-means

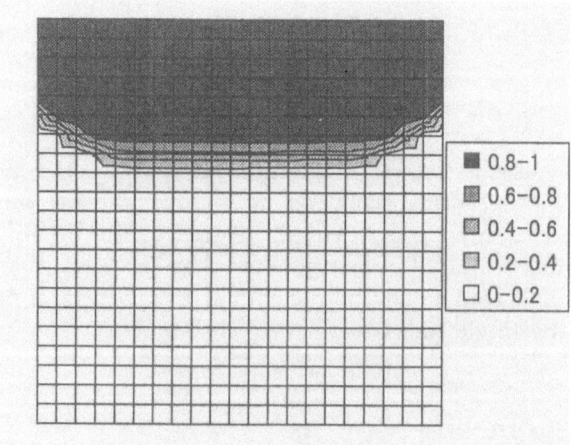

FIGURE 5.13. Contours of classification function for the 'north' cluster by the entropy method

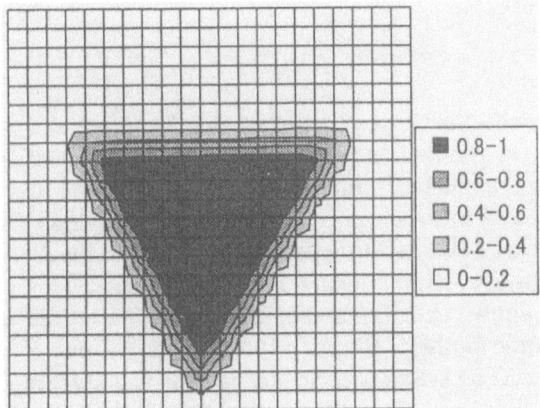

FIGURE 5.14. Contours of classification function for the 'central' cluster by the entropy method

(II) (symmetricity)

$$T(x, y) = T(y, x), \quad \text{for all } x, y \in X.$$

(III) (transitivity)

$$T(x, z) \geq \max_{y \in X}\{T(x, y), T(y, z)\} \quad \text{for all } x, z \in X.$$

As his term of *similarity relation* conflicts with the general term of the similarity measure in cluster analysis, we use the term of *fuzzy equivalence relation* instead, which satisfies **(I)**, **(II)**, and **(III)**.

The characteristic of fuzzy equivalence relation is that an arbitrary α-cut T_α is a crisp equivalence relation. Namely, T_α induces a partition such as (5.1).

Generally, the condition for a fuzzy equivalence relation is very strict. Nevertheless, one can generate a fuzzy equivalence relation when a reflexive and symmetric relation S is given.

Let us write the maximum operation as $\vee$ and the max-min composition as a multiplication, i.e.,

$$S^{k+1}(x, z) = \max_y \min\{S^k(x, y), S(y, z)\} = \bigvee_y \{S^k(x, y) \wedge S(y, z)\},$$

$$k = 1, 2, \ldots$$

We define the transitive closure of S by

$$S^* = S \vee S^2 \vee \cdots \vee S^n \vee \cdots$$

When X is finite,

$$S^* = S \vee S^2 \vee \cdots \vee S^n. \qquad (5.50)$$

Zadeh [48] mentions hierarchical classification using the transitive closure which is the fuzzy equivalence relation, since an α-cut of S^* is a partition and for $\alpha_1 \leq \alpha_2$, a cluster induced from $S^*_{\alpha_2}$ is included in a cluster induced from $S^*_{\alpha_1}$. Thus, a hierarchical classification is obtained from S^*. The resemblance between the dendrogram and the hierarchical classification generated from the transitive closure has been remarked by researchers, and Dunn [10] has shown that the method of the transitive closure is equivalent to the single link method. Remark that we should put $S = (s(x_i, x_j))$ for the equivalence. The transitive closure is thus the algebraic representation and the single link is an algorithmic representation of the same method, although the two appear quite different. We will show that a single geometric structure underlies these methods by describing a fuzzy graph.

Fuzzy graph and single link

A geometric representation of the same clustering has been studied using a fuzzy graph in which edges of a graph are with membership values.

Recall that a crisp graph G is a pair $G = (V, E)$ in which V is a finite set of vertices, and $E \subseteq V \times V$ is the set of edges of G. An edge $e \in E$ is written as $e = vw$ for $v, w \in V$ when e connects those two vertices v and w. When v and w are connected by an edge, we say v is adjacent to w or w is adjacent to v. In order to distinguish two or more graphs $G, G', \ldots$, the set of vertices, edges, and the memberships are indexed as $V(G)$, $E(G)$, $V(G')$, $E(G')$, and so on. Details of the general theory of graphs which are omitted here are given in standard textbooks (e.g., Chardrand and Lesniak [5]).

A fuzzy graph FG is a pair $FG = (V, \mu)$ in which V is a finite set of vertices, the set E of edges is assumed to be $E = V \times V$ and hence omitted, and μ is the membership defined on the set of edges $\mu \colon E \to [0, 1]$. (If the membership values are deleted, a fuzzy graph becomes a complete graph.) The symbols $V(FG)$ and μ_{FG} are moreover used.

Let us see an example.

Example 4. Figure 5.15 illustrates an example of a fuzzy graph FG. Although there are five vertices v_1, v_2, v_3, v_4, v_5 and we are assuming the complete graph, only six edges are shown with their memberships. This implies that we do not write the edges of zero membership values for simplicity:

$$\mu(v_1 v_4) = \mu(v_2 v_4) = \mu(v_2 v_5) = \mu(v_3 v_5) = 0.$$

Since a fuzzy graph should be regarded as a collection of its α-cuts as crisp graphs, $FG_\alpha = (V(FG), E(FG_\alpha))$ in which $V(FG)$ is the same for

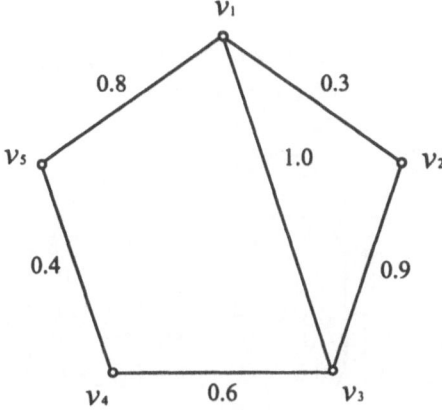

FIGURE 5.15. A fuzzy graph FG

all $0 \le \alpha \le 1$, whereas

$$E(FG_\alpha) = \{vw : v, w \in V(FG), \mu_{FG}(vw) \ge \alpha\}. \qquad (5.51)$$

Figure 5.16 illustrates FG_α in Example 4 for $\alpha = 0.85$. Notice that FG_α for $\alpha = 0.85$ is written as $FG_{0.85}$.

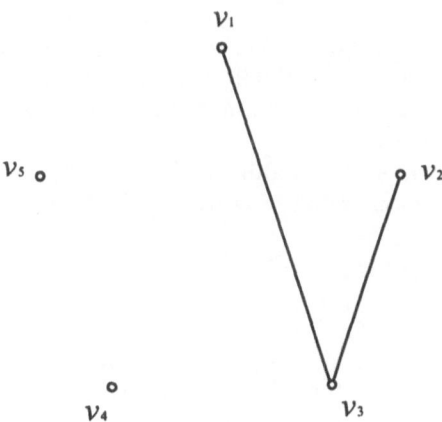

FIGURE 5.16. An α-cut of FG in Fig.5.5 with $\alpha = 0.85$.

A fuzzy graph FG naturally induces a fuzzy relation S on $V(FG)$, that is,

$$S(v, w) = \mu_{FG}(vw), \quad v, w \in V(FG), v \neq w$$
$$S(v, v) = 1 \quad v \in V(FG).$$

For the above example,

$$S = \begin{array}{c} \\ v_1 \\ v_2 \\ v_3 \\ v_4 \\ v_5 \end{array} \begin{array}{ccccc} v_1 & v_2 & v_3 & v_4 & v_5 \\ \begin{pmatrix} 1 & 0.3 & 1 & 0 & 0.8 \\ 0.3 & 1 & 0.9 & 0 & 0 \\ 1 & 0.9 & 1 & 0.6 & 0 \\ 0 & 0 & 0.6 & 1 & 0.4 \\ 0.8 & 0 & 0 & 0.4 & 1 \end{pmatrix} \end{array}.$$

It is obvious that $S(v, w)$ is the maximum value of the α-cut for which v is adjacent to w. For example, v_1 is adjacent to v_2 in $FG_{0.3}$ in Example 4, but for $\alpha > 0.3$ they are not connected by an edge.

We will consider the meaning of the transitive closure using fuzzy graphs. First, an walk in a crisp graph [5] should be introduced. A u-v walk of length $k(> 1)$ in G is an alternating sequence of vertices and edges:

$$u = v_0, e_1, v_1, e_2, v_2, \cdots, v_{k-1}, e_k, v_k = v$$

in which $e_j = v_{j-1}v_j \in E(G)$ or $v_{j-1} = v_j$, $j = 1, \ldots, k$. (Notice that the definition herein is different from the general one in [5]. Namely, we assume that there is always a u-u walk of any length.)

It is easy to see that $S^2(v, w)$ is the maximum value of α by which there is v-w walk of length 2 in FG_α. For example, the maximum value by which there is a v_1-v_2 walk of length 2 is $\alpha = 0.9$ in FG_α in Example 4. Readers can easily check that $S^2(v_1, v_2) = 0.9$. In the same manner, it is not difficult to see that $S^k(v, w)$ is the maximum value of α by which there is a v-w walk of length k in FG_α.

It is now obvious to see the meaning of the transitive closure S^*. Namely, $S^*(v, w)$ is the maximum value of α by which there is a v-w walk of any length in FG_α.

For Example 4,

$$S^* = \begin{array}{c} \\ v_1 \\ v_2 \\ v_3 \\ v_4 \\ v_5 \end{array} \begin{array}{ccccc} v_1 & v_2 & v_3 & v_4 & v_5 \\ \begin{pmatrix} 1 & 0.9 & 1 & 0.6 & 0.8 \\ 0.9 & 1 & 0.9 & 0.6 & 0.8 \\ 1 & 0.9 & 1 & 0.6 & 0.8 \\ 0.6 & 0.6 & 0.6 & 1 & 0.6 \\ 0.8 & 0.8 & 0.8 & 0.6 & 1 \end{pmatrix} \end{array}.$$

Readers can easily check that these values are the maximum α-cuts by which the corresponding pair of vertices are connected by a series of edges.

In FG_α, an α-cut of FG, we can observe clusters of vertices, namely, those vertices that are connected each other by walks form a cluster. In Fig. 5.16, we have three clusters: $\{v_1, v_2, v_3\}$, $\{v_4\}$, and $\{v_5\}$. It is also immediate to see that these clusters coincide with the equivalence classes derived from the equivalence relation S_α^*. The equivalence between the classes derived from the transitive closure and the groups of connected vertices in a fuzzy graph is thus observed.

Each α-cut of a fuzzy graph induces subgraphs: in each subgraph vertices are connected and between two subgraphs vertices are not connected. Such a subgraph is a connected component of FG_α and hence may be called a connected component of a fuzzy graph.

The single link algorithm sequentially generates the connected components of a fuzzy graph in which objects are vertices and similarity measures are the memberships. To see this, it is sufficient to remark that the updating formula

$$s(G_r, G_i) = \max\{s(G_p, G_i), s(G_q, G_i)\}$$

implies the level of α at which the two connected components will be connected by an edge vw, $v \in G_r$, $w \in G_i$.

In the above example, the sequence of connected components are as follows.

$$\alpha = 1.0: \quad \{v_1, v_3\}, \{v_2\}, \{v_4\}, \{v_5\}$$
$$\alpha = 0.9: \quad \{v_1, v_2, v_3\}, \{v_4\}, \{v_5\}$$
$$\alpha = 0.8: \quad \{v_1, v_2, v_3, v_5\}, \{v_4\}$$
$$\alpha = 0.6: \quad \{v_1, v_2, v_3, v_4, v_5\}.$$

Readers can understand that connected components are obtained from the single link algorithm.

We have now shown the equivalence of three concepts:

I. Clusters by the single link,

II. Classes by the transitive closure,

III. Connected components of a fuzzy graph.

Complete discussion on the equivalence of the above three methods as well as clusters derived from the minimum spanning tree is found in Miyamoto [24]. Formal proof is omitted here and an example has been shown instead. Emphasis on the fuzzy graph is important in understanding hierarchical clustering, since it is a fundamental structure in the single link method. In the proof [24], the key is a simplified version of Kruskal's old algorithm [19] for the maximum spanning trees, which generates connected components of the corresponding fuzzy graph.

Miyamoto [25] discusses a variation of the fuzzy graph in which vertices have membership values and its application to agglomerative clustering.

5.7   Conclusions

Nonhierarchical as well as hierarchical clustering using fuzzy and crisp techniques have been overviewed. In particular, recent methods using the concept of regularization have been emphasized. As mentioned above, variations of the regularization methods have not been tested in real applications.

Usefulness of fuzzy classification functions is in their capability of exhibiting features of the generated clusters, and hence further studies of them are necessary.

There are few studies nowadays about hierarchical fuzzy classifications, as the hierarchical methods are well-established, and it is recognized that there is a small room for further studies. Nevertheless, fuzzy relations are essential for agglomerative clustering, as shown above. If a new and theoretically sound hierarchical method is developed, it should be based on fuzzy set theory, somehow. However, due to space constraints, we are able to show only one numerical example here.

The contents of this chapter is different in many aspects from those of a recent book by Höppner *et al.* [13], which may be useful for readers who are interested in various other techniques in fuzzy clustering.

5.8   REFERENCES

[1] M.R. Anderberg, *Cluster Analysis for Applications*, Academic Press, New York, 1973.

[2] E. Backer, *Cluster Analysis by Optimal Decomposition of Induced Fuzzy Sets*, Delft Univ. Press, Delft, The Netherlands, 1978.

[3] J.C. Bezdek, *Pattern Recognition with Fuzzy Objective Function Algorithms*, Plenum, New York, 1981.

[4] L. Bobrowski, J.C. Bezdek, "c-means clustering with the ℓ_1 and ℓ_∞ norms," *IEEE Trans. on Syst., Man, and Cybern.*, Vol. 21, No. 3, pp.545-554, 1991.

[5] G. Chardrand, L. Lesniak, *Graphs and Digraphs*, 2nd ed., Wadsworth, Monterey, California, 1986.

[6] A.P. Dempster, N.M. Laird, D.B. Rubin, "MAXIMUM likelihood from incomplete data via the *EM* algorithm," *J. of the Royal Statistical Society, B.*, Vol.39, pp.1-38, 1977.

[7] R.O. Duda, P.E. Hart, *Pattern Classification and Scene Analysis*, Wiley, New York, 1973.

[8] J.C. Dunn, "A fuzzy relative of the ISODATA process and its use in detecting compact well-separated clusters," *J. of Cybernetics*, Vol.3, pp.32-57, 1974.

[9] J.C. Dunn, "Well-separated clusters and optimal fuzzy partitions," *J. of Cybernetics*, Vol.4, pp.95-104, 1974.

[10] J.C. Dunn, "A graph theoretic analysis of pattern classification via Tamura's fuzzy relation," *IEEE Trans., Syst., Man, and Cybern.*, Vol.4, pp.310-313, 1974.

[11] B.S. Everitt, *Cluster Analysis*, 3rd ed., Arnold, London, 1993.

[12] R.J. Hathaway and J.C. Bezdek, "Switching regression models and fuzzy clustering," *IEEE Trans. on Fuzzy Syst.*, Vol.1, pp.195-204, 1993.

[13] F. Höppner, F. Klawonn, R. Kruse, and T. Runkler, *Fuzzy Cluster Analysis*, Wiley, Chichester, 1999.

[14] A.K. Jain and R.C. Dubes, *Algorithms for Clustering Data*, Prentice Hall, Englewood Cliffs, NJ, 1988.

[15] K. Jajuga, "L_1-norm based fuzzy clustering," *Fuzzy Sets and Systems*, Vol. 39, pp.43-50, 1991.

[16] J.M. Keller, M.R. Gray, and J.A. Givens, Jr., "A fuzzy k - nearest neighbor algorithm," *IEEE Trans., on Syst., Man, and Cybern.*, Vol.15, pp.580-585, 1985.

[17] T. Kohonen, *Self-Organizing Maps*, 2nd Ed., Springer, Berlin, 1997.

[18] R. Krishnapuram and J.M. Keller, "A possibilistic approach to clustering," *IEEE Trans. on Fuzzy Syst.*, Vol.1, No.2, pp.98-110, 1993.

[19] J.B. Kruskal, Jr., "On the shortest spanning subtree of a graph and the traveling salesman problem," *Proc. Amer. Math. Soc.*, Vol.7, No.1, pp.48-50, 1956.

[20] R.-P. Li and M. Mukaidono, "A maximum entropy approach to fuzzy clustering," *Proc. of the 4th IEEE Intern. Conf. on Fuzzy Systems (FUZZ-IEEE/IFES'95)*, Yokohama, Japan, March 20-24, 1995, pp.2227-2232, 1995.

[21] J.B. MacQueen, "Some methods of classification and analysis of multivariate observations," *Proc. of 5th Berkeley Symposium on Math. Stat. and Prob.*, pp.281-297, 1967.

[22] F.Masulli, M.Artuso, P.Bogus, and A.Schenone, "Fuzzy clustering methods for the segmentation of multivariate medical images," *Proc. of the 7th International Fuzzy Systems Association World Congress (IFSA'97)*, June 25-30, 1997, Prague, Chech, pp.123-128, 1997.

[23] Z. Michalewicz, *Genetic Algorithms + Data Structures = Evolution Programs*, Springer, Berlin, 1992.

[24] S. Miyamoto, *Fuzzy Sets in Information Retrieval and Cluster Analysis*, Kluwer Academic Publishers, Dordrecht, 1990.

[25] S. Miyamoto, "Fuzzy graphs as a basic tool for agglomerative clustering and information retrieval," In: O.Opitz, *et. al.*, (Eds.), *Information*

and Classification: Concepts, Methods, and Applications, Springer-Verlag, Berlin, pp.268-281, 1993.

[26] S. Miyamoto and Y. Agusta, "An efficient algorithm for ℓ_1 fuzzy c-means and its termination," *Control and Cybernetics* Vol. 24, No.4, pp.421-436, 1995.

[27] S. Miyamoto and S. Katoh, "Metaheuristic methods for optimal clustering," *Computing Science and Statistics*, Vol.29, No.2, pp.439-443, 1997.

[28] S. Miyamoto and M. Mukaidono, "Fuzzy c - means as a regularization and maximum entropy approach," *Proc. of the 7th International Fuzzy Systems Association World Congress (IFSA'97)*, June 25-30, 1997, Prague, Chech, Vol.II, pp.86-92, 1997.

[29] S. Miyamoto and Y. Agusta, "Algorithms for L_1 and L_p fuzzy c-means and their convergence," in: C. Hayashi, *et. al.*, (Eds.), *Data Science, Classification, and Related Methods*, Springer-Verlag, Tokyo, pp.295-302, 1997.

[30] S. Miyamoto and S. Katoh, "Crisp and fuzzy methods of optimal clustering on networks of objects," *Proc. of KES'98*, April 21-23, 1998, Adelaide, Australia, pp.177-182, 1998.

[31] S. Miyamoto and K. Umayahara, "Fuzzy clustering by quadratic regularization," *Proc. of FUZZ-IEEE'98*, May 4-9, 1998, Anchorage, Alaska, pp.1394-1399, 1998.

[32] S. Miyamoto and K. Umayahara, "Two methods of fuzzy c-means and classification functions," *Proceedings of the 6th Conference of the International Federation of Classification Societies (IFCS-98)*, Rome, 21-24, July, 1998, pp.105-110, 1998.

[33] Y. Nakamori, M. Ryoke, and K. Umayahara, "Multivariate analysis for fuzzy modeling," *Proc. of the 7th International Fuzzy Systems Association World Congress (IFSA'97)*, June 25-30, 1997, Prague, Chech, Vol.II, pp.93-98, 1997.

[34] N.K. Pal, J.C. Bezdek, and E.C.-K. Tsao, "Generalized clustering networks and Kohonen's self-organizing scheme," *IEEE Trans. on Neural Networks*, Vol.4, No.4, pp.549-557, 1993.

[35] F.P. Preparata and M.I. Shamos, *Computational Geometry: An Introduction*, Springer, New York, 1985.

[36] V.V. Raghavan and K. Birchard, "A clustering strategy based on a formalism of the reproductive processes in natural systems," *Proceedings of the Second International Conference on Information Retrieval*, pp.10-22, 1979.

[37] R.A. Redner and H.F. Walker, "Mixture densities, maximum likelihood and the EM algorithm," *SIAM Review*, Vol.26, No.2, pp.195-239, 1984.

[38] K. Rose, E. Gurewitz, and G. Fox, "A deterministic annealing approach to clustering," *Pattern Recognition Letters*, Vol.11, pp.589-594, 1990.

[39] E.H. Ruspini, "A new approach to clustering," *Information and Control*, Vol. 15, pp.22-32, 1969.

[40] M. Sato and Y. Sato, "On a general fuzzy additive clustering model," *Intern. J. of Intelligent Automation and Soft Computing*, Vol.1, No.4, pp.439-448, 1995.

[41] R.N. Shepard and P. Arabie, "Additive clustering: representation of similarities as combinations of discrete overlapping properties," *Psychological Review*, Vol.86, No.2, pp.87-123, 1979.

[42] S. Tamura, S. Higuchi, and K. Tanaka, "Pattern classifications based on fuzzy relations," *IEEE Trans. on Syst., Man, and Cybern.*, Vol. 1, No. 1, pp.61-66, 1971.

[43] A.N. Tihonov, "Solutions of incorrectly formulated problems and the regularization method," *Dokl. Akad. Nauk. SSSR*, Vol.151, pp.1035-1038, 1963.

[44] A.N. Tihonov and V.Y. Arsenin, *Solutions of Ill-Posed Problems*, Wiley, New York, 1977.

[45] K. Umayahara and Y. Nakamori, "*N*-dimensional views in fuzzy data analysis," *Proc. of 1997 IEEE Intern. Conf. on Intelligent Processing Systems*, Oct.28-31, Beijing, China, pp.54-57, 1997.

[46] N. Wu, *The Maximum Entropy Method*, Springer, Berlin, 1997.

[47] L.A. Zadeh, "Fuzzy sets," *Information and Control*, Vol.8, pp.338-353, 1965.

[48] L.A. Zadeh, "Similarity relations and fuzzy orderings," *Information Sciences*, Vol.3, pp.177-200, 1971.

6

Soft-Competitive Learning Paradigms

Zhi-Qiang Liu
Michael Glickman
Yajun Zhang

6.1 Introduction

Learning is the ability to autonomously select, update, and store relevant information in memory; and the ability to predict and create based on what has been learned.

Natural *evolution* ensures that the species evolve and adapt to the environment. However, natural evolution is a slow process compared to the lifetime of the species, and therefore works only at the level of large groups of organisms. Learning provides *individuals* with the ability to respond to specific environmental demands. Through learning, individual human beings develop internal qualities in addition to what has been offered by Nature in the evolutionary process. Perhaps the most fundamental and important property of the human learning process is the ability to create new knowledge, to analyze and predict. This ability has enabled us to make quantum leaps in our insights into the universe.

In order to develop human-centered servant modules, we must instill advanced learning capabilities into these servant modules, which is important both from theoretical and practical points of view. Making machines learn and think has been our long standing obsession; in particular, in the last twenty years we have made significant progresses in the development of machine learning techniques. In general, machine learning refers to computer models that improve machine's performance and in some cases through learning the machine is able to demonstrate limited *creativity*.

An intelligent machine must have a structure for representing knowledge, rules for reasoning, and dynamics for adaptation. This means that the designer must either build explicitly the rules into the system for carrying out certain tasks, or the system must *learn* them. Learning may result in better and flexible performance and may solve the difficulties in programming the rules for complex intelligent systems. In classical expert systems a large number of explicit rules has to be formulated by the human programmer (or expert). If a learning system can be placed in the environment, it can be *trained* to program by analyzing examples in the environment. Such a system possesses the ability to produce rules on the basis of the examples. As a consequence, it can function without explicit formulation of the rules by the programmer. It can develop its own structure and continuously adapt to its environment without the need for detailed instructions (rules) from the human supervisor.

Over the last few decades researchers have developed many machine learning paradigms. In this chapter, we will first briefly present some basic learning methods then focus on some new techniques in competitive learning.

6.2 Learning by Neural Networks

Neural network is a network of interconnected units, called *neurons* or *nodes*. Information processing in neural networks takes the form of activating neurons and transmitting activations along weighted connections to other neurons. Each neuron processes incoming activations according to their values and weights producing an output signal passed to other units. The system learns according to given *samples*. Learning changes weights that adapt the neural network to the new pattern. The set of weights expresses the accumulated knowledge of the system. This knowledge can be obtained by the following learning methods:

- Supervised learning;
- Unsupervised learning;
- Reinforcement learning.

6.2.1 Supervised Learning

Supervised learning is a paradigm for learning a certain classification or behavior pattern, when the input-output pairs of training data sets are provided. This learning method is implemented by presenting training examples to the learning unit. The examples are labeled. The training set is used by the learning system to determine the pattern criteria for each class. A supervised learning scheme involves an algorithm that, for the given input data, computes the output and compares it with the *desired* output data. The objective of this algorithm is to tune the network so that the error between the desired and calculated outputs becomes as small as possible. When this happens, we say that the network has been *trained* to perform as what is expected by the teacher (supervisor). The first supervised learning scheme, the *Perceptron* model [15], dates back to late the 50s and deals with simulation of the human visual perceptual system. Later this model was extended to *multilayer perception models*, one of which is the back-propagation algorithm (see e.g., [16]) that is most widely used for solving classification problems. Although an important learning paradigm, supervised learning lacks the ability to *autonomously* organize, order and select relevant information from its input. The traditional supervised learning resembles rote learning that recalls only what's stored.

6.2.2 Unsupervised Learning

Unsupervised learning, on the contrary, deals with initially unlabeled data and is used when labeled training examples are not available or little known beforehand. No desired output is specified, so that regularities in input samples must be autonomously detected and stored by the system. The unsupervised learning system must be able to extract relevant properties from the training samples, find common patterns among them and formulate the discrimination criteria for object recognition. An impressive example of unsupervised learning is the ability of babies in recognizing faces: they are able to recognize faces after a short period of exposure to the face without being explicitly instructed. The technique for implementing unsupervised learning is clustering that is the process of grouping or classifying objects on the basis of shared characteristics. The objects are some physical or abstract entities and their characteristics are represented in terms of attribute values, relations between objects or both. Clustering can be characterized as discovery learning, since it involves finding similarities between objects and grouping them into clusters. A well-known paradigm is competitive learning in which the prototypes that represent different clusters compete for their right to represent an input sample. From the classic competitive learning paradigm [17] a variety of competitive learning methods have been developed, which are differentiated in their approaches to competition and learning (weigh update) rules.

6.2.3 Reinforcement Learning

Reinforcement learning (also called *reward-penalty learning*) in some sense combines the above two paradigms. Generally, a reinforcement learning system consists of the following elements: a policy, a reward function, a value function, and an optional environmental model [21]. Reinforcement learning does not prescribe an action for each situation. Instead, the learner has to select the action that produces the highest reward (reinforcement signal). In reinforcement learning, a sample vector is presented to the network, and the output is evaluated. If based on some measurable outcomes (e.g., a desirable move in chess playing), it is considered as good, a reward is given to the system resulting in strengthening the weight vectors. On the contrary, for a bad output signal the network is punished as 'not properly set'. Whereas supervised learning can be considered as learning with a teacher, reinforcement learning is *learning with a critic*. Reinforcement learning is a close model of certain human learning processes (e.g., secondary reinforcement) and received much attention in classical psychology [19, 23]. In biological learning, rewards or punishments may be provided by some instructors, but most often they are expressed as environmental changes (negative or positive) as the result of particular actions of the learner.

In artificial intelligence research, a classic system is the checker player developed in 1959 in which Samuel used the reinforcement principle [18]. More recently, researchers have proposed two reinforcement strategies: *immediate* reinforcement strategy (e.g., REINFORCE algorithms [28]) in which the reinforcement is determined by only the most recent input-output pair, and *delayed* reinforcement strategy in which the learning system observes the pairs in order to predict the expected reward. This approach is the basis of *temporal-difference methods* [20], such as Q-learning [29, 30].

Although there are different machine learning paradigms, they complement each other and play different roles in different application areas. In this chapter, we will consider soft-competitive learning techniques that have been developed recently and have shown a greater potential in their applications in human-centered systems. Section 6.3 discusses the competitive learning paradigm and its different realizations. This is followed by the introduction of some new competitive learning algorithms. These ideas are based on more flexible approaches to competition and learning rules, which have variable degree of frequency sensitivity and continuous winning counter. We will propose a new competitive learning paradigm: Compensated Competitive Learning (CCL). The CCL does not give heavy penalty to remote nodes that have weaker interference to competition.

To illustrate the effectiveness of our compensated competitive learning algorithm, we compare test results with some well-known competitive learning methods.

6.3   Competitive Learning Paradigm

Competitive learning can be described as a case of a *vector quantizer* [10], which is a clustering algorithm that provides a mapping of the p-dimensional Euclidean space $\Re^p$ into a *finite* subset W. We will refer to elements of W as *prototype vectors* or *weight vectors*. Any p-dimensional vector from $\Re^p$ can be attributed to a particular weight vector. Each weight vector can be associated with a *cluster*, which is the subset of $\Re^p$ consisting of all elements, attributed to this weight vector. Thus the set of weight vectors W divides the whole Euclidean space $\Re^p$ into n clusters, where n is the number of elements in W.

Normally the mapping is performed by the *nearest prototype rule*, according to which, for a vector $\vec{x}$ from $\Re^p$, we can choose a prototype $\vec{w_i}$ so that it gives the minimum distance from $\vec{x}$,

$$\vec{Q}(x) = \vec{w}_i, \text{ if } d(\vec{x}, \vec{w}_i) \leq d(\vec{x}, \vec{w}_j), j = 1, ...n, \tag{6.1}$$

where

$$d(\vec{x}, \vec{y}) = \|\vec{x} - \vec{y}\|, \tag{6.2}$$

is the distance between $\vec{x}$ and $\vec{y}$.

To evaluate how good the set of prototypes represents the clusters, the *distortion value* D is introduced as expectation $E[d(\vec{x}, \vec{Q}(x))]$ of the distances between a vector and its prototype,

$$D = E[d(\vec{x}, \vec{Q}(x))] = \int d(\vec{x}, \vec{Q}(x))p(\vec{x})dx, \tag{6.3}$$

where $p(x)$ is the probability density function. The main objective of a competitive learning scheme is to find the set of prototypes that produces the minimum total distortion. The total distortion can be expressed as the sum of subdistortions D_i for each cluster,

$$D_i = \int_{x \in C_i} d(\vec{x}, \vec{w}_i)p(\vec{x})dx, \tag{6.4}$$

where C_i denotes the cluster, corresponding to the i-th weight vector. Therefore

$$D = \sum_{i=1}^{W} D_i. \tag{6.5}$$

Hence the problem can be reduced to minimizing the subdistortion in each cluster. Theoretically, the minimum subdistortion for cluster i is reached when its prototype is located in the *centroid* of the cluster [8]. In the Euclidean space, the centroid can be evaluated as the mathematical expectation of vectors corresponding to the cluster,

$$\vec{z}_i = E[\vec{x}|\vec{x} \in C_i] = \int_{\vec{x} \in C_i} \vec{x}p(\vec{x})dx. \tag{6.6}$$

In practice, the centroid can be determined from a finite number of elements from $\Re_p$ called *sample vectors*. Assuming equiprobable distribution ($p(x) =$ constant), for a given set of S sample vectors, we have

$$X = \{\vec{x}_1, \vec{x}_2...\vec{x}_S\}, \ S \geq n,$$

the centroid location is given by

$$\vec{z}_i = \frac{1}{S} \sum_{\vec{x}_k \in C_i} \vec{x}_k. \tag{6.7}$$

Note that condition (6.7) is sufficient for *local* minimum of D, and only *necessary* for a global minimum. This approach is used in one of the first VQ algorithms, the LBG algorithm [13]. The LBG algorithm uses the nearest prototype rule (6.1) to adjust partitions and recalculate the centroid position using Equation (6.7) until a certain convergence criterion is met. Since Equation (6.7) does not guarantee the global minimum, the same problem exists for LBG algorithm. Therefore its success is largely dependent on the initial locations of prototypes.

6.3.1   Classic Competitive Learning

Consider a set of S sample vectors in the p-dimensional Euclidean space $\Re^p$ $X = \{\vec{x}_s\}_{s=1}^S$, where $\{\vec{x}_s\}_{s=1}^S$ is a concise notation of $\{\vec{x}_1, \vec{x}_2, \ldots, \vec{x}_S\}$. The algorithm of classic competitive learning can be described as follows.

Initialization Mark all sample vectors as active.
  Choose a reasonable number n of prototypes $\{\vec{w}_i\}_{i=1}^n$.
Learning For k from 1 to S perform the following *learning cycle*:
  1. Take an arbitrary active sample vector $\vec{x}_k$.
  2. Competition:
     Find the winning node $\vec{w}_c$, such as

$$\|\vec{x}_k - \vec{w}_c\| = min_i\|\vec{x}_k - \vec{w}_i\|, \qquad (6.8)$$

     where $\|\ldots\|$ denotes the Euclidean norm.
  3. Weight update:
     Adjust the winning prototype:

$$\begin{aligned} \Delta\vec{w}_c &= \alpha_c(\vec{x}_k - \vec{w}_c), \qquad (6.9) \\ \vec{w}_c' &= \vec{w}_c + \Delta\vec{w}_c. \end{aligned}$$

  4. Mark sample vector $\vec{x}_k$ as inactive.

We notice that

- The classic competitive learning uses the Euclidean distance as a measure in its competition.
- The purpose of competition is to find *one* prototype that is the closest to the sample vector (the minimum distance rule). This vector, *the winner*, is the **only** vector that is updated during current learning cycle.
- The winner is shifted toward the sample vector according to the weight update rule (6.9). The positive factor α_c is the *learning rate* which shows how strong the sample will affect the weight change. Normally $0 < \alpha_c \leq 1$.

The learning scheme, according to which the winner is found by the minimum-distance rule and is shifted *toward* the sample vector, while other prototypes remain unchanged, is referred to as the *winner-take-all* (WTA) learning scheme. In particular, the classic competitive learning can be characterized as the WTA learning scheme.[1]

The value of weight update, $\|\Delta \vec{w}_c\|$, (sometime also called *reward*) can be defined as follows,

$$\|\Delta \vec{w}_c\| = \alpha_c \|\vec{x}_k - \vec{w}_c\|. \tag{6.10}$$

That is, the reward is proportional to the distance between the winner and the sample vector. Hence we can expect it to be more significant at the beginning (i.e., for small numbers of learning cycle k). We use function $argmin_j\ F_j$ to denote the smallest value of index j producing the minimum of F_j. In other words, $i = argmin_j\ F_j$ means that for any valid argument j the following conditions hold:

1. $F_i \leq F_j$;
2. $F_i = F_j$ implies $i \leq j$.

Later, we will use argument i with an optional condition, separated by a vertical bar (|) that restricts the range of the argument; for example, $min_{j|j \neq c} F_j$ indicates that the scan is restricted by values j, other than specified parameter c. An important property of $argmin_j\ F_j$ is that it remains unaffected when any monotonically increasing function is applied to F_j, that is

$$argmin_j\ F_j = argmin_j\ G(F_j),$$

where function G satisfies the condition: $G(y) > G(x)$ whenever $y > x$. In particular, F_j can be multiplied by any constant positive factor, or square function can be applied to F_j. Consequently, rule (6.8) is equivalent to

$$c = argmin_i\ \|\vec{x}_k - \vec{w}_i\|^2. \tag{6.11}$$

Equation (6.11) is very efficient in computation, because it eliminates the need to evaluate square roots in each cycle. In addition, it offers generality.

Figure 6.1(a) illustrates typical behaviors of the classic competitive learning algorithm. In this figure, the sample data consists of three clusters, each is represented by 200 normally distributed sample vectors. We have assumed six prototypes in the vicinity of the samples' global mean ($\vec{x}_m = \frac{1}{S}\sum_{k=1}^{S}\vec{x}_k$) and set the learning rate, $\alpha = 0.5$. We can see that at the end of learning cycles, only three prototypes have been relocated close to the local mean of corresponding cluster, whereas other three prototypes were

---

[1]Some authors use winner-take-all to mean the classic competitive learning. In this chapter, it is used in a broader context.

not affected. Figure 6.1(b) demonstrates a serious problem with the classic competitive learning algorithm. In this case, we have placed the six prototypes far away from the clusters. As a result, the reward produced by the first sample vector is big enough to make the winning prototype considerably closer to *all* sample vectors than all other five prototypes. Therefore, the winner in the first learning cycle keeps winning in all the following cycles, and in fact, oscillates between the three clusters. All other prototypes have no effect on competition and are referred to as the *'dead nodes'*.

6.4   Overview of Competitive Learning Schemes

The early competitive learning model was proposed by Von der Malsburg [14] in his study of the visual cortex. The classic competitive learning algorithm [17] assumes competition mechanism based on the *winner-take-all* paradigm. Each time, when a sample vector is presented, the nearest prototype is considered to be *the winner*. The learning consists of shifting the winner toward current sample vector. It has been shown [25] that the classic competitive learning algorithm is comparable to the LBG algorithm discussed previously. Also, [25] proves that the classic competitive learning scheme can sometimes escape from unstable local minima, while LBG algorithm cannot. The convergence of the classic competitive learning algorithm depends on the number of prototypes and their original locations. Rumelhart and Zipser [17], who demonstrated the advantages of the competitive learning mechanism, pointed out the problem of *'dead nodes'* as shown in Figure 6.1. If initially prototypes are located close to each other and far away from the sample data, it may result in missing some important clusters, because some prototypes will never win whereas the 'successful' prototypes will take possession of all the sample data and oscillate among them. Rumelhart and Zipser proposed two ways to avoid the problem of dead nodes.

6.4.1   Winner-Take-Most (WTM) Paradigm

The problem of dead nodes can be avoided by using the *winner-take-most* paradigm that treats all prototypes, as winners to a *certain degree*, and updates their weights accordingly. The winner-take-most principle is used, for example, in the fuzzy competitive learning algorithm [6] and the soft-competition scheme [27].

The fuzzy competitive learning algorithm combines the idea of classic competitive learning with the fuzzy set theory [31]. In the fuzzy competitive learning algorithm there is no obvious winner, because each prototype is considered to be a winner only to a certain degree. Consequently, each prototype is adjusted proportionally according to a *fuzzy-scaling function*

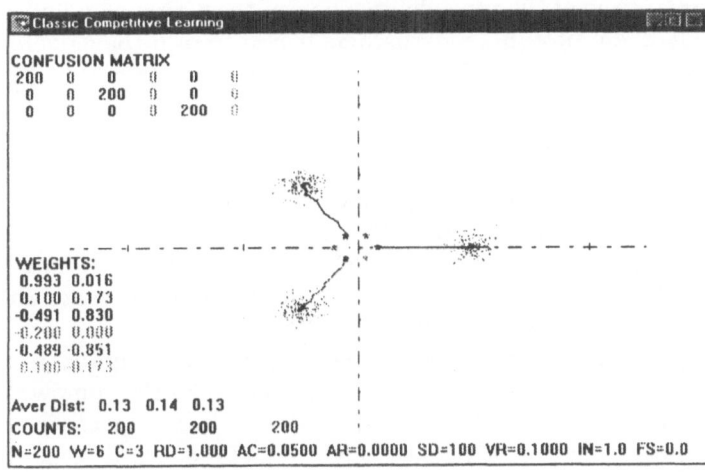

(a)

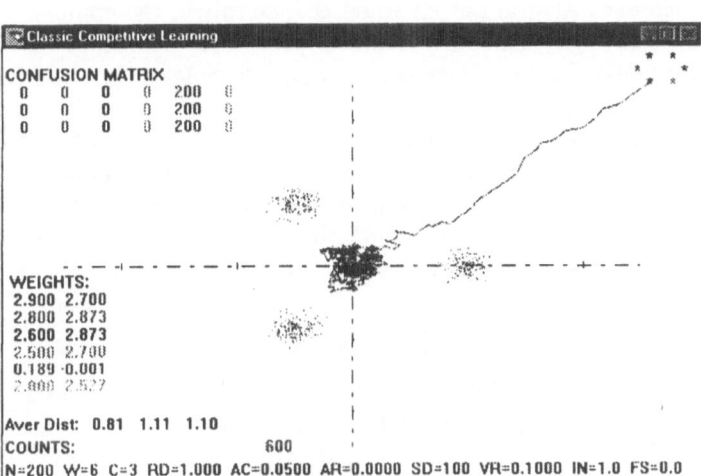

(b)

FIGURE 6.1. Classic competitive learning. In each case, six prototypes were used. (a) shows three winning prototypes, whereas the other three prototypes remain unchanged; (b) in this case, only one prototype that is closest to the three clusters has taken over all the three clusters leaving the other five prototypes unchanged.

that represents the prototype's contribution to the competition. The similar idea is used in the soft-competition scheme, but the learning rule is based on the Gibbs sampler [9].

The winner-take-most strategy overcomes the problem of dead nodes. However it has a significant side effect: since all prototypes are attracted to each sample, some of them are detracted from their corresponding clusters. As a result, the prototypes may become biased toward the global mean of the clusters.

6.4.2   Competitive Learning with Conscience

Another solution, suggested by Rumelhart and Zipser [17] is to modulate the 'sensitivity' of prototypes, so that less frequent winners can increase their chance to win the next time. deSieno [22] proposed a method that achieves this effect and introduced a *conscience* parameter that reduces the rate of frequent winners by making them 'guilty'.

Ahalt *et al.* [1] proposed the frequency sensitive competitive learning (FSCL) algorithm that uses conscience as a factor in the competition. Instead of comparing 'pure' Euclidean distances to determine the winner, the FSCL approach multiplies the distance by a factor which is responsible for the number of previous winnings of the current prototype. This algorithm, therefore, introduces an *equiprobable principle* in the competition, according to which each neuron has an equal chance to win the competition [25]. Frequency sensitivity can be combined with fuzzy learning, resulting in the so-called fuzzy frequency-sensitive competitive learning (FFSCL) algorithm [6].

However, the use of non-Euclidean distance may lead to a problematic situation when the solution is far from being optimal. In particular, when an inappropriate prototype is taken as the winner, one may encounter the problem of 'shared clusters' meaning that a number of prototypes is attracted by the same cluster.

6.4.3   Penalizing in Competitive Learning

Xu *et al.* considered the problem of shared clusters and proposed the rival penalized competitive learning (RPCL) algorithm [26]. The basic idea of RPCL is to shift the second place winner, the *rival*, away from the sample vector preventing its interference in the competition. Although the optimal number of prototypes is initially unknown, the penalizing mechanism can be used to push the redundant prototypes away from the set of sample data, so that these prototypes will have little effect on the competition. The authors suggest that these redundant prototypes determine the locations of reserved clusters that might be useful for further extension.

The authors demonstrated some improvements using RPCL over FSCL. However, as it will be shown later, RPCL has some new problems itself. In addition, it has a side effect: a prototype can be wrongly penalized or the penalizing effect may be too big. We will be show later that the problem of shared clusters is the consequence of non-Euclidean metric in the competition, and can be solved by modifying FSCL so that the effect of non-Euclidean distance can be reduced.

6.4.4   Learning Schemes with Variable Number of Prototypes

Some learning schemes suggest to start with a few number of prototypes and introduce new prototypes in the process of competition. In 1980, Linde *et al.* [13] suggested the *splitting method* for improving the LBG algorithm. According to this method only one prototype is initially considered. At any time when the convergence criterion is met, each prototype is replaced by two prototypes. As a result, we can finish the competition with 2^m winning prototypes, where m can be any nonnegative number.

Goldberg [7] proposed to use the genetic algorithm for training neural networks. He introduces a *fitness measure*: the neurons with high fitness values are reproduced, whereas neurons with small fitness values are eliminated. Genetic algorithm has been used in competitive and selective learning (CSL) algorithms [24, 25]. The CSL algorithm adds *selection mechanism* to competition. The fitness measure of a cluster is defined as a function of its subdistortion defined by Equation (6.4). Therefore, the selection mechanism used by CSL is based on the *equidistortion* principle, according to which a larger number of prototypes have to be assigned to subregions with larger subdistortion. The prototypes with bigger distortions are split. Eventually all prototypes have approximately equal distortion. Thus, in cases when sample distribution is close to normal, the equidistortion principle means that the clusters containing a larger number of sample data will have a bigger number of prototypes, which makes this paradigm applicable only to applications where clusters are expected to have similar number of elements.

6.4.5   Fuzzy Clustering Algorithms

Instead of considering an element as a member of a definite cluster, this element can be viewed as belonging to a cluster to a *certain degree*. In fact, this approach considers clusters as fuzzy sets and is referred to as *fuzzy clustering*. With fuzzy clustering, the membership function u_{ij} is evaluated for any element i that shows the extent to which this element belongs to a particular cluster j. This is the main idea of the well-known Fuzzy C-means algorithm (FCM) developed in the late 70s by Bezdek [2, 3, 4, 5].

More recently, Krishnapuram and Keller [11] have developed a possibilistic C-Means Algorithm (PCM) which is a modification of the classic FCM algorithm. In PCM, the membership function represents the *degree of typicality*, i.e., u_{ij} shows the possibility of element i to be a member of cluster j, whereas in FCM membership functions represent the degree of sharing defined on a convex hull which requires that $\sum_{i=1}^{C} u_{ij} = 1$. In the PCM approach, however, clusters may overlap and $\sum_{i=1}^{C} u_{ij}$ is not necessarily 1. The PCM algorithm defines learning and membership update based on minimizing the following cost function,

$$J_i(\vec{\beta}_i, U_i; X) = \sum_{j=1}^{n} (u_{ij})^m \, \|\vec{x}_j - \vec{\beta}_j\|^2 + \eta_i \sum_{j=1}^{n} (1 - u_{ij})^m, \qquad (6.12)$$

where $\vec{\beta}_j$ is the parameter vector to be estimated and η_i is a constant.

6.5   Fuzzy Competitive Learning and Soft Competition

Dead prototypes can be eliminated if the winner is not given the *absolute* right to possess the sample data. The fuzzy competitive learning (FCL) [6] suggests that each prototype should learn according to its place in the competition. For each prototype, the algorithm uses a *fuzzy-scaling function* that shows the share of this prototype in the competition. For the ith prototype, the fuzzy-scaling function Z_i is evaluated as follows,

$$Z_i = \left(\frac{F_i}{\sum_{l=1}^{W} F_l} \right)^m, \qquad (6.13)$$

where

$$F_i = \left(\frac{1}{\|\vec{x}_k - \vec{w}_i\|} \right)^{\frac{1}{m-1}}. \qquad (6.14)$$

The quantity m, appearing as the exponent in (6.14) is greater than 1 and responsible for fuzziness. As m increases, the clustering becomes fuzzier. The weight-update rule for FCL is defined as follows,

$$\|\Delta \vec{w}_i\| = \alpha \, Z_i \, \|\vec{x}_k - \vec{w}_i\|. \qquad (6.15)$$

Note, however, that the learning rule given by (6.13)-(6.15) can be applied only when the sample data is far enough from its nearest prototype. As the prototypes approach their clusters (and therefore the corresponding sample data), the denominator in (6.14) will become small enough to cause overflow. The closer m is to 1, the sooner overflow will occur. For this reason it is necessary to restrict the use of the fuzzy-scaling function by a particular distance threshold.

Yair *et al.* proposed a soft-competition scheme (SCS) [27] that also assumes learning according to the location of the prototype in the competition. The share P_i^k of a prototype in the step is evaluated according to the following rule,

$$P_i^k = \frac{e^{-\beta(k)\|\vec{x}_k - \vec{w}_i\|^2}}{\sum_{l=1}^{n} e^{-\beta(k)\|\vec{x}_k - \vec{w}_l\|^2}}, \qquad (6.16)$$

where the factor $\beta(k)$ increases with each step, which results in

$$\lim_{k \to \infty} \beta(k) = \infty. \tag{6.17}$$

The authors suggest $\beta = C \bullet ln(n)$. The weight update rule in the soft-competition scheme is given as follows,

$$\|\Delta \vec{w}_i\| = \alpha_i^k P_i^k \|\vec{x}_k - \vec{w}_i\|. \tag{6.18}$$

The learning rate α_i^k is defined for each prototype and is reduced with each competition. The authors used the following formula for updating α_i^k,

$$\alpha_i^k = 1/n_i^k, \tag{6.19}$$

where n_i^k is determined recursively,

$$n_i^k = n_{i-1}^k + P_i^k. \tag{6.20}$$

n_i^k increases according to the result of the competition for the ith prototype at the kth learning cycle, which causes the reduction of α_i^k. As a result, more successful prototypes will be less rewarded in subsequent steps.

Both fuzzy competitive learning and soft-competition scheme can be characterized as the *winner-take-most* (WTM) learning, in that the nearest prototype to the sample vector receives a bigger reward. For this type of learning schemes, we can express the weight update rule as follows,

$$\|\Delta \vec{w}_i\| = \alpha_i^k S_i^k \|\vec{x}_k - \vec{w}_i\|, \tag{6.21}$$

where the scaling factor S_i^k may change with learning, but for each learning cycle k it satisfies the following conditions:

$$\sum_{i=1}^{W} S_i^k = 1, \tag{6.22}$$

$$S_i^k \geq S_j^k \ , \ \text{if} \ \ \|\vec{x}_k - \vec{w}_i\| \leq \|\vec{x}_k - \vec{w}_j\|. \tag{6.23}$$

The classic weight update rule can be considered as a special case of (6.21) when S_i^k is one for the winner and zero for the other prototypes. A scheme using the fuzzy competitive learning paradigm can be void of dead nodes, but usually converges much slower than that for the classic competitive learning algorithm. For this reason, the initial learning rate α should be much bigger than for the classic competitive learning scheme.

There is another problem in the 'winner-take-most' paradigm. Since after each learning cycle, all prototypes get a shift toward the sample, this behavior has a negative effect: a prototype may be detracted from its optimal location and move toward a sample vector belonging to a different cluster. As a result, prototypes may become *biased toward the global means*

6.5.1 Conscience and Frequency Sensitive Competitive Learning

The concept of *conscience* was introduced by deSieno [22] as a way to reduce ability of frequent winners to win in successive learning cycles. This
can be achieved by providing each prototype $\vec{w}_i$ with a *winning counter* n_i
that is initially set to 0 and increased by 1 whenever prototype i wins. This
winning counter can contribute to the competition and make the frequent
winners 'guilty'. The conscience mechanism [22] makes the distinction between competition and adjusting prototypes. The winning prototype is still
determined according to the closest Euclidean distance to the sample. de-
Sieno introduces a *bias* that depends on the number of times a prototype
has won the competition. The bias p_i for the ith prototype is given by the
following recurrent formula,

$$p_i^{new} = p_i^{old} + B(y_i - p_i^{old}), \qquad (6.24)$$

where

$$y_i = \begin{cases} 1, & \text{if } \vec{w}_i \text{ is the winner} \\ 0, & \text{if } \vec{w}_i \text{ is not the winner} \end{cases} \qquad (6.25)$$

and B is a constant chosen so that $0 < B \ll 1$. In his experiments, deSieno
used $B = 0.0001$.

With the introduction of bias, the winner $\vec{w}_i$ is determined as follows

$$\min_k \{ \|\vec{x}_k - \vec{w}_i\|^2 - b_i \}, \qquad (6.26)$$

where $b_i = C(1/n - p_i)$, C is the bias factor (conscience is disabled when
$C = 0$), and n is the number of prototypes.

The conscience introduced by deSieno is *additive* conscience, since the
bias is added to the distance. In some other competition schemes, the conscience is represented as a *factor* of a distance in competition. In other
words, the distance between sample vector $\vec{x}_k$ and prototype $\vec{w}_i$ involves a
conscious factor F_i that increases, or at least does not decrease, with the
growth of winning counter n_i,

$$D(\vec{x}_k, \vec{w}_i) = F_i \|\vec{x}_k - \vec{w}_i\|^2, \qquad (6.27)$$

so that

$$F_i \geq F_j \text{ whenever } n_i > n_j. \qquad (6.28)$$

Function F_i depends not only on the winning counter, but also on some
other parameters related to prototype $\vec{w}_i$. By choosing an appropriate representation of F_i, we can make the conscious factor more or less effective
in competition. In particular, taking $F_i \equiv 1$ eliminates conscience causing

the algorithm to be identical to the classic competitive learning algorithm. Since conscience affects only the competition process and is not concerned with weight update, conscience mechanism can be used in either WTA or WTM learning. The frequency sensitive competitive learning (FSCL) algorithm [1] uses the conscious factor proportional to the winning counter

$$\gamma_i = \frac{n_i}{\sum_{j=1}^{n} n_i},$$
(6.29)

where $\sum_{j=1}^{n} n_i$ is always equal to the number of learning cycle k, hence the expression for γ can be rewritten simply as [2]

$$\gamma_i = \frac{n_i}{k}.$$
(6.30)

Therefore, the distance used by FSCL in competition is defined as

$$FSSD(\vec{x}_k, \vec{w}_i) = \gamma_i \|\vec{x}_k - \vec{w}_i\|^2,$$
(6.31)

and is referred to as the *frequency sensitive square distance (FSSD)*.

The authors of FSCL assume a 'winner-take-all' paradigm with the weight update rule (6.9) being equivalent to that used in the classic competitive learning algorithm. However, we can replace (6.9) by (6.15), or in general by (6.21) to include fuzziness in learning [6]. This modification results in the fuzzy frequency sensitive competitive learning (FFSCL).

Figure 6.2 shows the behavior of the FSCL algorithm scheme. In this figure we used the same set of sample data used in the classic competitive learning algorithm. The learning rate is 0.5. Clearly as shown in Figure 6.2, the FSCL scheme solves the dead-node problem nicely and performed well when the classic competitive learning algorithm failed (compare Figure 6.2(b) with Figure 6.1(b)). Although FSCL enables all prototypes to participate in the competition, Figure 6.2 shows another problem: the classic competitive learning algorithm uses appropriate number of prototypes to represent the clusters and leaves the redundant prototypes "dead", whereas the FSCL causes some prototypes to *share* the same cluster. This is also shown in the *confusion matrix* displayed in the upper left corner of each figure, in which the rows correspond to the clusters, and the columns to the prototypes. An element e_{ij} in the matrix represents the number of sample vectors from the ith class that have been shared by prototype j using the nearest prototype rule; for instance, in Figure 6.2(a) cluster 1 is shared by prototype 1 with 131 sample data and prototype 2 with 69 sample data. This is in sharp contrast to that obtained from the classic competitive learning algorithm shown in Figure 6.1. This phenomenon will

---

[2]Since constant factor $\frac{1}{k}$ does not have any effect on comparison, it can be left out; so anywhere in the following formulae, including next section, γ_i can be replaced by n_i.

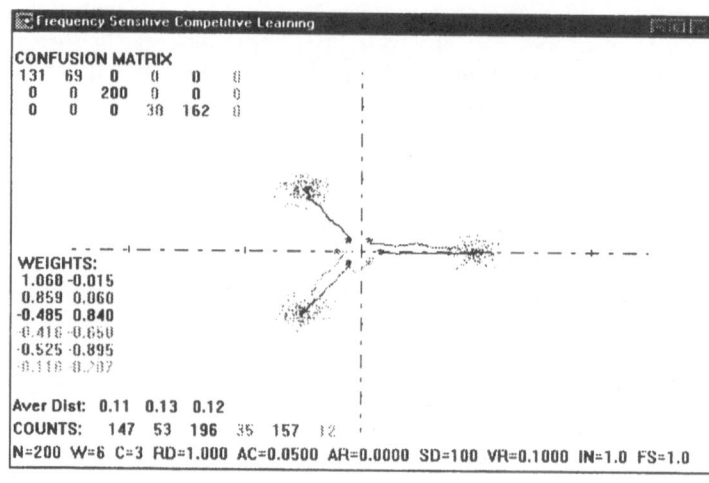

(a)

(b)

FIGURE 6.2. Frequency sensitive competitive learning: in both (a) and (b) all six prototypes have participated in the competition.

cause significant difficulties in decision making in many real-world applications. It will be shown later that shared clusters is the result of using the non-Euclidean distance in the competition. We also consider ways to reduce this effect.

6.5.2 Rival Penalized Competitive Learning

Xu *et al.* [26] proposed an improvement of the WTA learning rule by introducing a *penalizing* mechanism and developed the rival penalized competitive learning (RPCL) algorithm.

In this section, we use the term *penalizing* in a broad context, where as the result of competition, one or more prototypes are moved *away* from the sample vector. That is, the penalty is applied to a prototype $\vec{w}_r$ by sample vector $\vec{x}_k$, which means that $\vec{w}_r$ is shifted so that $\Delta\vec{w}_r$ and $\vec{x}_k - \vec{w}_r$ are collinear and have *opposite* directions. Whereas learning can be considered as a positive reinforcement that encourages the winner to represent the sample vector, penalizing on the contrary can be considered as a negative reinforcement, that would detract a prototype from the sample. The basic penalizing mechanism can be described as follows: once a cluster has been represented by a prototype, another prototype that may potentially share the cluster is *pushed* away from the cluster, preventing it from interfering in the competition.

The RPCL algorithm uses frequency-sensitive distance defined by (6.31) in the competition, and penalty is applied in each learning cycle to a *single* prototype that wins the second place in the competition. The authors claim that the second place winner, the *rival*, is the most probable prototype to share the cluster with the winner. In other words, RPCL establishes the winner $\vec{w}_c$ and rival $\vec{w}_r$ according to the following rules:

$$c = argmin_i \; \gamma_i \|\vec{x}_k - \vec{w}_i\|^2,$$
$$r = argmin_{i|i\neq c} \; \gamma_i \|\vec{x}_k - \vec{w}_i,\|^2,$$

where c and r stand for the winner and the rival, respectively; and $\vec{x}_k$ is the sample vector at the learning cycle k.

As the result of competition, RPCL assigns weight updates for the winner and its rival (*reward* and *penalty* respectively), according to the following rules:

$$\Delta\vec{w}_c = \alpha_c(\vec{x}_k - \vec{w}_c), \qquad (6.32)$$
$$\vec{w}_c' = \vec{w}_c + \Delta\vec{w}_c,$$

$$\Delta\vec{w}_r = -\alpha_r(\vec{x}_k - \vec{w}_r), \qquad (6.33)$$
$$\vec{w}_r' = \vec{w}_r + \Delta\vec{w}_r,$$

where $0 < \alpha_r \ll \alpha_c \le 1$ are the learning rates for the winner and the rival, respectively. The positive factor α_r (6.33) gives the *penalizing rate* for the rival. The negative sign on the right-hand side of the equation specifies that weight update for the rival is directed *from* the sample vector. Since penalizing may substantially slow down the convergence process, it is necessary to choose α_r to be *much* less than α_c: $0 < \alpha_r \ll \alpha_c \le 1$.

In [26], the advantage of RPCL algorithm to FSCL is demonstrated. The authors compare the performance of FSCL and RPCL on different number of prototypes using sample data from 4 clusters, each represented by a set of 100 normally distributed sample vectors with variance 0.1. The learning and penalizing rates α_c and α_r were set as constants equal to 0.05 and 0.002, respectively. However, the penalizing rule, defined by (6.33) appears to us inconsistent with the idea of penalizing, since the value of penalty does not really depend on how this neuron interferes in the competition. In fact, it will be shown (Section 6.6) that a bigger penalty may be imposed on the prototypes that less deserve it. As a consequence, an inappropriate prototype can be substantially penalized (*over-penalized*), or on the other hand, under-penalized. To illustrate these drawbacks, consider the data in Figure 6.3 that consist of 5 clusters. The centers of clusters were chosen on a circumference with radius 1.5:

$$C_n^1 = 1.5 \; cos\frac{2\pi n - 1}{5},$$
$$C_n^2 = 1.5 \; sin\frac{2\pi n - 1}{5},$$

where $n = 1 \ldots 5$ is the cluster number. The clusters are numbered from 1 to 5 counter clockwise, where zero is assigned to a cluster whose center is located on the positive direction of the x-axis (coordinates (1.5,0)). A set of 500 sample vectors were chosen (normal distribution with variance 0.1).

Figures 6.3 and 6.4 show the performance of RPCL with the learning rate $\alpha_c = 0.05$, and penalizing rates equal to 0, 0.001, 0.0021, and 0.003, respectively. The case where $\alpha_r = 0$ corresponds to the frequency sensitive competitive learning (FSCL) and shows the familiar problem of shared clusters. We can see that the problem of shared clusters persists for $\alpha_r = 0.001$ (underpenalizing, see Figure 6.3(b)). With the increase of α_r, the penalizing effect becomes more significant. When $\alpha_r = 0.0021$ (Figure 6.4(a)) it appears to be "ideal" and produces a perfect confusion matrix. When $\alpha_r = 0.003$, we see the effect of overpenalizing. Figure 6.4(b) shows that only 4 out of the 7 prototypes approached the clusters. Other prototypes drifted away as the result of penalizing. Since there are 5 clusters, one prototype oscillates between two clusters (in the lower part of Figures 6.4(a) and (b))

In the following section, we propose the Compensated Competitive Learning CCL) that is also a penalizing scheme and solves the problems associated with the RPCL discussed above.

6.6   Compensated Competitive Learning

In Section 6.5.2, we discussed the rival penalized competitive learning (RPCL) algorithm. Penalizing does not simply deactivate redundant pro-

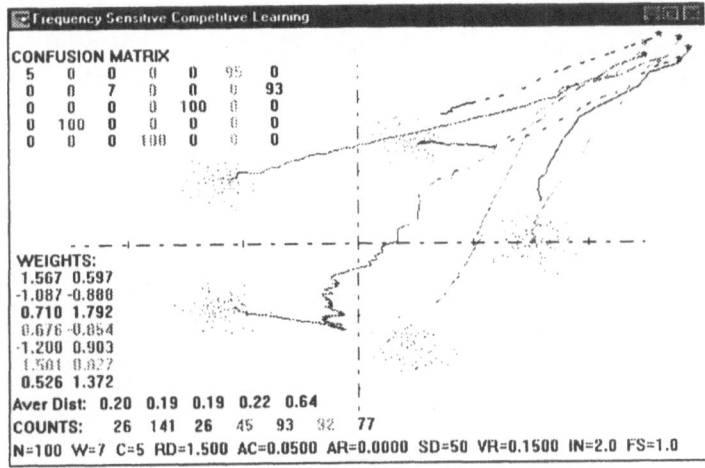

(a)

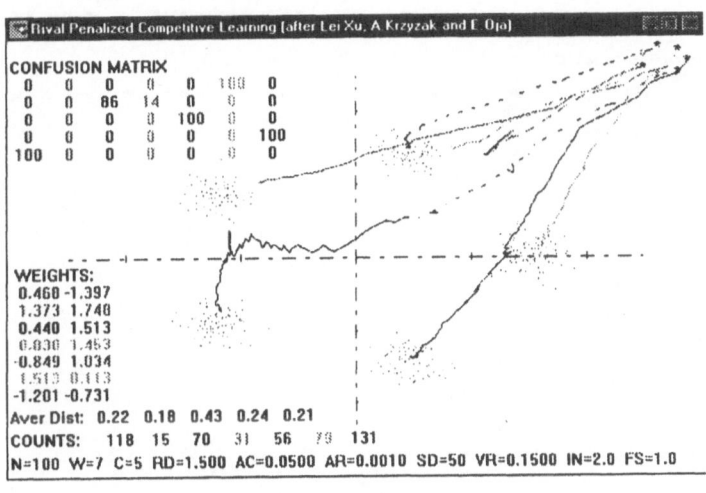

(b)

FIGURE 6.3. Comparing frequency sensitive and rival penalized CL schemes: (a) FSCL ($\alpha_r = 0$) - cluster 2 is shared; (b) RPCL ($\alpha_r = 0.001$) - still cluster 2 is shared (underpenalizing).

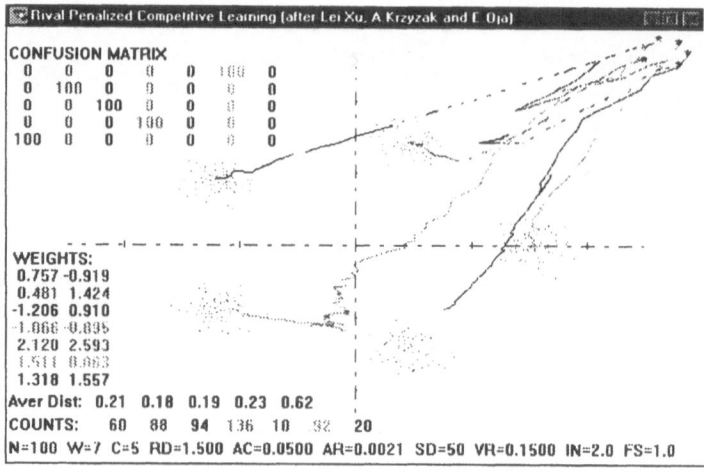

(a)

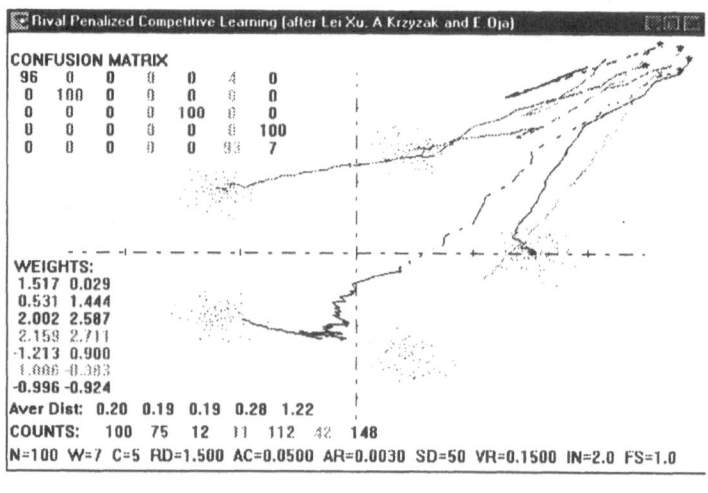

(b)

FIGURE 6.4. Problems with rival penalized competitive learning: (a) $\alpha_r = 0.0021$ - no shared clusters, but cluster 5 is badly represented; (b) $\alpha_r = 0.003$ - clusters 4 and 5 share the prototype (overpenalizing).

totypes, but also pushes these vectors away from the "competition arena". However, as Figures 6.3 and 6.4 show, rival penalizing in RPCL can be insufficient or excessive within a small range of penalizing constants. Even with the appropriate use of learning and penalizing rates, prototypes may be unnecessarily repelled from their clusters. These negative effects arise as the following drawbacks in RPCL:

- RPCL assumes that penalizing is applied to exactly one prototype, no matter how close this prototype is and other prototypes are to the cluster. If two or more prototypes tend to share the cluster, only one of them is penalized. Other rival(s) may not be penalized until another sample vector, belonging to the corresponding cluster, is presented.

- The shift of the rival prototype $\Delta\vec{w}_r$ does not reflect the strength of this prototype's interference in the competition. Therefore rule (6.33) gives

$$|\Delta\vec{w}_r| = \alpha_r\|\vec{x}_k - \vec{w}_r\|; \qquad (6.34)$$

that is, the prototypes located farther away from the cluster receive bigger penalty. The fact, however, is that the actually distant rival prototypes do not cause substantial interference in the competition and do not deserve such a heavy penalty. It is the near, rival prototypes that should be penalized accordingly. Therefore, the level of penalty imposed on the rival prototype by the RPCL is inproportional to the interference produced by the penalized prototype.

6.6.1 The Concept of Compensated Competitive Learning

Our goal is to develop a penalizing scheme that provides a negative reinforcement according to the ability of a prototype to obstruct the competition. As an alternative to the rival penalized scheme, we introduce the concept of *compensation*, assuming that the negative update of a prototype is intended to compensate its possibility to share the cluster. This weight update is applied not only to the second place winner, but also to those vectors that tend toward a cluster already represented by the winner. This new penalizing scheme is called Compensated Competitive Learning (CCL) and is based on the following principles:

- More than one prototype (neuron) can be penalized;
- The shift for a penalized prototype depends on how strong it interferes in the competition.

In the compensated competitive learning algorithm, α_r is the learning rate for non-winners. As for RPCL, $\alpha_r \ll \alpha_c$; however, as it will be shown later, the convergence is not as sensitive to the choice of α_r as that in the RPCL. The CCL algorithm can be summarized as follows.

CCL Algorithm For any pattern $\vec{x}_k$:

    1. Find the winning prototype $\vec{w}_c$, giving the minimum of

$$D(\vec{x}_k, \vec{w}_i) = \gamma_i\|\vec{x}_k - \vec{w}_i\|^2. \qquad (6.35)$$

(This is the same as for FSCL and RPCL).

2. Adjust the weight of winning prototype:

$$\Delta \vec{w}_c = \alpha_c (\vec{x}_k - \vec{w}_c), \qquad (6.36)$$

$$\vec{w}_c' = \vec{w}_c + \Delta \vec{w}_c. \qquad (6.37)$$

3. Determine which prototypes have to be penalized. The number of penalized prototypes may change, or it could be constant. Let P be the number of prototypes to be penalized.

4. Adjust the weight of each penalized prototype, according to the formula

$$\Delta \vec{w}_i = \alpha_r \mu_i (\vec{x}_k - \vec{w}_i)/P, \qquad (6.38)$$

where

$$\mu_i = \left[\frac{D(\vec{x}_k, \vec{w}_c)}{D(\vec{x}_k, \vec{w}_i)} \right]^q \qquad (6.39)$$

and $D(\vec{x}_k, \vec{w}_i)$ is the distance used in competition, see (6.35).

As suggested by CCL's weight update rule (6.38), the penalty depends also on the new factor μ_i given by (6.39), which reflects the *ratio of distances* (multiplied by the relative frequency) from the sample to the winner and to the current prototype. Due to the definition of the winning prototype, the ratio is never greater than 1, i.e., $\mu_i \leq 1$ (in case $\|\vec{x}_k - \vec{w}_c\| = \|\vec{x}_k - \vec{w}_i\| = 0$, we just set μ_i to 1). Using the exponent q we can adjust the effect of μ_i. Factor $\frac{1}{P}$ in (6.38) is introduced in order to reduce the total penalizing effect with a large number of penalized prototypes. However, we need to set some parameters in some special situations:

- Set $\Delta \vec{w}_i$ to 0 (no penalizing) when $P = 0$;
- Set μ_i to 1 when denominator is 0.

6.6.2 *Varying the Number of Penalized Vectors in CCL*

RPCL can be considered as a special case of CCL when P is set to 1 and $q = 0$. Alternatively, P can be set as a constant that is one less than the number of prototypes, meaning that all non-winners are penalized. However, to ensure that penalizing does not slow down the convergence rate at its later stage, it seems to be a good idea to start with penalizing all non-winners and gradually reduce P. A reasonable choice for P would be

$$P = \text{int} \frac{n-1}{1 + \ln k}, \qquad (6.40)$$

where n is the number of prototypes, S is the sample number (1,2...), and int for integer part.

To demonstrate the effect of exponent q, we consider the cases of q equal to 0, 1, and 2. If $q = 0$, then μ_i is always 1, and the distances do not have any effect (as in RPCL). If $q = 1$, the equations in the CCL algorithm yield

$$\|\Delta \vec{w}_i\| = \alpha_r \bullet \frac{1}{P} \bullet \frac{n_c}{n_i} \bullet \|\vec{x}_k - \vec{w}_c\|, \qquad (6.41)$$

where n_i is number of winning.

As Equation (6.41) shows, the shift is now proportional to the distance between the sample and the *winning* prototype rather than to the distance between the sample and the current prototype. Therefore, we can expect the reduction of penalizing effect when the winner is close to the sample. Another factor that contributes to the shift is the ratio $\frac{n_c}{n_i}$ which takes into account the 'activity' of prototype $\vec{w}_i$. Thus, for an active prototype the number of winning is comparable to that of the winning prototype. On the other hand, n_i will be much smaller than n_c, if $\vec{w}_i$ is a dead node, hence the ratio $\frac{n_i}{n_c}$ will be high enough to cause considerable penalty.

If $q = 2$, we have

$$\|\Delta \vec{w}_i\| = \alpha_r \bullet \frac{1}{P} \bullet \left(\frac{n_c}{n_i}\right)^2 \bullet \|\vec{x}_k - \vec{w}_c\|. \qquad (6.42)$$

Now the penalizing effect depends not only on the distance of the sample to the winning prototype, but also on the ratio between the distance from the sample to the winning prototype and the distance from the sample to the current prototype.

Figure 6.5 demonstrates the performance of the compensated competitive learning algorithm on the set of sample data used previously (Section 6.5.2). Figures 6.3 and 6.4 show that the rival penalized competitive learning scheme works successfully only in a limited range of penalizing constant α_r. Now, let's consider the behavior of the compensated competitive learning scheme with $\alpha_r = 0.01$ (Figure 6.5(a) and $\alpha_r = 0.002$ (Figure 6.5(b)). The penalizing constant $\alpha_r = 0.01$ is big enough to give substantial penalty, but only at the initial stage. As prototypes are approaching the clusters, the penalizing effect is being reduced. When $\alpha_r = 0.002$, the penalty is considerably small from the initial steps, but in this case we again obtain a good confusion matrix. We see that, unlike RPCL, CCL works reliably in a bigger range of penalizing constants.

Figures 6.6 to 6.9 further compare the performances of the classic competitive learning, the soft-competitive learning, rival penalized learning, and the compensated completive learning algorithms. In these figures, we generated three clusters; each cluster has 150 samples and the Gaussian distribution with variance 0.05. The initial positions of the prototypes were chosen randomly. For the classic competitive learning algorithm we set the

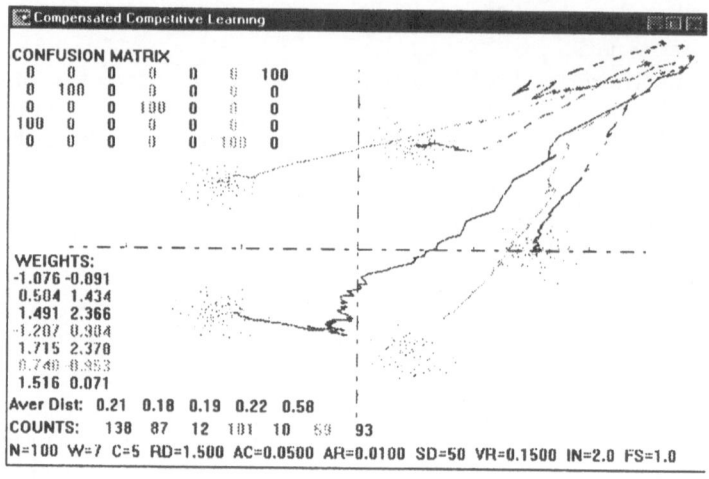

(a)

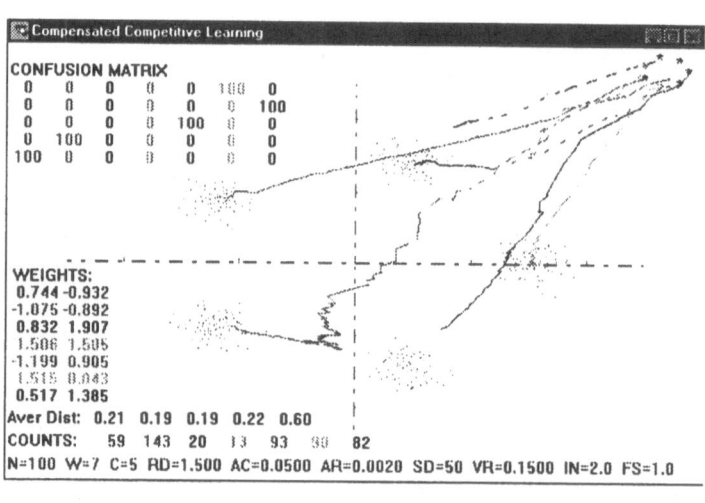

(b)

FIGURE 6.5. Compensated Competitive Learning (CCL) scheme ($q = 1$): (a) Penalizing constant is 0.01; (b) Penalizing constant is 0.002

learning rate to 0.1. In the soft competitive scheme (6.16), we set $C = 0.8$ and $\alpha_i^k = 0.1$. For the RPCL algorithm, we set the learning rate to 0.2 and the penalty parameter to -0.02. In the compensated competitive learning algorithm, only one rival will be penalized at a time. The compensation exponent, $q = 1$, the learning rate was set to 0.2, and the penalty -0.02.

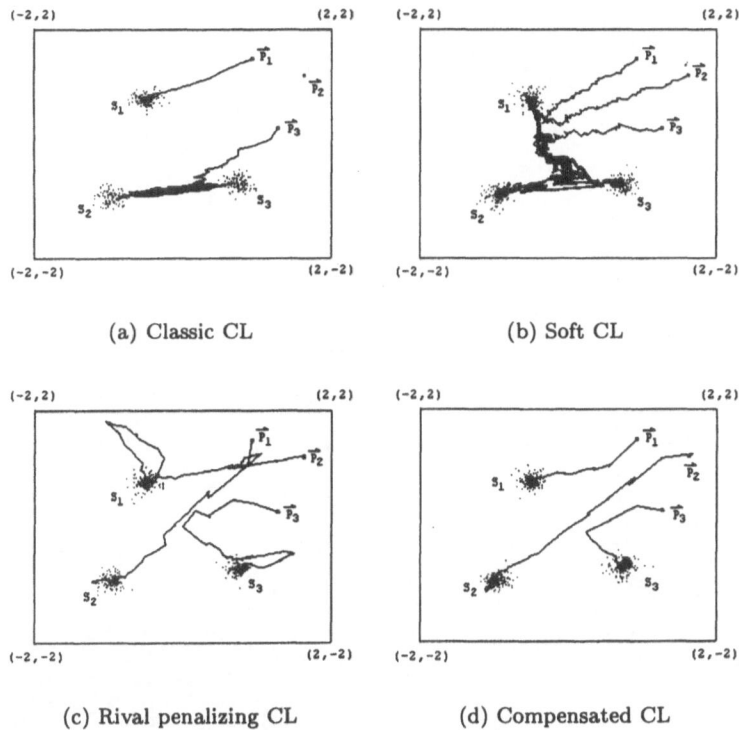

(a) Classic CL   (b) Soft CL

(c) Rival penalizing CL   (d) Compensated CL

FIGURE 6.6. Three clusters with three prototypes: (a) P_1 has settled in S_1; P_2 is a dead node; and P_3 is oscillating between S_2 and S_3. In (b), (c), and (d), all prototypes were able to settle to their corresponding clusters: P_1 has settled in S_1, P_2 in S_2, and P_3 in S_3.

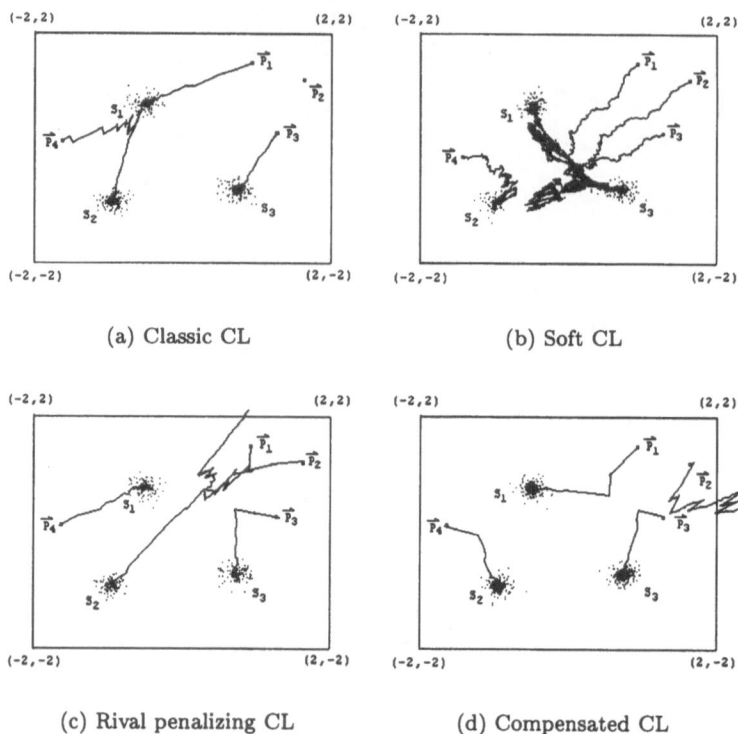

(a) Classic CL   (b) Soft CL

(c) Rival penalizing CL   (d) Compensated CL

FIGURE 6.7. Three clusters with four prototypes: (a) P_1 has taken S_1; P_2 is a dead node; P_3 is in S_3; and P_4 in S_4. (b) P_1 has settled in S_1; P_2 and P_3 now share S_3; P_4 has taken S_2. (c) P_1 has been penalized and pushed away; P_2, P_3, and P_4 get S_2, S_3, and S_1, respectively. (d) P_2 has been penalized and pushed away; P_1, P_3, and P_4 get S_1, S_3, and S_2, respectively.

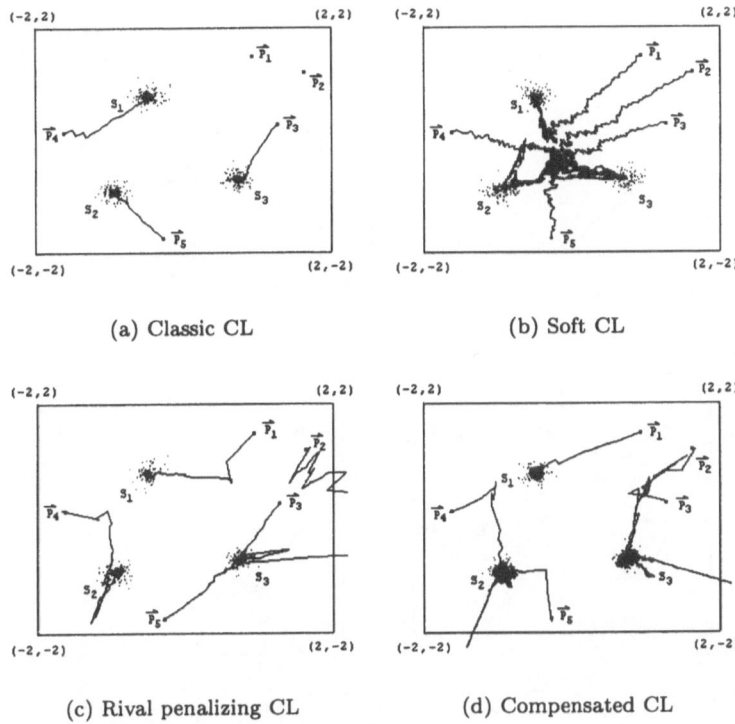

(a) Classic CL

(b) Soft CL

(c) Rival penalizing CL

(d) Compensated CL

FIGURE 6.8. Three clusters with five prototypes: (a) Both P_1 and P_2 are dead nodes; P_3, P_4, and P_5, have settled to S_3, S_1, S_2, respectively. (b) P_1 gets S_1; P_2 and P_3 share S_3; and P_4 and P_5 share S_2. (c) P_1 is in S_1; both P_2 and P_3 have been penalized and pushed away; P_4 and P_5 are in S_2 and S_3, respectively. (d) P_1 is in S_1; both P_2 and P_4 have been penalized and pushed away; P_3 and P_5 are in S_3 and S_2, respectively.

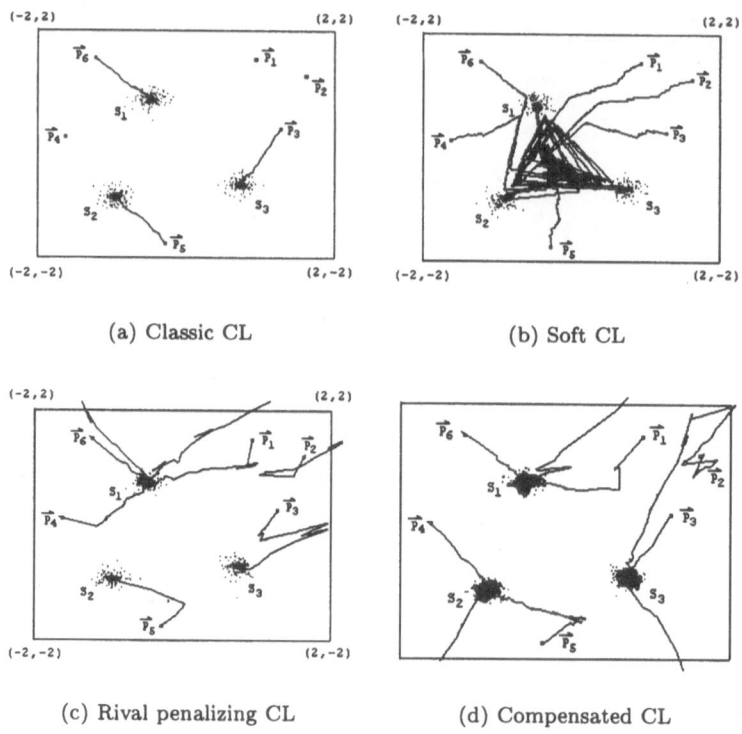

(a) Classic CL

(b) Soft CL

(c) Rival penalizing CL

(d) Compensated CL

FIGURE 6.9. Three clusters with six prototypes: (a) three dead nodes: P_1, P_2, and P_4; P_3, P_5, and P_6 have settled to S_3, S_2, and S_1, respectively. (b) P_1, P_2, and P_3 share S_3; P_4 and P_6 share S_2; and P_5 gets S_1. (c) P_1, P_2, and P_6 have been penalized and pushed away; P_3, P_4, and P_5 have settled in S_3, S_1, and S_2, respectively. (d) P_1, P_2, and P_4 have been penalized and pushed away; P_3, P_5, and P_6 have settled in S_3, S_2, and S_1, respectively.

We can see from these figures that in addition to a proper number of prototypes, the classic competitive learning algorithm is very sensitive to the placement of the initial prototypes as compared to all other competitive learning algorithms. Although the soft-competitive learning algorithm does not have the problem of dead nodes, it suffers from the problem of shared nodes if there are more prototypes than there are clusters, which is shown in Figures 6.7 to 6.9. Both rival penalizing competitive learning (RPCL) and compensated competitive learning (CCL) algorithms are successful in removing the redundant prototypes, thus solving the problems of dead nodes and shared nodes. It has been demonstrated that CCL has a balanced penalizing effect to the prototypes, therefore it is more robust. However, we have to stress that in order for both RPCL and CCL to perform properly the number of initial prototypes must be equal or greater than that of the clusters. This can be a significant problem in many real applications, especially, in such applications as web mining and dynamic database query when the number of clusters vary dramatically. We have recently developed a new algorithm, self-spawning competitive learning (SSCL), which is far more flexible and robust and solves the problems associated with these algorithms. The theory of SSCL and experimental results will be presented in detail in another paper.

6.7   Conclusions

We presented a brief review of some major existing competitive learning schemes. The attention is given to competitive learning schemes involving soft-competition and penalizing. We presented a more general approach to conscience and penalizing than that proposed previously such as frequency sensitive competitive learning (FSCL) and rival penalized competitive learning (RPCL). In particular, we proposed an improved penalizing scheme, Compensated Competitive Learning, which is based on the concept that the value of negative reinforcement (penalizing) should reflect the possibility of a prototype to share the cluster with current winner. Therefore non-winners, located closer to the sample vector, get a bigger shift away form the sample than those remote prototypes that practically do not cause significant interference in the competition. Another advantage of this approach is that the negative reinforcement can be applied to as many prototypes as needed in order to ensure that one and only one prototype represents each cluster. The ideas presented in this chapter are general and can be used in any practical tasks where competitive learning is applicable. With the growing popularity of competitive learning algorithms, we expect that our new approach will find many applications in the development of intelligent servant modules.

6.8   REFERENCES

[1] S.C. Ahalt, A.K. Krishnamurty, P. Chen, and D.E. Melton, "Competitive learning algorithms for vector quantization," *Neural Networks*, Vol.3, No.3, pp.277-291, 1990.

[2] J.C. Bezdek, *Pattern Recognition with Fuzzy Objective Function Algorithms*, Plenum Press, New York, 1981.

[3] J.C. Bezdek, E. C.-K. Tsao, and N. Pal, "Kohonen clustering networks," *Proceedings of IEEE 1st International Conference on Fuzzy Systems*, pp.1035-1043, 1992.

[4] J. C. Bezdek and S. K. Pal, "Fuzzy models for pattern recognition background, significance, and key points," in *Fuzzy Models for Pattern Recognition*, J. C. Bezdek and S. K. Pal, editors, IEEE Press, pp.1-27, 1992.

[5] J. C. Bezdek, J.M. Keller, R. Krishnamupram, and S. K. Pal, *Fuzzy Models and Algorithms in Pattern Recognition and Image Processing*, Kluwer, Norwell, MA, 1999.

[6] F.L. Chung and T. Lee, "Fuzzy competitive learning," *Neural Networks*, Vol.7, No.3 pp.539-551, 1994.

[7] D.E. Goldberg, "Genetic algorithms in search, optimization in machine learning," Addison-Wesley, Cambridge MA, 1989.

[8] A. Gersho, "Asymptotically optimal block quantization," *IEEE Transactions on Information Theory*, Vol.25, No.4, pp.373-380, 1979.

[9] S. German and D. German, "Stochastic relaxation. Gibbs distributions and Bayesian restoration of images," *IEEE Transactions on Pattern Analysis and Machine Intelligence*, vol PAMI-6, pp.721-741, 1984.

[10] T. Kohonen, *Self-Organizing Maps*, 2nd Edition, Springer-Verlag, 1997.

[11] R. Krishnapuram and J. M. Keller, "A Possibilistic Approach to Clustering," *IEEE Transactions on Fuzzy Systems*, Vol.1, No.1, pp.98-110, 1993.

[12] R. Krishnapuram and J. M. Keller, "The possible C-Means Algorithm: Insights and Recommendations," *IEEE Transactions on Fuzzy Systems*, Vol.4, No.2, pp.385-393, 1996.

[13] Y. Linde, A. Buzo, R.M. Gray, "An algorithm for vector quantizer design," *IEEE Transactions on Communications*, Vol.28, No.1, pp.84-95, 1980.

[14] G. von der Malsburg, "Self Organization of Orientation Sensitive Cells in the Striate Cortex," *Kybernetic*, Vol.14 pp.85-100, 1972.

[15] F. Rosenblatt, "The Perceptron: A probabilistic model for information storage and organization in brain," *Psychology Review*, Vol.65, pp.386-408, 1958.

[16] D.E. Rumelhart, G.E. Hinton and R.J.Williams, "Learning internal representation by error propagation" in in D.E. Rumelhart et al *Parallel Distributing Processing vol 1*, MIT Press, Cambridge, MA, pp.318-362, 1986.

[17] D.E. Rumelhart and D. Zipser, "Feature discovery by competitive learning," *Cognitive Science*, Vol.9, pp.75-112, 1985.

[18] A.L. Samuel, "Same studies in machine learning using the game of checkers," IBM Journal on Research and Development, Vol.3, 210-229, 1959.

[19] B.F.Skinner, *The Behavior of Organisms: an experimental analysis*, Appleton-Century-Crofts. New York, 1938.

[20] R.S. Sutton, "Learning to predict by the method of temporal differences," *Machine Learning*, Vol.3, pp.9-44, 1988.

[21] R.S. Sutton and A.G. Barto, *Reinforcement Learning*, MIT Press, Cambridge, Massachusetts, 1998.

[22] D. deSieno, "Adding consciousness to competitive learning," *Proceedings of the 2nd IEEE International Conference on Neural Networks*, Vol.1, No.1, pp.117-124, 1988.

[23] E.L. Thorndike, "Animal Intelligence. Experimental studies," MacMillan, New York, 1911.

[24] N. Ueda and R. Nakano, "A competitive and selecting learning method for designing optimal vector quantizers," *Proceedings of IEEE International Conference on Neural Networks*, pp.1444-1450, 1993.

[25] N. Ueda and R. Nakano, "A new competitive learning approach based on an equidistortional principle for designing optimal vector quantizers." *Neural Networks*, Vol.7, No.8, pp.1211-1227, 1994.

[26] L. Xu, A. Krzyzak, and E.Oja, "Rival penalised competitive learning," *IEEE Transactions on Neural Networks*, Vol.4, No.4, pp.636-648, 1993.

[27] E. Yair, K. Zeger, and A.Gersho, "Competitive learning and soft competition for vector quantizer design," *IEEE Transactions on Signal Processing*, Vol.40, No.2, pp.294-309, 1992.

[28] R. J. Williams, "Simple statistical gradient-following algorithms for connectionist reinforcement learning," *Machine Learning*, Vol.8, pp.229-256, 1992.

[29] C.J.C.H. Watkins, *Learning from delayed rewards*, PhD. dissertation, Cambridge University, 1989.

[30] C.J.C.H. Watkins and P. Dayan, "Q-learning," *Machine Learning*, Vol.8, pp.279-292, 1992.

[31] L.A. Zadeh, "Fuzzy sets," *Information and Control*, Vol.8, pp.338-353, 1965.

7

Aggregation Operations for Fusing Fuzzy Information

Ronald R. Yager

7.1 Introduction

Fuzzy sets can play an important role in the construction of human centered systems. They provide a mechanism for representing information in way that is compatible with human cognition. Particularly notable here is the idea of a linguistic variable [1]. In many of these applications we are faced with the problem of fusing or combining fuzzy subsets. For example, in multi-criteria decision making we need to combine the criteria to form an overall decision function. Since each criteria can be very naturally expressed as a fuzzy subset we are faced with the problem of combining fuzzy subsets. Fuzzy querying of data bases, information retrieval and pattern recognition also require us to face the problem of combining requirements expressed as fuzzy subsets. The construction of fuzzy systems models [2], which have been particularly successful in intelligent controllers, often require the aggregation of fuzzy subsets in the antecedents of the rules. Another class of problems which involves the fusion of fuzzy subsets arises in domains in which we obtain information from multiple sources and we desire to forge this information into one unified value. For example, if we search the Internet for information about the future exchange rate between the dollar and the yen we would get many estimates. Often these estimates are expressed in linguistic terms requiring the use of fuzzy subsets to translate them. Data mining [3] is another area where the fusion of fuzzy information is required.

We briefly describe the general framework for the fusion of fuzzy information [4, 5, 6]. Assume $A_i, i = 1$ to n, are a collection of fuzzy subsets over the space X. The fusion of these fuzzy subsets is indicated as $A = A_1 \oplus A_2 \oplus \ldots \oplus A_n$. Here A is a fuzzy subset of X such that

$$A = \left\{ \frac{\mathrm{U}(A_1(x_1), A_2(x_2), \ldots, A_n(x_n)) \wedge \mathrm{R}(x_1, x_2, \ldots, x_n)}{\mathrm{L}(x_1, x_2, \ldots, x_n)} \right\}$$

where $\wedge$ is the Min [4]. In the above L is an operation that takes a collection of objects in X, $x_1, x_2, \ldots, x_n$, and returns another element in X, it is the <u>basic fusion function</u>. Generally L is an operation that reflects

the way we are going to aggregate or fuse the fuzzy sets. For example, if
the A_j are fuzzy numbers, then L could be the average of its arguments,
this would indicate an averaging of the fuzzy sets. In the case in which
the space X is not numeric the formulation of L requires very special attention. However in general, whatever the structure of X the L operator
is *idempotent*, if $x_j = x$ for all j then $L(x_1, x_2, \dots, x_n) = x$. R is called
the compatibility function, it returns a number in the unit interval which
measures the degree to which it reasonable to combine the objects in its argument to form new objects. It measures how compatible are the elements
in the collection $(x_1, x_2, \dots, x_n)$. While R is generally context dependent
it is clear that if $x_i = x_j$ for all i and j then $R(x_1, x_2, \dots, x_n) = 1$. Two
extreme examples of this compatibility function are: strongly compatible,
$R(x_1, x_2, \dots, x_n) = 1$ for all arguments and strongly incompatible, here
$R(x_1, x_2, \dots, x_n) = 0$ except in the case when all arguments are the same.
A typical example of this strongly incompatible case is one in which the
x_j are people, here we cannot combine people. In multiple criteria decision
making if the x_j are distinct alternative actions they are generally incompatible. On the other hand if X is the space of numbers than R can be
strongly compatible; however as noted in [4] often even in this case we only
have partial compatibility, this allows us to handle conflicts intelligently.
The function U reflects the imperative used to determine how strongly a
tuple $(x_1, x_2, \dots, x_n)$ is in the overall set, if $A_j(x_j)$ is the degree to which
x_j is in the set A_j. The function U aggregates the memberships in the
individual fuzzy subsets to find its membership in the fused fuzzy set, we
shall denote U as a membership aggregation function.

It is our intention here to focus on this operation U of aggregating membership functions. We note that while it appears here only as a part of the
general problem of fusion of fuzzy subsets there is an important special case
where it is the sole operation involved. Consider the special case where R
is strongly incompatible, $R(x_1, x_2, \dots, x_n) = 0$ except when all the x_j are
the same, where it equals one. Because of the required idempotency of L
we get

$$A = \left\{ \frac{U(A_1(x), A_2(x), \dots, A_n(x))}{x} \right\}$$

that is for each x, $A(x) = U(A_1(x), A_2(x), \dots, A_n(x))$. Thus in this case
our only concern is with the imperative used to decide how we aggregate
elements contained in the different sets to form a new set. The intersection
and union of fuzzy sets introduced by Zadeh in his pioneering paper [7]
provide a prototypical example of this kind of fusion of fuzzy sets. With
the intersection we are requiring that an element x be in all the A_j for it
to be in the fused set A, while with the union we are only requiring that it
be in at least one of A_j for it to be in the fused set A.

In this chapter we provide a number of different methodologies for aggregating fuzzy subsets. We first look at the intersection and the union

operation. We discuss two classes of aggregation operators, the t-norm and t-conorm, which generalize the intersection and union operations. We then turn to the issue of weighted intersections and unions. Next a generalization of the t-norm and t-conorm called the uninorm is introduced. We then move to mean operators. We provide a characterization for the mean operator and discuss one class called the generalized mean. We next turn to the Ordered Weighed Aggregation (OWA) operators which are also a type of mean operator. We then use these OWA operators to help make a connection between the natural language stipulation of aggregations in terms of linguistic quantifiers and the mathematical realization of the stipulated aggregation. Finally we discuss the idea of a fuzzy measure show how it can be used in aggregation operators.

7.2 Intersection and Union of Fuzzy Sets

In Zadeh's original work the union and intersection of fuzzy sets were defined in terms of the Max and Min operations: $F = A \cup B$ where $F(x) = \text{Max}[A(x), B(x)]$ and $E = A \cap B$ where $E(x) = \text{Min}[A(x), B(x)]$. The Max and Min operations are not the only choice for defining union and intersection of fuzzy sets, although they are very special choices and the ones most often used.

Assume A and B are two fuzzy subsets defined over the set X. We shall denote their intersection as a fuzzy subset $E = A \cap B$ and their union as a fuzzy subset $F = A \cup B$. In defining these operators a primary consideration is that they must reduce to the ordinary classic set operators when the subsets are crisp sets. Implicit in this is the requirement that these operators be defined in a **pointwise** manner: $E(x)$ and $F(x)$ must solely depend on $A(x)$ and $B(x)$. The formalization of this requirement is that $E(x) = \text{T}(A(x), B(x))$ and $F(x) = \text{S}(A(x), B(x))$. An important implication of this pointwiseness is that we can simply concentrate on the the structure of T and S for the definition of intersection and union.

One property generally expected of these operators is that of **commutativity**:

$$E = A \cap B = B \cap A \text{ and } F = A \cup B = B \cup A.$$

This requirement reflects into the conditions:

$$\text{T}(A(x), B(x)) = \text{T}(B(x), A(x))$$

$$\text{S}(A(x), B(x)) = \text{S}(B(x), A(x)).$$

Commutativity brings with it the property that the ordering of the sets aggregated is unimportant.

A second property that is generally required is that of **associativity**. This property is

$$(A \cap B) \cap C = A \cap (B \cap C) \text{ and } (A \cup B) \cup C = A \cup (B \cup C).$$

It can be expressed as

$$T(A(x), T(B(x), C(x))) = T(T(A(x), B(x)), C(x))$$

$$S(A(x), S(B(x), C(x))) = S(S(A(x), B(x)), C(x)).$$

Associativity, an extremely useful property, allows us to provide a definition of union (intersection) in terms of two sets and then extend it to any number of sets.

Another property required is **monotonicity**. Assume $A(x) \geq A(y)$ and $B(x) \geq B(y)$ then monotonicity requires $T(A(x), B(x)) \geq T(A(y), B(y))$ and $S(A(x), B(x)) \geq S(A(y), B(y))$.

The three conditions, commutativity, associativity and monotonicity are shared by both the intersection and union operations. A fourth condition is the existence of some fixed identity set for that operation. The identity set is such the union (intersection) of any set A with this identity results in A. The operations of intersection and union are distinguished by their respective identity sets. For the intersection the identity is the whole space, X, while for the union the identity is the null set, $\varnothing$. The introduction of these identities implies that $T(A(x), 1) = A(x)$ and $S(A(x), 0) = A(x)$.

The four conditions, commutativity, associativity, monotonicity and respective identities have been used to characterize the T and S operators which define the general class of intersection and union operators. An operator $T : [0, 1] \times [0, 1] \rightarrow [0, 1]$ is called a **t-norm** operator [8, 9, 10, 11] if

(1) $T(a, b) = T(b, a)$ Commutativity
(2) $T(a, T(b, c)) = T(T(a, b), c)$ Associativity
(3) $T(a, b) \geq T(c, d)$ if $a \geq c$ and $b \geq d$.   Monotonicity
(4) $T(a, 1) = a$ One Identity

We note that from monotonicity and the one identity that we have $T(a, 0) = 0$, this feature which we call the *dismissive* property implies that if there exists just one set which does not contain an element x then x is not in the aggregated set. More generally, this dismissiveness means that any component in the aggregation can unilaterally expel an element. Formally the effect of this is that the lower valued arguments are more important then the higher ones in formulating $T(a, b)$.

Some important examples of this operator are **Min**: $T(a, b) = \text{Min}(a, b)$, **Product**: $T(a, b) = ab$ and **Lukasiewicz**: $T(a, b) = \text{Max}(0, a + b - 1)$. The Min operator is the largest of all possible t-norms, $T(a, b) \leq \text{Min}(a, b)$.

The union operation is generalized by the t-conorm operator. An operator $S : [0, 1] \times [0, 1] \rightarrow [0, 1]$ is called a **t-conorm** operator if

(1) $S(a, b) = S(b, a)$ Commutativity
(2) $S(a, S(b, c)) = S(S(a, b), c))$ Associativity
(3) $S(a, b) \geq S(c, d)$ if $a \geq c$ and $b \geq d$.   Monotonicity
(4')$S(a, 0) = a$ Zero Identity

We note that from monotonicity and the zero identity that we have
$T(a, 1) = 1$, this feature which we call the *inclusiveness* property implies
that if there exists just one set which contains an element x then x is
in the aggregated set. More generally this inclusiveness means that any
component in the aggregation can unilaterally include an element. Formally
the effect of this is that the higher valued arguments are more important
then the lower ones in formulating $S(a, b)$.

Important examples of this are **Max**: $S(a, b) = \text{Max}[a, b]$, **Probabilistic
Sum**: $S(a, b) = a + b - ab$ and **Bounded Sum**: $S(a, b) = \text{Min}(1, a + b)$.
The Max operator is the smallest of all the t-conorms, $\text{Max}(a, b) \leq S(a, b)$.

The only distinction between the T and S operators is in the fourth
condition: 4 implies that the smallest argument is the most influential in the
formulation of the intersection while 4' implies that the biggest argument
is most influential in the formulation of the union.

An important relationship called De Morgan's law exists between these
two classes of operators: if T is a t-norm then $1 - T(1 - a, 1 - b)$ is a t-conorm
and if S is a t-conorm then $1 - S(1 - a, 1 - b)$ is a t-norm. t-norms and t-
conorms connected by this relationship are called duals. We note that Min
and Max are duals, the product and the probabilistic sum are also duals.

A number of other properties can be associated with the union and intersection operators we shall here briefly touch upon some of these properties.
Idempotency requires that for any a, $T(a, a) = a$ and $S(a, a) = a$. The
Min and Max are the only idempotent operators. At the other extreme from
idempotency is the **archimedean** property which is defined as $\forall a \in (0, 1)$
it is the case that $T(a, a) < a$ and $S(a, a) > a$. While many t-norms and
t-conorms have this property the prototypical example of operators having
this property are $T(a, b) = ab$ and $S(a, b) = a + b - ab$. A property closely
related to the above is that of **nilpotency**. For any sequence of $a_i \in (0, 1)$,
where $i \in N$ (the set of natural numbers), every nilpotent operator satisfies

$$\exists n_o < +\infty, T(a_1, a_2, \dots, a_{n_o}) = 0$$

$$\exists n_o < +\infty, S(a_1, a_2, \dots, a_{n_o}) = 1.$$

The prototypical example of operators having this property are $T(a, b) =
\text{Max}(0, x + y - 1)$ and $S(a, b) = \text{Min}(1, x + y)$.

One very active area of research has been the development of parameterized families of intersection and union. One example of these families is
the one proposed by Yager [12]:

$$T(a, b) = 1 - \text{Min}[1, ((1 - a)^p + (1 - b)^p)^{1/p}],$$

and

$$S(a, b) = \text{Min}[1, (a^p + b^p)^{1/p}],$$

where $p \in (0, \infty)$. Another interesting class was introduced by Frank [13]: these are such that $T(a, b) + S(a, b) = a + b$. For this Frank class $S(a, b) = 1 - \log_p \left[1 + \frac{(p^{(1-a)}-1)(p^{(1-b)}-1)}{p-1} \right]$ and $T(a, b) = \log_p \left[1 + \frac{(p^a-1)(p^b-1)}{p-1} \right]$, here again $p \in (0, \infty)$.

7.3   Weighted Unions and Intersections

Assume we have a collection of fuzzy subsets $A_1, A_2, \ldots, A_n$ and associated with each fuzzy subset is a weight $\alpha_j \in [0, 1]$ corresponding to the importance of the fuzzy subset, the larger the weight the more important the fuzzy subset in the aggregation. Let $\mathcal{R}$ indicate an aggregation operation, either intersection or union, our problem is to find the weighted aggregation [14, 15, 16, 17]

$$D = \mathcal{R}((A_1, \alpha_2), (A_1, \alpha_2), \ldots, (A_n, \alpha_n)).$$

The general procedure for accomplishing this task is to first transform, based upon their associated weights, the original fuzzy subsets A_j into modified fuzzy subsets, denoted B_j, and then perform the appropriate aggregation on these modified fuzzy subsets. Letting f be the transformation function we obtain new fuzzy subsets B_j, where $B_j(x) = f((A_j(x), \alpha_j))$ and then calculate $D = \mathcal{R}(B_1, B_2, \ldots, B_n)$. If $\mathcal{R}$ is the intersection and T is the t-norm operation being used then $D(x) = T(B_1(x), B_2(x), \ldots, B_n(x))$ and if $\mathcal{R}$ is the union operation and then $D(x) = S(B_1(x), B_2(x), \ldots, B_n(x))$.

Our focus in this section will be on the construction of the importance transformation function f. While the form of f will depend upon whether the operation $\mathcal{R}$ is a union or intersection the required characteristics of f is the same for both cases. In the following we describe these basic characteristics.

1. The ordering relationship between membership grades should be maintained:

    $$\text{if } A_j(x) \geq A_j(y) \text{ then } B_j(x) \geq B_j(y).$$

2. Importance of one is the nominal case: $A_j(x) = f(A_j(x), 1)$

3. Zero importance should result in a modified fuzzy subset that does not effect the aggregation.

4. $f(A_j(x), 0)$ should move monotonically to $f(A_j(x), 1)$:

    $$\frac{\partial f(A_j(x), \alpha_j)}{\alpha_j} \geq 0 \text{ if } f(A_j(x), 1) \geq f(A_j(x), 0)$$

$$\frac{\partial f(A_j(x), \alpha_j)}{\alpha_j} \leq 0 \text{ if } f(A_j(x), 1) \leq f(A_j(x), 0)$$

It is in the process of satisfying the third and fourth requirements that mandates different forms of f for the union and intersection. We recall for the t-norm one is the identity, $T(a, 1) = a$, thus a membership grade of one does not effect the intersection operation. Because of this we require that when $\mathcal{R}$ is an intersection we should have zero importance transform any membership grade to one, $f(A_i(x), 0) = 1$. For the case of union , t-conorm aggregations, zero is the identity, $S(a, 0) = a$, thus a membership grade of zero doesn't effect the union operation. Because of this we require that when R is a union we should have zero importance transform any membership grade to zero, $f(Ai(x), 0) = 0$. This distinction requires us to have different importance transformation operators for union and intersection. We now shall introduce some importance transformation operators which can be used to transform A_i into B_i. In [15] Yager suggested a general class of importance transformation operators that can be used for the <u>intersection</u> operation,

$$B_i(x) = S(A_i(x), \bar{\alpha}_i)$$

where S is any t-conorm and $\bar{\alpha}_i = 1 - \alpha_i$. Among this class of transformation operators are:

1. $B_i(x) = Max[\bar{\alpha}_i, A_i(x)]$ (here $S(a, b) = \text{Max}[a, b]$)
2. $B_i(x) = \bar{\alpha}_i + \alpha_i A_i(x)$ (here $S(a, b) = a + b - ab$)
3. $B_i(x) = (1 + A_i(x) - \alpha_i) \wedge 1$ (here $S(a, b) = \text{Min}(a + b, 1)$)

An importance transformation [14] which is not from this class is $B_i(x) = (A_i(x))^{\alpha_i}$.

If we used Min as our intersection operator and $\text{Max}[\bar{\alpha}_i, A_i(x)]$ as our transformation then $D = \mathcal{R}((A_1, \alpha_1), (A_2, \alpha_2), \ldots, (A_n, \alpha_n))$, where $\mathcal{R}$ is intersection would have

$$D(x) = \text{Min}_j[Max[\bar{\alpha}_j, Aj(x)]].$$

For the <u>union</u> operation a general class of importance transformation operators is

$$B_i(x) = T(A_i(x), \alpha_i)$$

where T is any t-conorm. Among this class of transformation operators are:

1. $B_i(x) = \alpha_i \wedge A_i(x)$ (here $T(a, b) = \text{Min}(a, b)$)
2. $B_i(x) = \alpha_i A_i(x)$ (here $T(a, b) = ab$)
3. $B_i(x) = (A_i(x) + \alpha_i - 1) \vee 0$ (here $T(a, b) = \text{Max}(a + b - 1, 0)$).

An additional transformation operator not from this class is $B_i(x) = 1 - (\bar{A}_i(x))^{\alpha_i}$

If we used Max as our union operator and $\text{Min}[\alpha_i, A_i(x)]$ as our transformation then $E = \mathcal{R}((A_1, \alpha_1), (A_2, \alpha_2), \ldots, (A_n, \alpha_n))$, where $\mathcal{R}$ is intersection would have

$$D(x) = \text{Max}_j[\text{Min}(\alpha_i, A_i(x))]$$

7.4 Uninorms

The t-norm and t-conorm can be seen as extreme types of aggregation operators. We note that for the t-norm from the monotonicity and the fact that $T(a, 1) = a$ we must have $T(a, b) \le a$. The implication of this is that the addition of a fuzzy subset to a t-norm based aggregation cannot result in an increase in the membership grade of any element. On the other since $S(a, 0) = a$ then monotonicity requires that $S(a, b) \ge a$. Here the implication is that the addition of a fuzzy subset to a t-conorm based aggregation can't result in a decrease in the membership grade of any element.

In an attempt to address this issue in [18] Yager and Rybalov introduced a generalization of the t-norm and t-conorm which they called uninorms. Fodor and DeBaets [19, 20, 21] have studied the structure of these operators in considerable detail.

Definition 36 *A **uninorm** R is a mapping $RI \times I \to I$ having the following properties*

1. ***Commutativity:*** *$R(x, y) = R(y, x)$*
2. ***Monotonicity:*** *$R(x, y) = R(u, v)$ for $x \ge u, y \ge v$*
3. ***Associativity:*** *$R(x, R(y, z)) = R(R(x, y), z)$*
4. ***Identity:*** *There exists some fixed element $g \in [0, 1]$, called the identity, such that we have $R(x, g) = x$.*

If $A_j, j = 1$ to n, are a collection of fuzzy subsets we can implement a uninorm based aggregation of these fuzzy subsets as $A = \text{Uni}(A_1, A_2, \ldots, A_n)$ such that

$$A(x) = R(A_1(x), A_2(x), \ldots, A_n(x))$$

where R is a uni-norm operator.

The uni-norm operator has the same first three of the properties of the t-norm and t-conorm but the fourth condition is more general in that it allows for any identity. In introducing the uni-norm we are essentially allowing the identity to be anywhere in the unit interval rather than just at the extremes of one and zero as in the cases of the t-norm and t-conorms. Since $R(a, g) = a$ then the monotonicity condition implies that $R(a, b) \le a$

for $b < g$ and $R(a, b) \geq a$ for $b > g$. Thus g determines a boundary which separates those values which can cause a decrease in the aggregated value from those which can cause an increase in the aggregated value. With this property we see that addition of a fuzzy subsets to a uninorm based aggregation can cause an increase or decrease in overall membership grade. We note that t-norms are special cases of uninorms in which $g = 1$ while t-conorms have $g = 0$.

Having introduced the concept of a uni-norm we next turn to the question of the existence or construction of such operators. For $g = 1$ or 0 there exists a large class of such uni-norms corresponding to the t-norms and t-conorms respectively. In [19] Fodor, Yager and Rybalov introduced a general class of uni-norms for any g. In the following we describe this class.

Let T and S be any t-norm and t-conorm respectively. The mappings R defined by I and II below are two general classes of uni-norm operators with identity g.

I.
(a) $R(x, y) = gT(\frac{x}{g}, \frac{y}{g})$ if $0 \leq x, y \leq g$
(b) $R(x, y) = g + (1 - g)S(\frac{x-g}{1-g}, \frac{y-g}{1-g})$ if $g \leq x, y \leq 1$
(c) $R(x, y) = \text{Min}(x, y)$ if $\text{Min}(x, y) \leq g \leq \text{Max}(x, y)$
II.
(a) $R(x, y) = gT(\frac{x}{g}, \frac{y}{g})$ if $0 \leq x, y \leq g$
(b) $R(x, y) = g + (1 - g)S(\frac{x-g}{1-g}, \frac{y-g}{1-g})$ if $g \leq x, y \leq 1$
(c) $R(x, y) = \text{Max}(x, y)$ if $\text{Min}(x, y) \leq g \leq \text{Max}(x, y)$

The difference between the two classes are in item (c), in the first case when one of the arguments is above the identity and the other below the identity we take the Min while in the second case we take the Max. In particular we note that for the first class $R(0, 1) = 0$ and for the second class $R(0, 1) = 1$. The following Figure 7.1 provides visualization of the above structure.

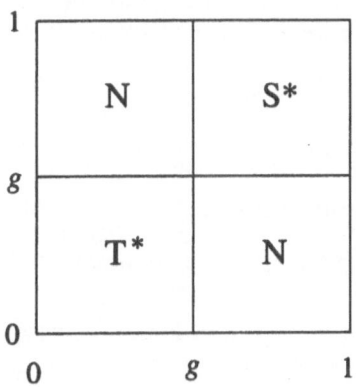

FIGURE 7.1. Structure of uni-norm class in $[0, 1] \times [0, 1]$

In the above

$$T^*(x, y) = g T(\frac{x}{g}, \frac{y}{g})$$

$$S^*(x, y) = g + (1 - g) S(\frac{x - g}{1 - g}, \frac{y - g}{1 - g})$$

and N(x, y) is either the Max or the Min.

Let us look at the form of T* and S* for some special cases of T and S.

$T(x, y) = Min[x, y]$ $T^*(x, y) = Min[x, y]$
$S(x, y) = Max[x, y]$ $S^*(x, y) = Max[x, y]$
$T(x, y) = xy$ $T^*(x, y) = \frac{xy}{g}$
$S(x, y) = x + y - xy$ $S^*(x, y) = \frac{x + y - xy - g}{1 - g}$
$T(x, y) = Max[0, x + y - 1]$ $T^*(x, y) = Max[0, x + y - g]$
$S(x, y) = Min[1, x + y]$ $S^*(x, y) = Min[1, x + y - g]$

In [18] we provided an example of a uninorm not of the above class, called the **three-Π operator**. For this operator for all x and y in $[0, 1]$ $R(x, y) = \frac{xy}{\bar{x}\bar{y} + xy}$ where $\bar{x} = 1 - x$. In this uninorm the identity $g = 0.5$. Using the associativity property we get that

$$R(x_1, x_2, \dots, x_n) = \frac{\prod_{j=1}^{n} x_j}{\prod_{j=1}^{n} \bar{x}_j + \prod_{j=1}^{n} x_j}$$

7.5   Mean Aggregation Operators

In this section, we shall introduce a another type of aggregation operators which are called the **mean aggregation operators**. A mapping $M : [0, 1]^n \to [0, 1]$ is called a mean aggregation operator of dimension n if it satisfies the following axiomatic conditions

M1. **Commutativity:** The aggregation is indifferent to the ordering of the arguments.

M2. **Monotonicity:** $M(a_1, a_2, \dots, a_n) \geq M(b_1, b_2, \dots, b_n)$ if $a_i \geq b_i \forall i$.

M3. **Bounded:** $Min_i[a_i] \leq M(a_1, a_2, \dots, a_n) \leq Max_i[a_i]$

It can be shown that this operator is idempotent, $M(a_1, a_2, \dots, a_n) = a$ if $a_i = a$ $\forall i$.

The simple average $\hat{a} = \frac{1}{n} \sum_{j=1}^{n} a_j$ is the prototypical example of a mean operator. The Min, Max and median are also mean operators.

Mean operators can be used to provide for an aggregation of fuzzy sets. If $A_j, j = 1$ to n, are a collection of fuzzy subsets we can implement a mean based aggregation of these fuzzy subsets as $A = \mathcal{M}(A_1, A_2, \ldots, A_n)$ such that

$$A(x) = M(A_1(x), A_2(x), \ldots, A_n(x))$$

where M is a mean operator of dimension n.

Associativity, except in the special cases of Min and Max, is generally not available for mean operators [11]. The lack of associativity has some important implications. One is that the definition for the mean operators of n arguments is not implicit in the definition of the operator for two arguments. Any aggregation operator G can be seen as a family of aggregation operators $G = [G_2, G_3, \ldots, G_j, \ldots,]$ where G_j indicates the form of the operator used when the number of arguments is j. With uninorms, we have the associativity property, this mandates the form of G_j based upon the form of G_2. In the case of mean operators, which do not have associativity, our definition of the operator G must be such that it specifies, either implicitly or explicitly, the form for all argument sizes. Furthermore, since all that is required is that each G_j satisfy the conditions M1 to M3 we are quit free in specifying the form of the aggregation at the different dimensions. This situation requires us to introduce some imperative for assuring that the definition of G at different dimensions are consistent. In [22] Yager suggested one criteria for assuring this. A mean operator is said to be **consistent** if the following condition is satisfied by G_n for all n's:

$$\text{if } a_{n+1} \geq G_n(a_1, \ldots, a_n) \text{ then } G_{n+1}(a_1, \ldots, a_n, a_{n+1}) \geq G_n(a_1, \ldots, a_n)$$

$$\text{if } a_{n+1} \leq G_n(a_1, \ldots, a_n) \text{ then } G_{n+1}(a_1, \ldots, a_n, a_{n+1}) \leq G_n(a_1, \ldots, a_n)$$

We see that this consistency condition can be seen as effectively requiring that if we add some element to our aggregation our new aggregated value should move in the direction of this element.

A closely related property to this is the property which Yager [23, 24] called **self identity**. An operator G is said to possess the self-identity property if

$$G_{n+1}(a_1, \ldots, a_n, a_{n+1}) = G_n(a_1, \ldots, a_n), \quad \text{if } a_{n+1} = G_n(a_1, \ldots, a_n).$$

The self identity condition says that adding an element to an aggregation whose value is the same as the aggregation leads to no change in the aggregate value. It can be shown that if we impose the self identity condition on a aggregation that is monotonic we attain the consistency condition described above.

A further property that one can associate with a mapping G is **naturalness**. By this we mean the condition that $\forall a, G_1(a) = a$. Thus naturalness implies that one argument aggregations evaluate to the one argument.

We see that self identity and naturalness implies idempotency. From naturalness $G_1(a) = a$ in addition self identity implies $G_2(a, a) = a$. Continuing to apply self identity we get idempotency.

We now can define a special class of mean operators which we call **self consistent mean operators**. Before introducing these operators we discuss the concept of a bag. A bag associated with [0, 1] like a set is any collection of elements drawn from [0, 1]. A bag is different from a set in that it allows multiple copies of the same element. We shall use $\mathcal{B}^{[0,1]}$ to indicate the set off all bags associated with the interval [0, 1]. Essentially we can view $\mathcal{B}^{[0,1]}$ to consist of all the possible arguments that can be applied to our aggregation. For a given bag B $\in \mathcal{B}^{[0,1]}$ we indicate the dimension of B, Dim(B), as the number of elements in B. We emphasize that $\mathcal{B}^{[0,1]}$ contains bags of all different dimensions. A mapping

$$H : \mathcal{B}^{[0,1]} \to [0, 1]$$

is called a **self consistent mean operator** if H satisfies the following conditions:

1. **Naturalness**: $H(a) = a$

2. **Commutativity**: The ordering of the arguments does not matter.

3. **Monotonicity**: For bags of the same dimension condition M2 applies.

4. **Self Identity**.

It can be easily shown that for any fixed dimension n a self consistent mean operator satisfies **M1-M3** and is thus a mean operator.

In [25] a parameterized family of mean operators, called the *generalized mean operator*, is described. The generalized mean operator is defined as

$$H^\alpha(a_1, \ldots, a_n) = \left(\frac{a_1^\alpha + a_2^\alpha + \cdots + a_1^\alpha}{n} \right)^{\frac{1}{\alpha}}$$

where $\alpha \in (-\infty, \infty)$ is a parameter. It can be shown that for any α, H^α is a self consistent mean.

This family reduces to a number of interesting special cases for particular values of the parameter α:

α	$H^\alpha(a_1, \ldots, a_n)$
∞	$\text{Max}_i(a_1, \ldots, a_n)$
1	$\frac{1}{n} \sum_{j=1}^n a_j$
0	$(a_1, a_2, \ldots, a_n)^{1/n}$
-1	$\frac{n}{\frac{1}{a_1} + \frac{1}{a_2} + \cdots + \frac{1}{a_n}}$ (harmonic mean)
$-\infty$	$\text{Min}_i(a_1, \ldots, a_n)$

7.6   Ordered Weighted Averaging Operators

In Yager [26] introduced a family of mean operators called the Ordered Weighted Averaging (OWA) operators [27]. We shall see that the OWA aggregation operators allow us to easily adjust the degree of *anding* and *oring* implicit in an aggregation.

Definition 37 *An OWA operator of dimension n is a mapping $f : R^n \to R$ that has an associated n vector W*

$$W = \begin{bmatrix} w_1 \\ w_2 \\ \vdots \\ w_n \end{bmatrix}$$

such that

1. $w_i \in [0,1]$

2. $\sum_i w_i = 1$.

where $f(a_1, \ldots, a_n) = \sum_j w_j b_j$ with b_j the jth largest of the a_i.

It can be shown that the OWA operator is a mean operator, it is commutative, monotonic and bounded. The distinguishing feature of this operation is the re-ordering step, in particular an argument a_i is <u>not</u> associated with a particular weight w_i but rather a weight is associated with a particular ordered position of aggregate. This ordering step introduces a nonlinearity into the aggregation process. The OWA operator can be viewed as a special kind of inner product $f(a_1, \ldots, a_n) = W^T B$ here B is an n vector whose j component b_j is the jth largest of the a_i.

It is noted that different OWA operators are distinguished by their weighting vector W. Three important special cases of OWA aggregations can be pointed out

1. F^*: In this case $w_1 = 1$ and $w_j = 0$ for all other j and $F^*(a_1, \ldots, a_n) = \text{Max}_i[a_i]$

2. F_*: In this case $w_n = 1$ and $w_j = 0$ for all other j and $F_*(a_1, \ldots, a_n) = \text{Min}_i[a_i]$

3. F_A: In this case $w_j = \frac{1}{n}$ for all j and $F_A(a_1, \ldots, a_n) = \frac{1}{n} \sum_i a_i$.

Another example of OWA operators is the median aggregator. We define W_{med} as follows: if n is odd, $w_{\frac{n+1}{2}} = 1$ and $w_j = 0$ for all other j; if n is even, $w_{\frac{m}{2}} = w_{\frac{m}{2}+1} = 0.5$ and $w_j = 0$ for all other j.

An interesting class of OWA operators is defined by the following weighting vector $w_k = 1$ and $w_j = 0$ for $j \neq k$. This form of aggregation selects the kth largest element in the argument as its aggregated value. OWA operators, like other mean operators, can be used to provide for an aggregation of fuzzy sets. If $A_j, j = 1$ to n, are a collection of fuzzy subsets we can implement an OWA based aggregation of these fuzzy subsets as $A = \mathcal{OWA}(A_1, A_2, \ldots, A_n)$ such that

$$A(x) = F_w(A_1(x), A_2(x), \ldots, A_n(x))$$

where F_w is an OWA operator with weighting vector W.

The OWA operator allows us, by appropriate choice of the weighting vector, to smoothly move from the Min to the Max type aggregation. In order to classify these OWA operators in regards to their location on this continuum a measure of **orness**, associated with any vector W, can be introduced. The **orness** measure is defined as

$$\mathbf{orness}(W) = \frac{1}{n-1} \sum_{i=1}^{n} ((n-1)w_i).$$

We note that for any W the **orness**(W) is always in the unit interval. Furthermore it is noted that the nearer W is to an "or" the closer its measure is to one while the nearer it is to an "and" the closer its measure is to zero. Supporting this it can be easily shown that

1. **orness**$(W^*) = 1$

2. **orness**$(W_*) = 0$

3. **orness**$(W_F) = .5$.

Generally it should be appreciated that an OWA operator with much of the non-zero weights at near the top will be an *orlike* operator, **orness**$(W) \geq .5$. At the other extreme when the weights are clustered near the bottom the OWA operator will be *andlike*, **orness**$(W) \leq .5$.

Except for the extreme cases of **orness**$(W) = 1$ and 0 which are obtained by W^* and W_* respectively numerous different vectors W can attain a given measure of orness. To help further distinguish amongst the different OWA aggregators a measure, called the dispersion (or entropy) of an OWA vector W is defined as $\text{Disp}(W) = -\sum_i w_i ln(w_i)$.

The measure of dispersion can be used to indicate the degree to which we use all the information contained in the argument, $(a_1, \ldots, a_n)$ when calculating the aggregated value $F(a_1, \ldots, a_n)$. In particular the larger $\text{Disp}(W)$ the more of the information provided by the argument we use.

The great generality that the OWA operators provide us brings with it the issue of selecting the weights that are to be used in a particular application. A number of approaches are available to accomplish this task. Before mentioning the more "objective/formal" approaches the most straightforward approach is to subjectively select the weights used in the vector W. This subjective selection is not to be done blindly but should be guided by information about the application at hand as well as knowledge about functioning of the OWA operator.

A very useful approach to the generation of the OWA weights was suggested by O'Hagan [28, 29]. This approach makes use of the measures of *orness* and *entropy* introduced earlier.

O'Hagan suggested using the following procedure for calculating the weights associated with an OWA aggregation. The first step is to choose a desired level of *orness*, α associated with the aggregation. The second step is to select from among the weighting vectors that attain this level of *orness* the one that has the maximum dispersion (entropy). As suggested by O'Hagan we can obtain the weights by solving the following mathematical programming problem:

Max: $-\sum_i w_i ln(w_i)$
Subject to:
$$\alpha = \frac{1}{n-1} \sum_i^n (n-i) w_i$$
$$\sum_i^n w_i = 1, \quad w_i \in [0,1].$$

We note that by just specifying one parameter, α, we get the whole weighting vector W.

Another approach to obtaining the OWA weights is to learn the weights from observations on the argument and aggregated value. Our data will be a collection of m observations. Each observation is comprised of an n-tuple of values $(a_{k1}, a_{k2}, \ldots, a_{kn})$, called the arguments, and an associated single value called the aggregated value, which we shall denote as d_k. Our goal is to find an OWA operator, a weighting vector W, that *best* models the process of aggregation used in that data set. We denote the reordered objects of the k-th sample by $b_{k1}, b_{k2}, \ldots, b_{kn}$ where b_{kj} is the jth largest element of the argument collection $a_{k1}, a_{k2}, \ldots, a_{kn}$. Using these ordered arguments the problem of modelling the aggregation process is to find the vector of OWA weights $W = [w_1 w_2 \ldots w_n]^T$ that best satisfies

$$b_{k1} w_1 + b_{k2} w_2 + \cdots + b_{kn} w_n = d_k$$

for every $k = 1$ to m.

Applying the learning technique used in neural networks we look for a weighting vector $W = [w_1 w_2 \ldots w_n]^T$ that minimizes the instantaneous errors e_k where,

$$e_k = (b_{k1} w_1 + b_{k2} w_2 + \cdots + b_{kn} w_n - d_k)^2$$

The solution to this problem [30, 31] is not the simple application of the Widrow-Hoff rule that it appears to be. The situation is complicated by the fact that the minimization problem is a constrained optimization problem, since the w_i have to satisfy $\sum_{i=1}^n w_i = 1$ and $w_i \in [0,1], i = (1,n)$.

To circumvent the constraints on the weights we express the w_i's as

$$w_i = \frac{e^{\lambda_i}}{\sum_{j=1}^n e^{\lambda_j}}, i = (1,n)$$

Using the gradient descent method we obtain the following rule for updating the parameters

$$\lambda_i(l+1) = \lambda_i(l) - \beta w_i(l)(b_{ki} - \hat{d}_k)(\hat{d}_k - d_k)$$

where β denotes the learning rate ($0 \leq \beta \leq 1$) and $\lambda_i(l+1)$ is our new estimate for λ_i. For each $i, w_i(l) = \frac{e^{\lambda_i(l)}}{\sum_{j=1}^n e^{\lambda_j(l)}}$ is the estimate of w_i after the lth iteration and $\hat{d}_k = \sum_{j=1}^n b_{kj} w_j(l)$. The process of updating the λ_i continues until we get small changes of parameter estimates $\Delta_i = |\lambda_i(l+1) - \lambda_i(l)|, i = (1, n)$.

7.7 Linguistic Quantifiers and OWA Operators

As indicated, the OWA operators provides for aggregations which lie between the intersection (and) and the union (or). If $A_j, j = 1$ to n, are a collection of fuzzy subsets and F_W is an OWA operator we can construct a fuzzy subset D of X such that $D(x) = F_W(A_1(x), A_2(x), \ldots, A_n(x))$. By appropriately selecting W we can implement different aggregations between *anding* and *oring* of the fuzzy sets.

There exists in natural language a class of words (concepts) which convey aggregation policies that have as its extremes the Min/Intersection/and and the Max/Union/or operators. These extremes are conveyed respectively by the terms "for all" and "at least one." Zadeh [32] called these concepts *linguistic quantifiers*. Examples of these words are *almost all, few, many, most, nearly half, at least twenty percent* as well as the the extreme cases of *all and their exists* (at least one). We note that these words are conveying information about proportions. Zadeh [32] suggested that linguistic quantifiers which say something about proportion can be represented as fuzzy subsets of the unit interval. Thus if Q is a linguistic quantifier, such as *most*, then Q can be represented as a fuzzy subset Q of I where for each $r \in I, Q(r)$ indicates the degree to which the proportion r satisfies the concept denoted by Q. Thus if Q is the fuzzy subset representing *most* then if $Q(.8) = 1$ we are saying that 80% is completely compatible with the idea conveyed by the linguistic quantifier *most* while if $Q(.5) = 0.7$ we are indicating that the proportion 50% is only 0.7 compatible with the concept of *most*.

Having provided for a representation of linguistic quantifiers in terms of fuzzy sets we characterize some special classes of linguistic quantifiers. We shall say a linguistic quantifier Q is regular non-decreasing if: $Q(0) = 0, Q(1) = 1$ and when $r_1 > r_2$ we have $Q(r_1) \geq Q(r_2)$. These linguistic quantifiers are characterized by the fact that the satisfaction to the quantifier increases as the proportion increases. Among this class of linguistic quantifiers are concepts like *at least* α, *most* and *all*.

We shall call a quantifier Q regular non-increasing if: $Q(0) = 1, Q(1) = 0$ and when $r_1 < r_2$ we have $Q(r_1) \geq Q(r_2)$. These quantifiers are characterized by the fact that the satisfaction to the quantifier decreases as the proportion increases. Among this class of linguistic quantifiers are concepts like *at most* α and *few*.

We shall call a quantifier Q regular unimodal if: $Q(0) = Q(1) = 0, Q(r) = 1$ for $a \leq r \leq b$, when $r_2 \geq r_1 \leq a$ then $Q(r_1) \leq Q(r_2)$ and when $r_2 \geq r_1 \geq b$ then $Q(r_2) \leq Q(r_1)$. Among this class of quantifiers are concepts such as *about* α.

Linguistic quantifiers can be used in the aggregation of fuzzy subsets in a process called *quantifier guided aggregation*. Assume $A_1, A_2, \ldots, A_n$ are a collection of fuzzy subsets and let Q be a proportion quantifier of the type we just described. We shall use the notation $D = Q(A_1, A_2, \ldots, A_n)$ to indicate the fuzzy subset D is formed in a manner that the membership grade $D(x)$ is obtained by requiring that x is a member of Q of the A_i. This type of aggregation is particularly useful in applications of fuzzy set theory to multi-criteria decision making, information retrieval and pattern recognition. In these applications we use the fuzzy subset A_i to represent a condition associated with an object being a solution to the problem. Rather then requiring all the conditions to be satisfied we only need Q of the conditions satisfied. As we shall see the OWA aggregation operators play a central role in implementing this type of aggregation.

To calculate $D = Q(A_1, A_2, \ldots, A_n)$ when Q is regular non-decreasing we proceed as follows:

1. Obtain the fuzzy subset Q representing the quantifier Q.

2. Calculate the OWA weighting vector W associated with the fuzzy subset Q by

$$w_i = Q\left(\frac{i}{n}\right) - Q\left(\frac{i-1}{n}\right) \text{ for } i = 1 \text{ to } n,$$

3. For each $x \in X$ calculate $D(x) = F_w(A_1(x), A_2(x), \ldots, A_n(x))$.

Figure 7.2 illustrates the process of obtaining the weights.

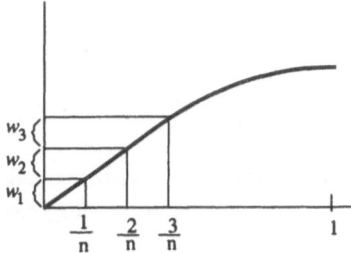

FIGURE 7.2. Obtaining Weights from a Quantifier

The regularity and the increasing nature of Q guarantees the satisfaction of required conditions of an OWA vector W.

We now turn to the implementation of quantifier guided aggregation when the quantifier Q is regular nonincreasing. In order to do this we need introduce the idea of an antonym. If Q is a fuzzy subset defined on the unit interval the fuzzy subset $\hat{Q}$ defined such that $\hat{Q}(r) = Q(1-r)$ is called the antonym of Q. It can be shown that when Q is a regular non-increasing quantifier $\hat{Q}$ is a regular non-decreasing quantifier. Having introduced this antonym of Q we can now provide a procedure for evaluating $D = \mathcal{Q}(A_1, A_2, \ldots, A_n)$ when $\mathcal{Q}$ is represented by a regular non-increasing quantifier Q.

1. Obtain the fuzzy subset Q representing the quantifier $\mathcal{Q}$

2. Calculate the OWA weighting vector $\hat{W}$ associated with the fuzzy subset $\hat{Q}$, the antonym of Q

$$\hat{w}_i = \hat{Q}\left(\frac{i}{n}\right) - \hat{Q}\left(\frac{i-1}{n}\right) \text{ for } i = 1 \text{ to } n,$$

3. For each $x \in X$ calculate $D(x) = F_{\hat{W}}((1-A_1(x)), (1-A_2(x)), \ldots, (1-A_n(x)))$.

We now turn to the case of unimodal quantifiers. We first note that any unimodal quantifier Q can be expressed in terms of a regular nondecreasing quantifier Q_1, and a regular nonincreasing quantifier Q_2. Assume a and $b \in I$ are the beginning and the end of the interval where Q is one. We define Q_1 as $Q_1(x) = Q(x)$ for $0 \leq x < a$ and $Q_1(x) = 1$ for $a \leq x \leq 1$. We define Q_2 as $Q_2(x) = 1$ for $0 \leq x < b$ and $Q_2(x) = Q(x)$ for $b \leq x \leq 1$. We see $Q = Q_1 \cap Q_2$ where $Q(x) = \text{Min}[Q_1(x), Q_2(x)]$. We can use these two quantifiers to evaluate $D = \mathcal{Q}(A_1, A_2, \ldots, A_n)$ when $\mathcal{Q}$ is a regular unimodal quantifier Q.

1. Calculate $D_1 = Q_1(A_1, A_2, \ldots, A_n)$
2. Calculate $D_2 = Q_2(A_1, A_2, \ldots, A_n)$
3. $D = D_1 \cap D_2$

7.8   Aggregation Using Fuzzy Measures

Another class of fuzzy set aggregation operators is based upon the concept of a fuzzy measure, a concept introduced by Sugeno [33, 34].

Definition 38 *Assume X is a finite set of elements. A set function μ : $2^X \to [0,1]$, having the properties:*

1. $\mu(\varnothing) = 0$

2. $\mu(X) = 1$

3. If $A \subset B$ then $\mu(A) \leq \mu(B)$.

is called a fuzzy measure on X.

The fuzzy measure provides a generalization of the probability measure which plays such a central role in probability and integration theory. We recall that a probability measure P is characterized by the property of additivity; for all sets disjoint sets A and B, $A \cap B = \varnothing$, $P(A \cup B) = P(A) + P(B)$. In the fuzzy measure this property is replaced by the more general one of monotonicity as captured in property three above. We note that probability measures are special cases of fuzzy measures since additivity implies monotonicity.

Sugeno introduced a class of fuzzy measure which he called the μ_λ-fuzzy measure, these are sometimes called Sugeno measures. These measure have the additional property that for all disjoint subsets A and B we have

$$\mu_\lambda(A \cup B) = \mu_\lambda(A) + \mu_\lambda(B) + \lambda\mu_\lambda(A)\mu_\lambda(B),$$

where $\lambda > -1$. We note that for the case where $\lambda = 0$ we get the probability measure.

Another class of fuzzy measures are based upon t-conorms [35]. If we define a measure μ on X such that $\mu(\varnothing) = 0$ and $\mu(X) = 1$ and for any two disjoint subsets A and B of X we define $\mu(A \cup B) = S(\mu(A), \mu(B))$ where S is a t-conorm then μ is a fuzzy measure. We note in this case $\mu(A) = \underset{x_i \in A}{S}[\mu(\{x_i\})]$, this implies that $\mu(A)$ is completely determined from the values $\mu(\{x_i\})$. In the case of the Max t-conorm, for any disjoint A and B we get $\mu(A \cup B) = \underset{x \in X}{\text{Max}}[\mu(A), \mu(B)]$. Using this t-conorm we have $\mu(A) = \underset{x \in A}{\text{Max}}[\mu(x)]$. If we use the bounded sum t-conorm then $\mu(A) = \text{Min}[1, \sum_{x \in A} \mu(\{x\})]$. We note that if $\sum_{x \in X} \mu(\{x\}) = 1$ then this is simply the additive case.

We can also define a class of fuzzy measures based upon the t-norm operators [35]. If we define a measure μ on X such that $\mu(\varnothing) = 0$ and $\mu(X) = 1$ and for any two subsets A and B for which $A \cup B = X$, we define $\mu(A \cap B) = T(\mu(A), \mu(B))$ where T is a t-norm then μ is a fuzzy measure. Let us look at these t-norm based measures in more detail. Consider the family of subsets $F_i = X - \{x_i\}$. We first note that for any i and $j, i \neq j$, it is the case that $F_i \cup F_j = X$. Next we note that any subset A of X can be expressed as an intersection of a collection from this family $A = \underset{x_i \notin A}{\bigcap} F_i$. Therefore we see that for any A, $\mu(A) = \underset{x_i \notin A}{T}(\mu(F_i))$. Thus we can express μ for any A by simply having μ for all F_i. The condition that $\mu(\varnothing) = 0$ is satisfied if $\mu(F_i) = 0$ for some F_i.

We now turn to the role of the fuzzy measures in the aggregations of fuzzy sets. Let $\mathcal{A} = \{A_1, \ldots, A_n\}$ be a collection of fuzzy subsets defined over the space X. Our goal is to aggregate these fuzzy subsets. We let the aggregation be specified as $D = \mathcal{G}(A_1, \ldots, A_n)$, $\mathcal{G}$ is called the aggregation imperative and carries the information of how D is formed from the A_i. $\mathcal{G}$ will be formally expressed in terms of a function μ defined on $2^{\mathcal{A}}$ which will have the properties of a fuzzy measure.

For each subset E of $\mathcal{A}$ we let $\mu(E) \in [0, 1]$ indicate how satisfied we would be if we formed D by just considering the subsets A_i that lie in E. Here we observe that if $E = \varnothing$, we would be completely dissatisfied, $\mu(\varnothing) = 0$ and if $E = \mathcal{A}$ we would be completely satisfied, $\mu(\mathcal{A}) = 1$. Furthermore, since larger E the more satisfied, we have $\mu(A) \geq \mu(B)$ if $B \subset A$. Thus μ is a fuzzy measure.

If $A_i(x)$ indicates the degree to which the element x satisfies the fuzzy subset A_i then $E(x) = \text{Min}_{A_i \in E}[A_i(x)]$ indicates the degree to which x satisfies all the elements in E. Since we have already established that $\mu(E)$ is a measure of how satisfied we would be if we only considered the elements in E in formulating D, then $\text{Min}[E(x), \mu(E)]$ is the degree to which we are satisfied with x based on using the elements in E.

Using this we can now describe a general method of constructing $D = \mathcal{G}(\mathcal{A})$. In this method D is the weighted union over all subsets E of $\mathcal{A}$, that is $D = \bigcup_{E \subseteq \mathcal{A}} (E, \mu(E))$. Implementing $(E, \mu(E))$ using the Min operator we get as the membership function of D

$$D(x) = \text{Max}_{E \subset \mathcal{A}}[E(x) \wedge \mu(E)].$$

This formulation is an example of a fuzzy integral [33].

Thus we see that the membership of an element x in D is obtained as follows. For any subset of criteria we find the degree to which we are satisfied with these, $\mu(E)$, and we find out the degree to which the element x is in the intersection of these criteria, $E(x)$, we then maximize this over all subsets of $\mathcal{A}$. If we denote $D_E(x) = E(x) \wedge \mu(E)$ then we have $D(x) = \text{Max}_{E \subset \mathcal{A}} D_E(x)$

The resulting form of aggregation strongly depends upon the structure of μ. Let us look at some special cases. The first case is one in which we shall assume that the measure is defined by $\mu^*(\{A_i\}) = 1$ for all $i = 1, \ldots, n$. Assume E is any nonnull subset of $\mathcal{A}$ hence there exists at least one $A_j \in E$ and thus $\{A_j\} \subset E$, since μ^* is a fuzzy measure, $\mu^*(E) \geq \mu^*(E_j) = 1$. In this case $\mu^*(E) = 1$ for all non-null subsets. From this it can be shown that $D(x) = \text{Max}_i[A_i(x)]$. Thus we see that the union operation is a special case of the preceding in which $\mu(E) = 1$ for all $E \neq \varnothing$.

Consider next the case in which $\mu_*(\mathcal{A}) = 1$ and $\mu_*(E) = 0$ for all $E \neq \mathcal{A}$. In this case it can be shown that $D(x) = \text{Min}_i A_i(x)$, thus this corresponds to the intersection of the A_i.

It is interesting to note that μ^* and μ_* form the extreme fuzzy measures, for any other measure $\mu_*(E) \leq \mu(E) \leq \mu^*(E)$. Thus we anticipate all the aggregations base on fuzzy measures to lie between these two extremes.

Assume we define a measure μ on $2^{\mathcal{A}}$ by the max t-conorm, $\mu(A \cup B) = \text{Max}(\mu(A), \mu(B))$, with the understanding that $\mu(\varnothing) = 0$ and $\mu(\mathcal{A}) = 1$. If we denote $\alpha_i = \mu(\{A_i\})$ then we get

$$D(x) = \text{Max}_i[A_i(x) \wedge \alpha_i]$$

which is case of the weighted union, here $\mu(\{A_i\})$ is the importance of A_i.

Assume we define a measure μ on $2^{\mathcal{A}}$ by the Min t-norm $\mu(A \cup B) = \text{Min}(\mu(A), \mu(B))$. This also has boundary conditions $\mu(\varnothing) = 0$ and $\mu(\mathcal{A}) = 1$. In this case we can uniquely express μ for any subset of $\mathcal{A}$ in terms of $\mu(Fi)$, where $F_i = A - \{A_i\}$. In particular, for any subset A

$$A = \bigcap_{x_i \notin A} F_i.$$

It can be shown that in this case $D(x) = \text{Min}_i[A_i(x) \vee \mu(F_i)]$. If we denote $\mu(F_i) = \bar{w}_i = 1 - w_i$ then we see that this is a formulation for weighted intersection with the importance of A_i equal to $1 - \mu(F_i)$.

We shall consider one other case of fuzzy measure, $\mu(E) = F(\text{Card}(E))$. Here the fuzzy measure just depends upon the number of elements in E, of course we again require $\mu(\varnothing) = 0$ and $\mu(\mathcal{A}) = 1$. Here all subsets A and B of $\mathcal{A}$ having the same cardinality have the same μ value. If $\text{Card}(A) = n$ then all we need to define μ are a collection of values $w_j = F(j)$ for $j = 0$ to n such that $w_0 = 0$, $w_n = 1$ and $w_j \geq w_i$ if $j > i$. Consider now $D(x) = \text{Max}_{E \subset \mathcal{A}}[E(x) \wedge \mu(E)]$. If $M(j)$ is the collection of subsets of $\mathcal{A}$ have j elements then $D(x) = \text{Max}_j\{\text{Max}_{E \subset M(j)}[E(x) \wedge w_j]\}$. It can be shown that in this case $D(x) = \text{Max}_j[b_j \wedge w_j]$ where b_j is the jth largest of the $A_i(x)$.

Recently considerable attention has been given to use of fuzzy measures with the Choquet integral to form aggregation functions [36, 37, 38, 39, 40]. Again assume we have a collection of fuzzy subsets $\mathcal{A} = \{A_1, \ldots, A_n\}$ defined over the space X and our goal is to aggregate these fuzzy subsets. We let the imperative for our aggregation be expressed in terms of a fuzzy measure μ. Using the Choquet integral our aggregated fuzzy subset is a fuzzy subset D with

$$D(x) = \mathbf{G}(A_1(x), \ldots, A_n(x)).$$

where $\mathbf{G}$ is defined as follows. Assume b_j is the jth largest of the $A_i(x)$ and let $H(j)$ be the subset of $\mathcal{A}$ consisting of the fuzzy subsets having the j highest membership grades for x then

$$\mathbf{G}(A_1(x), \ldots, A_n(x)) = \sum_{j=1}^{n} b_j \Delta_j$$

where $\Delta_j = \mu(H(j)) - \mu(H(j-1))$. (We note $H(j) = \varnothing$).

While any fuzzy measure μ can be used in this aggregation procedure one interesting measure will be considered in the following. Let us associate with each subset A_i a weight α_i and assume these weights lie in the unit interval and sum to one. Let $F(j)$ for $j = 1$ to n be a mapping into the unit interval such that $F(0) = 0$, $F(n) = 1$ and $F(j) \geq F(\hat{j})$ if $j > \hat{j}$. We now define a measure μ on $\mathcal{A}$ such that

$$\mu(E) = F(\text{Card}(E)) \sum_{A_j \in E} \alpha_j.$$

Let us first see that this is a fuzzy measure. If $E = \varnothing$ then $\text{Card}(E) = 0$ and since $F(0) = 0$ then $\mu(E) = 0$. If $E = \mathcal{A}$ then $\sum_{A_j \in E} \alpha_j = 1$ and $F(\text{Card}(E)) = 1$ hence $\mu(E) = 1$. Assume $A \subset B$ then $\text{Card}(B) \geq \text{Card}(A)$ and $\sum_{A_j \in B} \alpha_j \geq \sum_{A_j \in A} \alpha_j$ and hence $\mu(B) \geq \mu(A)$. Using this measure it appears that we can capture the importance of the individual A_j by α_j and we can use the function F to indicate the type of aggregation. For example if $F(n) = 1$ and $F(j) = 0$ for all $j \neq n$ then we have an intersection type aggregation and if $F(j) = 1$ for all $j \geq 1$ we have a union.

7.9   Conclusion

We have discussed a number of different methodologies for aggregating fuzzy subsets. We looked first at the intersection and the union operation and discussed the t-norm and t-conorm, which generalized the intersection and union operations. Weighted intersections and unions were next discussed. A generalization of the t-norm and t-conorm called the uninorm was then introduced. Mean operators were considered and we provided a characterization for these operator and discussed one class called the generalized mean. We introduced the Ordered Weighed Aggregation (OWA) operators. It was shown how these operators make a connection between the natural language stipulation of aggregations and the mathematical realization of the stipulated aggregation. Finally we discussed the fuzzy measure showed how it can be used in aggregation operators.

7.10   REFERENCES

[1] L.A. Zadeh, "Toward a theory of fuzzy information granulation and its centrality in human reasoning and fuzzy logic," *Fuzzy Sets and Systems*, Vol.90, pp.111-127, 1997.

[2] R.R. Yager and D.P. Filev, *Essentials of Fuzzy Modeling and Control*, John Wiley: New York, 1994.

[3] R.R. Yager, "On linguistic summaries of data," in *Knowledge Discovery in Databases*, Piatetsky-Shapiro, G. and Frawley, B. (eds.), Cambridge, MA: MIT Press, pp.347-363, 1991.

[4] R.R. Yager, "A general approach to the fusion of imprecise information," *International Journal of Intelligent Systems* Vol.12, pp.1-29, 1997.

[5] R.R. Yager and A. Kelman, "Fusion of fuzzy information with considerations for compatibility, partial aggregation and reinforcement," *International Journal of Approximate Reasoning* Vol.15, pp.93-122, 1996.

[6] D. Dubois, H. Prade, and R.R. Yager, "Merging fuzzy information," Technical Report# MII-1910 Machine Intelligence Institute, Iona College, New Rochelle, NY, 1999.

[7] Zadeh, L. A., "Fuzzy sets," *Information and Control* Vol.8, pp.338-353, 1965.

[8] U. Hohle, "Probabilistic uniformization of fuzzy topologies," *Fuzzy Sets and Systems* vol.1, 1978.

[9] E.P. Klement, "Characterization of fuzzy measures constructed by means of triangular norms," *J. of Math. Anal. and Appl.* Vol.86, pp.345-358, 1982.

[10] C. Alsina, E. Trillas, and L. Valverde, "On some logical connectives for fuzzy set theory," *J. Math Anal. and Appl.* Vol.93, pp.15-26, 1983.

[11] D. Dubois and H. Prade, "A review of fuzzy sets aggregation connectives," *Information Sciences* Vol.36, pp.85-121, 1985.

[12] R.R. Yager, "On a general class of fuzzy connectives," *Fuzzy Sets and Systems* Vol.4, pp.235-242, 1980.

[13] M.J. Frank, "On the simultaneous associativity of $F(x,y)$ and $x + y - F(x,y)$," *Aequat. Math.* Vol.19, pp.194-226, 1979.

[14] R.R. Yager, "Fuzzy decision making using unequal objectives," *Fuzzy Sets and Systems* Vol.1, pp.87-95, 1978.

[15] R.R. Yager, "A note on weighted queries in information retrieval systems," *J. of the American Society of Information Sciences* Vol.38, pp.23-24, 1987.

[16] D. Dubois and H. Prade, "Weighted minimum and maximum operations in fuzzy sets theory," *Information Sciences* Vol.39, pp.205-210, 1986.

[17] E. Sanchez, "Importance in knowledge systems," *Information Systems* Vol.14, pp.455-464, 1989.

[18] R.R. Yager and A. Rybalov, "Uninorm aggregation operators," *Fuzzy Sets and Systems* Vol.80, pp.111-120, 1996.

[19] J.C. Fodor, R.R. Yager, and A. Rybalov, A., "Structure of uni-norms," *International Journal of Uncertainty, Fuzziness and Knowledge-Based Systems* Vol.5, pp.411-427, 1997.

[20] B. De Baets, and J. Fodor, "On the structure of uninorms and their residual implicators," *Proc. Eighteenth Linz Seminar on Fuzzy Set Theory*, Linz, Austria, pp.81-87, 1997.

[21] B. De Baets, "Uninorms: The known classes," in *Fuzzy Logic and Intelligent Technologies for Nuclear Science and Industry*, D. Ruan, H.A. Abderrahim, P. D'hondt, and E.E. Kerre (Eds.), World Scientific: Singapore, Vol.21-28, 1998.

[22] R.R. Yager, "MAM and MOM operators for aggregation," *Information Sciences* Vol.69, pp.259-273, 1993.

[23] R.R. Yager, "Toward a unified approach to aggregation in fuzzy and neural systems," *Proceedings World Conference on Neural Networks*, Portland, Vol.II, pp.619-622, 1993.

[24] R.R. Yager, "A unified approach to aggregation based upon MOM and MAM operators," *International Journal of Intelligent Systems* Vol.10, pp.809-855, 1995.

[25] Dyckhoff, H. and Pedrycz, W., "Generalized means as model of compensative connectives," *Fuzzy Sets and Systems* Vol.14, pp.143-154, 1984.

[26] R.R. Yager, "On ordered weighted averaging aggregation operators in multi-criteria decision making," *IEEE Transactions on Systems, Man and Cybernetics* Vol.18, pp.183-190, 1988.

[27] R.R. Yager and J. Kacprzyk, *The Ordered Weighted Averaging Operators: Theory and Applications*, Kluwer: Norwell, MA, 1997.

[28] M. O'Hagan, "A fuzzy neuron based upon maximum entropy-ordered weighted averaging," in *Uncertainty in Knowledge Bases*, B. Bouchon-Meunier, R.R. Yager, and L.A. Zadeh, (Eds.), Springer-Verlag: Berlin, pp.598-609, 1990.

[29] M. O'Hagan, "Using maximum entropy-ordered weighted averaging to construct a fuzzy neuron," *Proceedings 24th Annual IEEE Asilomar Conf. on Signals, Systems and Comp uters*, Pacific Grove, Ca, pp.618-623, 1990.

[30] D.P. Filev and R.R. Yager, "Learning OWA operator weights from data," *Proceedings of the Third IEEE International Conference on Fuzzy Systems*, Orlando, pp.468-473, 1994.

[31] D.P. Filev and R.R. Yager, "On the issue of obtaining OWA operator weights," *Fuzzy Sets and Systems* Vol.94, pp.157-169, 1998.

[32] L.A. Zadeh, "A computational approach to fuzzy quantifiers in natural languages," *Computing and Mathematics with Applications* Vol.9, pp.149-184, 1983.

[33] M. Sugeno, *Theory of fuzzy integrals and its applications*, Doctoral Thesis, Tokyo Institute of Technology, 1974.

[34] M. Sugeno, "Fuzzy measures and fuzzy integrals: a survey," in *Fuzzy Automata and Decision Process*, Gupta, M.M., Saridis, G.N. and Gaines, B.R. (Eds.), Amsterdam: North-Holland Pub, pp.89-102, 1977.

[35] D. Dubois and H. Prade, "A class of fuzzy measures based on triangular norms," *International Journal of General Systems* Vol.8, pp.43-61, 1982.

[36] T. Murofushi and M. Sugeno, "An interpretation of fuzzy measures and the Choquet integral as an integral with respect to fuzzy measure," *Fuzzy Sets and Systems* Vol.29, pp.201-227, 1989.

[37] T. Murofushi and M. Sugeno, "Theory of fuzzy measures: representations, the Choquet integral and null sets," *J. Math Anal. Appl* Vol.159, pp.532-549, 1991.

[38] D. Denneberg, *Non-Additive Measure and Integral*, Kluwer Academic: Norwell, MA, 1994.

[39] F. Modave, F. and M. Grabisch, "Preference representation by Choquet integral: the commensurability hypothesis," *Proceedings of the Seventh International Conference on Information Processing and Management of Uncertainty in Knowledge-based Systems*, Paris, pp.164-171, 1998.

[40] H.T. Nguyen and M. Sugeno, *Fuzzy Systems: Modeling and Control*, Kluwer Academic Press: Norwell, Ma, 1998.

8

Fuzzy Gated Neural Networks in Pattern Recognition

Zhi-Qiang Liu
Venketachalam Chandrasekaran

8.1 Introduction

Pattern recognition is important in virtually all intelligent systems, in particular, in human centered systems, one major development is the *intelligent servant modules* (ISMs) that can react and interact with humans. To effectively respond to human's request ISMs must be able to recognize gestures, voice patterns, facial features and so on. Making systems to achieve such capabilities is a challenging task due to uncertainties which arise from incomplete or imprecise knowledge of what is being perceived together with data corruption due to inherent noise in sensors. Furthermore, the recognition system must be able to generalize from the "seen" samples to "unseen" patterns that are from the same population. In addition, the system will have to reject "unknown" patterns or to update the knowledge base, in some instances, to alert the user.

A general paradigm for recognition is to constrain the pattern recognition problem within the known classes. In this paradigm, the confidence with which we can identify an unknown pattern as one within the known classes or reject it as an "alien" depends on two important factors: the geometry of the feature space partitioning and the expected level of noise perturbations. Therefore, a proper choice of a partitioning strategy is critical to the pattern recognition task. Usually, we maximize the maximum *class frequency* in each of the partition and minimize the risk of misclassification to maintain a sufficient "cover" for noise perturbation.

In this chapter, we present a Fuzzy Gated Neural Network (FGNN) for pattern recognition [3, 4]. We propose a *selective* competitive learning paradigm for dynamic feature space partitioning that retains the robustness of the larger partitions and achieves accurate classification [1]. Furthermore, it eliminates the need for training the synaptic weights to a set of nodes at its input layer to represent the cluster centroids. This property makes it easy and fast to train and simple to implement in many practical applications, in particular, in human-centered systems which require near real-time response and intensive interactivity. We introduce a generalized

Gated Neuron Model (GGN) with multiple outputs, which is an integral part of the FGNN. We define the axonal outputs in terms of prediction strengths for time-dependent class labels. In the FGNN, the number of outputs equals that of the class labels. By combining the prediction strengths, we can accurately estimate the true class probability densities and obtain reliable class labels.

8.2 Generalized Gated Neuron Model

Computational intelligence aims at modeling the information content of the nervous system at the biophysical, circuit, and system levels. This is based on the conjecture–although controversial–that the brain *computes* in the sense of representing, processing, and storing information [5]. A complete understanding of information processing at the level of an individual cell requires the detailed knowledge of its anatomical structure, physiological properties of the neuron and its synapses, and a model of the cell that faithfully embodies this information. Unfortunately, we are far from being able to model the brain neurons even if we incorporated all known biophysical details [5]; can we ever? One of the strongest criticisms of the simplified artificial neuronal models is that these are not biological and are irrelevant to understanding the nervous system. Our objective, however, is to find a simple model that has the structure to flexibly represent continuous and discrete nonlinear systems and that can serve our purpose of developing intelligent servant modules for specific applications.

Over the last few decades researchers developed various models and applied these models in many applications, for instance, McCulloch and Pitts Model (1943) [6], Rosenblatt's Perceptron (1957) [7], Widrow's ADA-LINE (1962) [8], Oguztoreli's Discrete Neuron (1972) [9], Simplified Neuron [10]Lee and Lee's MP Fuzzy Neuron (1975) [11], Higher Order Correlation Neuron (1986) [12], Chandrasekaran *et al.*'s Gated Neuron (1993) [13], and Chandrasekaran *et al.*'s High Order Gated Neuron (1993) [14], Kwan *et al.*'s Generalized Fuzzy Neuron (1994) [15]. Fig. 8.1 shows two important models that are relevant to the following discussions.

Fuzzy Neuron: A Fuzzy Neuron (FN) has N weighted inputs $w_i * x_i$, where $w_i \in [0,1], i = 1, 2, \cdots, N$ are the weights and M outputs y_j. Each output is associated with a fuzzy membership value to express the degree the pattern **X** belongs to a fuzzy set. The following equations describe the fuzzy neuron:

$$z = h(w_1 x_1, w_2 x_2, \cdots, w_n x_n), \qquad (8.1)$$

$$s = f(z - T), \qquad (8.2)$$

$$y_j = g_j(s), \quad \forall j = 1, 2, \cdots, M, \qquad (8.3)$$

where z is the net input of FN, $h(\cdot)$ is the aggregating function, s is the

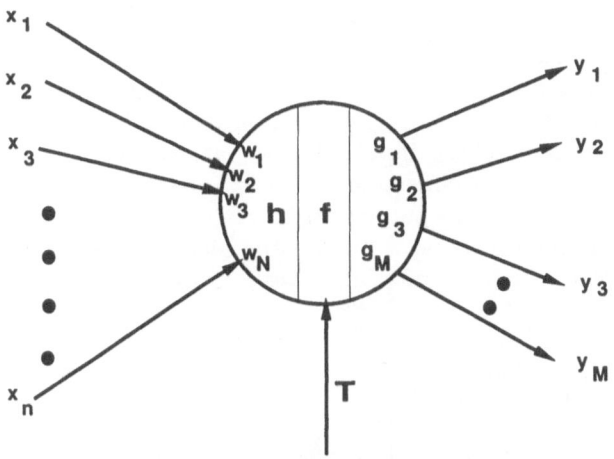

(a) Fuzzy Neuron

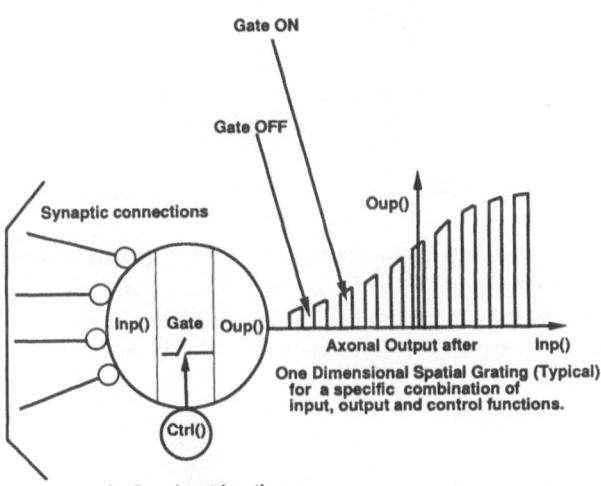

Inp() = Input function.
Ctrl() = Neuronal Gate Control function
Oup() = Output Non-linear function

(b) Gated Neuron

FIGURE 8.1. Two Recent Neuron Models

state of the FN, $f(\cdot)$ is the activation function, T is the activating threshold and $g_j(s)$ is the jth output function of FN representing the membership grade of the input pattern $\mathbf{X}$.

Gated Neuron: The Gated Neuron (GN) proposed in [13] has the ability to choose different input and output functions and to control the output via a neuronal gate.

Generalized Gated Neuron: We extend the above gated neuron model to a generalized gated neuron model that has multiple outputs. This model structurally decouples the input, the output, and the gate functions to facilitate many types connections; for instance, the gate can be placed anywhere within the input-output configuration or can be controlled totally from an external output. Fig. 8.2 shows the generalized gated neuron model that is capable of representing all other known models. In addition, this

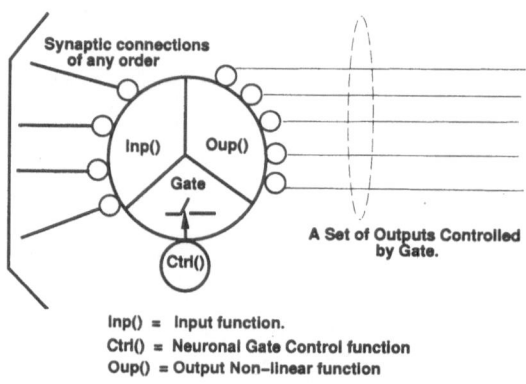

FIGURE 8.2. Generalized Gated Neuron Model representing an nonlinear system

model can perform a multi-dimensional spatial grating at its outputs. When the outputs are chosen to represent the fuzzy class membership values of the input pattern $\mathbf{X}$, the model becomes a fuzzy gated neuron model. In the following sections, we demonstrate its use in FGNNs for pattern recognition.

8.3   Fuzzy Gated Neural Networks

In the following section we discuss FGNN's architecture, functions, knowledge aggregation and class label prediction strategies.

8.3.1   System Structure

Fig. 8.3 shows the basic structure of FGNN which has two layers: the FGN layer and simple neuron layer. The input layer consists of $n + 1$ neurons for classifying n-dimensional input patterns. The "Virtual Neuron" (VN) (whose index is $n + 1$) provides class label predictions at its output only when all other neurons are "knocked out" of competition, i.e., when the input pattern resides in non-competition region and all neuronal gates are open. When an input pattern is presented, the network generates a sequence of winning node indexes which form a "spatio-temporal" *signature*. Each winning node has a set of class label predictions based on the class frequency data gathered at each winning node during training. The output classification layer performs knowledge aggregation by combining the predictions from time to time. At the end of the discrete interval indexed by T, the class label that has the maximum prediction strength is declared to be the predicted label of the input pattern.

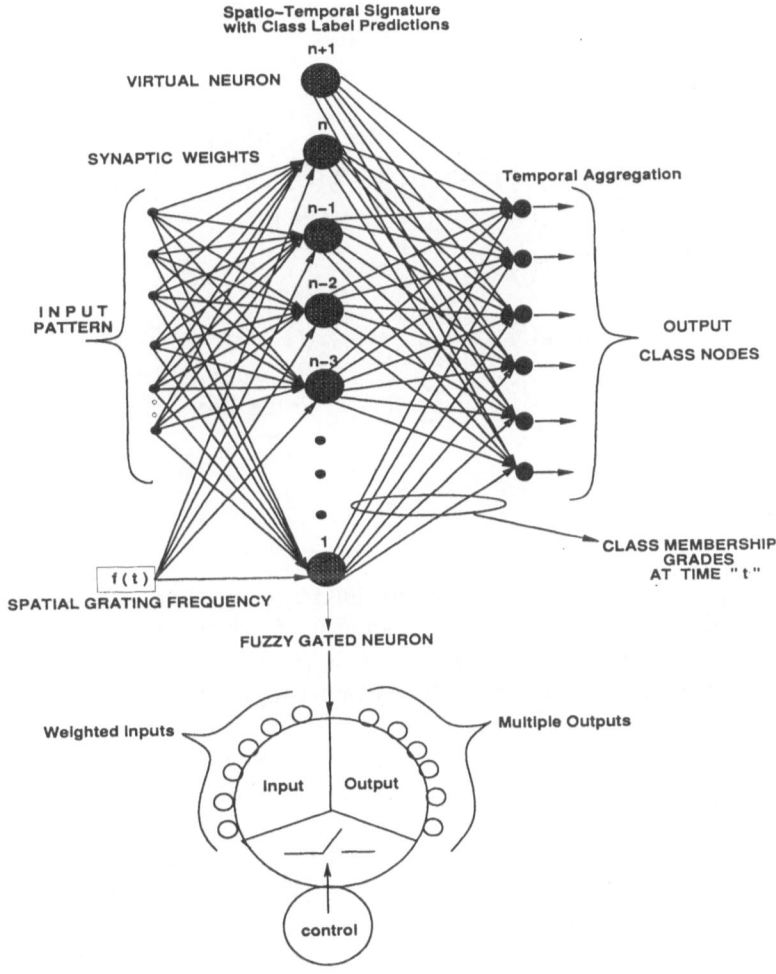

FIGURE 8.3. General Organization of FGNN

8.3.2 *Input, Gate, and Output Functions*

INPUT FUNCTION
The input function for neuron j computes the distance between the input pattern X and the weight vector W_j. We have chosen the city-block distance metric for its simplicity in computation.

GATE CONTROL
The j-th neuronal gate control function is constructed over several functions:

$$\text{gate}_j(\text{ctrl}_j(t)) = \begin{cases} \text{ON} & \text{if ctrl}_j(t) > 0, \\ \text{OFF} & \text{otherwise}; \end{cases} \tag{8.4}$$

$$\text{ctrl}_j(t) = \gamma_j(t) * \beta_j(t); \tag{8.5}$$

$$\gamma_j(t) = cos(2\pi f(t)d_j); \tag{8.6}$$

$$f(t) = f_{\max}(1 - \frac{t}{\text{T}}); \tag{8.7}$$

$$d_{\min}(t) = \overset{\min}{k}\ \{d_k|\gamma_k(t) > 0\}; \tag{8.8}$$

$$\beta_j(t) = \begin{cases} 1 & \text{if } d_j = d_{\min}(t), \\ 0 & \text{otherwise}; \end{cases} \tag{8.9}$$

where $f_{\max}$ is the maximum spatial grating frequency. For a given distance measure d_j, $f(t)$ controls the time sequence of jth neuronal competition over the discrete time interval T. $d_{\min}(t)$ is the minimum distance among the competing neurons at time t. The competing neuronal set is the set of neurons for which $\gamma_j(t) > 0$. When $\text{ctrl}_j(t) > 0$, it indicates that the neuron j has been *selected* in the competition and become the "winner", i.e., its $\gamma_j(t) > 0$ and $\beta_j(t) = 1$. Only the winning neuron has its gate switched on and outputs a set of predictions in terms of fuzzy class labels. When there is no competition between the neurons, VN takes control and provides the prediction as if it were the winner. In effect, VN represents the class label distributions within the regions of non-competition.

FUNCTIONS DEFINING MULTIPLE NEURONAL OUTPUTS
The fuzzy set of class label prediction strengths $C_i(t)$ at the winning neuron

i over m classes is given by

$$C_i(t) = \{g_{1i}(t)/1, g_{2i}(t)/2, \cdots, g_{mi}(t)/m\},$$

$$= \sum_{k=1}^{m} g_{ki}/k \quad \text{where } \sum \text{ is a fuzzy representation;} \quad (8.10)$$

$$g_{ki}(t) = \begin{cases} \frac{n_{ki}(t)}{n_i(t)} & \text{when node } i \text{ is winner,} \\ 0 & \text{otherwise;} \end{cases} \quad (8.11)$$

$$n_i(t) = \sum_{k=1}^{m} n_{ki}(t), \quad (8.12)$$

where $g_{ki}(t)$ is the k-th class membership grade associated with winning node i at time t. $n_{ki}(t)$ is the total number of training patterns of class k associated with the winning node i at time t and $n_i(t)$ is the total number of training patterns of all classes associated with the winning node i at time t.

The outputs of the winning neurons at the FGN layer are conveyed to the classification layer via unit weights.

8.3.3   Temporal Aggregation

In this section, we introduce two basic temporal aggregation methods in the classification layer: Mean aggregation and Bayesian aggregation. Mean aggregation considers averaging the prediction strengths $C_i(t)$ over $t \in \{0, 1, \cdots, T\}$. The Bayesian approach assumes that the class label predictions are statistically independent and obtains an overall prediction value by computing the product of probabilities. The final class label prediction is based on the class node that has the maximum average output. When the input patterns are noise-free, the two methods will produce similar results.

Let the winning node variable at any time t be denoted by $J(t)$ and the winning node sequence obtained for an input pattern p be $\{J_p(0), J_p(1), \cdots, J_p(T)\}$. These winning nodes provide a fuzzy set of class label predictions $C_{J_p(t)}(t)$. These temporal predictions are combined at the output layer of the class nodes to generate the overall prediction strength for each class. The class node that has the maximum strength is declared to be the label of the input pattern p.

MEAN AGGREGATION
The following equation defines the outputs at the class nodes based on the Mean aggregation strategy:

$$o_k^{(1)} = \frac{\sum_{t=0}^{T} g_{kJ_p(t)}(t)}{\sum_{k=1}^{m} \sum_{t=0}^{T} g_{kJ_p(t)}(t)}. \quad (8.13)$$

BAYESIAN AGGREGATION

The winning node sequence for a specific input pattern depends on the ordering of spatial grating frequency $f(t)$ in the set $\mathcal{F}$. Therefore the winning events and the associated predictions are statistically independent. If we treat $g_{kJ_p(t)}(t)$ as probabilities, these can be conveniently combined using the statistical independence assumption. The following equation estimates the conditional probabilities over the time-sequence of winning nodes,

$$o_k^{(2)} = \frac{\prod_{t=0}^{T} g_{kJ_p(t)}(t)}{\sum_{k=1}^{m} \prod_{t=0}^{T} g_{kJ_p(t)}(t)}. \tag{8.14}$$

Unlike other popular networks, FGNN does not require iterative weight adaptation schemes. This greatly reduces network setup time and enables on-line learning for any task space modifications. However, weight initialization is important to the success of this algorithm. We have shown that the necessary and sufficient condition for weight initialization is that *the weights need to be constrained within the convex hull of the input space and to be well spread to have representative samples from all regions* [16].

8.4   Comparison between FGNN and STFM

This section presents the dynamic feature space partitioning strategy of FGNN and discusses its similarities with STFM [1]. Furthermore, we extend the theoretical foundations of STFM to FGNN architecture.

8.4.1   FGNN's Operational Characteristics

The dynamic feature space partitioning strategy uses all possible combinations of the decision planes between the neurons in order to generate uniquely addressable fragments of the feature space. This is achieved in three steps:

- Under a selective competition criterion, a set of competing neurons is chosen and the closest node to the input pattern is declared as the winner.
- The selection criterion is changed in the time interval $[0, T]$ to produce a time-sequence of winning nodes. Each winning node represents a decision region based on the decision planes between the competing set of neurons.
- The intersection of the decision regions of the winning nodes results in a fragment of the feature space that contains the input pattern.

SELECTIVE COMPETITION

We implement the selective competition by first performing a spatial grating transformation $\mathcal{G}$ of the feature space and uniquely labeling the regions as competition and non-competition regions for each neuron. In this process, each neuron modulates the frequency $f(t)$ by its distance d_j to the input pattern. If the input pattern p resides in any of the competition regions of a neuron, then the neuron is selected to be in the competition set. Spatial grating for each neuron is implemented by its gate control function $\gamma_j(t)$ with parameters $f(t)$ and d_j. Therefore, the decision for any neuron j to compete is a "local" decision. Fig. 8.4 illustrates the neuronal selection process.

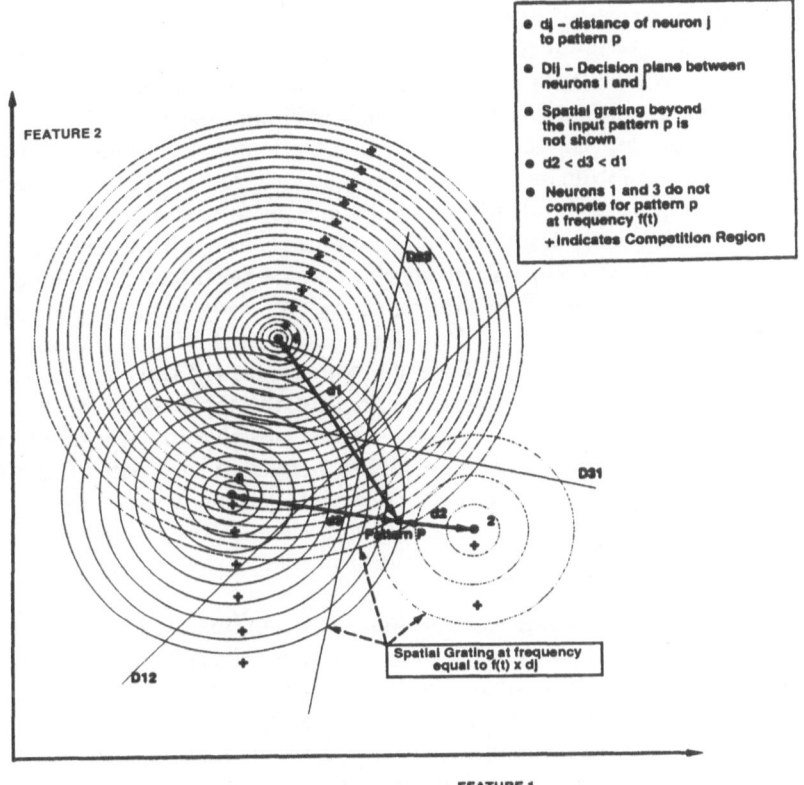

FIGURE 8.4. Spatial Grating and Selective Neuronal Competition

SPATIO-TEMPORAL SIGNATURE

As the grating frequency is decreased monotonically to zero in a fixed time interval T, the selection criterion is changed over time. This deterministically changes the competing set. A typical spatio-temporal signature generation is shown in Fig. 8.5.

The top part of Fig. 8.5 shows the winning node indexes that constitute the spatio-temporal signature generated in FGNN for an input pattern p.

A typical set of class label predictions $C_{J_p(t)}(t)$ associated with the winning nodes $J_p(t)$ is shown in the bottom part of Fig. 8.5.

REGION SHARING

The selective competition strategy usually results in a competition set of size smaller than n for all frequencies greater than zero. This means that a smaller number of participants share the input pattern space. The neurons that are close to the ones *knocked out* of competition subdivide the feature space "owned" by those eliminated from competition, as shown in Fig. 8.6. Elimination of a node enables its decision region to be partitioned and shared by its immediate neighbors. Such a partition is possible only when the decision planes between the immediate neighboring nodes cross the input space owned by the node that is eliminated from the competition. For example, the cluster "A" in Fig. 8.6 is not partitioned by any of the decision planes during the node elimination process. Those nodes that are outside the region will have no influence in further partitioning. It is therefore evident that the decision region associated with a winning node is dependent on the competition set chosen at a particular frequency. The dynamic nature of the region sharing activity provides a method to obtain different sets of class membership values associated with a winning node at various time instances. Such membership values aid the computation of the class possibilities in the neighborhood of the input pattern, resulting in an increased prediction accuracy. This is possible in view of FGNN's natural ability to use the decision planes between every pair of neuronal nodes to divide the input pattern space into uniquely addressable fragments. FGNN represents a new type of *soft*-competitive learning paradigm that has been demonstrated to out-perform many traditional winner-take-all learning techniques.

Each of the modules described in FGNN is functionally identical to that STFM except that FGNN uses a one-dimensional array of gate neurons. In addition, computation of predictions and the temporal aggregation at the class nodes are also the same. FGNN is similar to STFM in respect of selective competition among the neurons, dynamic feature space partitioning and knowledge aggregation followed by class label prediction. The main difference lies in the training of weights W at the input layer. In contrast to iterative weight adaptation using SOM algorithms, in STFM, FGNN selects from the training pattern set directly. Since the weight initialization in FGNN does not have any relation to the topology preserving concept, it is topology constraint-free. The two major theoretical foundations of STFM described in [1] can be readily extended to FGNN and are given as follows.

Theorem 39 *The spatio-temporal signature represented by a time-sequence of winning nodes generated in a FGNN is invariant over a small neighborhood of the input pattern.*

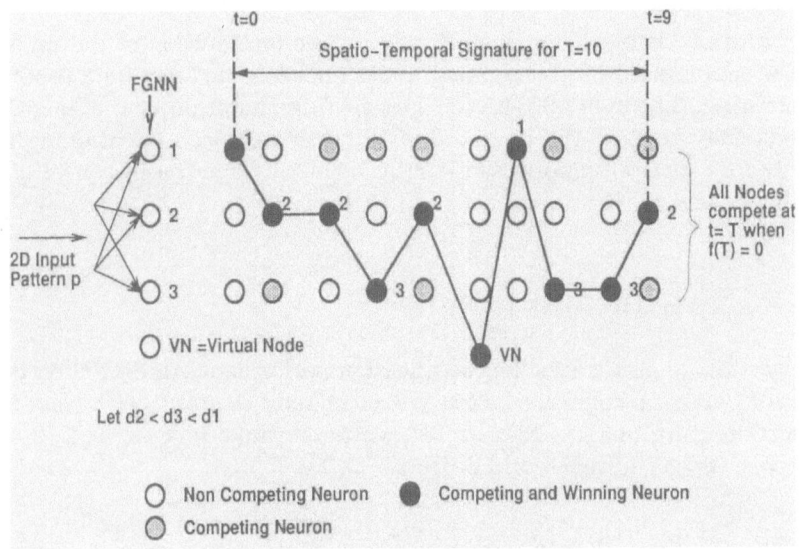

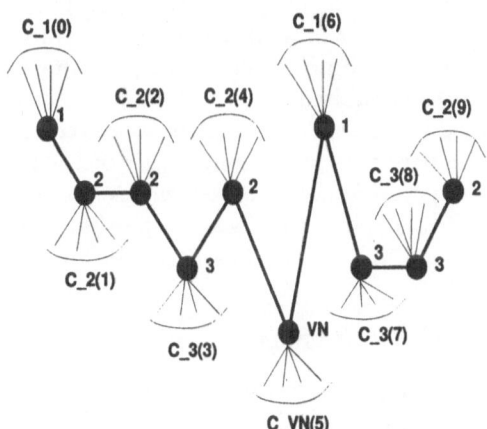

An Example of Spatio-Temporal Signature For An Input Pattern p together
with its class membership grades on a 4-Class Problem

FIGURE 8.5. Typical Spatio-Temporal Signature Generation with Class Membership Grades

Theorem 40 *The classification performance of FGNN will always be greater than or equal to SOFM if the class label prediction is based on the true label of the decision regions represented by the spatio-temporal signature.*

The proof of Theorem 39 depends only on the intersection of the decision regions represented by the winning nodes and does not require a topological ordering. Theorem 40 relies on further fragmentation of the partitions at frequency zero by the unique dynamic feature space partitioning represented by spatio-temporal signatures. Refer to Chandrasekaran *et al.* [1] for the details.

8.5   Experimental Results

This section demonstrates the classification performance of FGNN on three types of pattern recognition tasks. We extracted features from 8-class Brodatz 2D texture images, 12-class 3D synthetic range images, and 12-class 3D range images of real world objects.

8.5.1   2D Real World Texture Data

In this experiment we used a set of eight Brodatz texture images. The size of original texture image is 512×256 pixels. For feature extraction, we divide each image into 128 blocks of 32×32. Subimages of 16×16 are then extracted from the center of each 32×32 block. Each subimage is then convolved with 3×3 pixel masks of eight different types as described by Law [17]. For each subimage, we obtained a vector of eight real-valued elements; each value corresponded to a feature extracted by the convolution process. The complete data set had eight sets of 128 feature vectors; one set for each texture type. We used the following eight types of texture images in this experiment: i) raffia, ii) grass, iii) pigskin, iv) wool, v) leather, vi) sand, vii) water, and viii) wood. These are shown in Fig. 8.7

8.5.2   3D Synthetic Images

In this test, we restricted our implementation to 3D CAD models and view-dependent range images generated using the CAD tools. We generated a set of objects constructed by selecting subsets of a common pool of pieces which could assume different states (type of cylinder,etc), relative positions and intersections. Fig. 8.8 shows an example view of each object(top) and a full set of five views for one object(bottom) generated for this test. One of the main reasons for using such view-dependent input samples is that the computation of surface properties is in general restricted to what is visible.

We computed unary and binary part features with respect to the full 3D properties of the range data. For feature extraction, we need to segment the

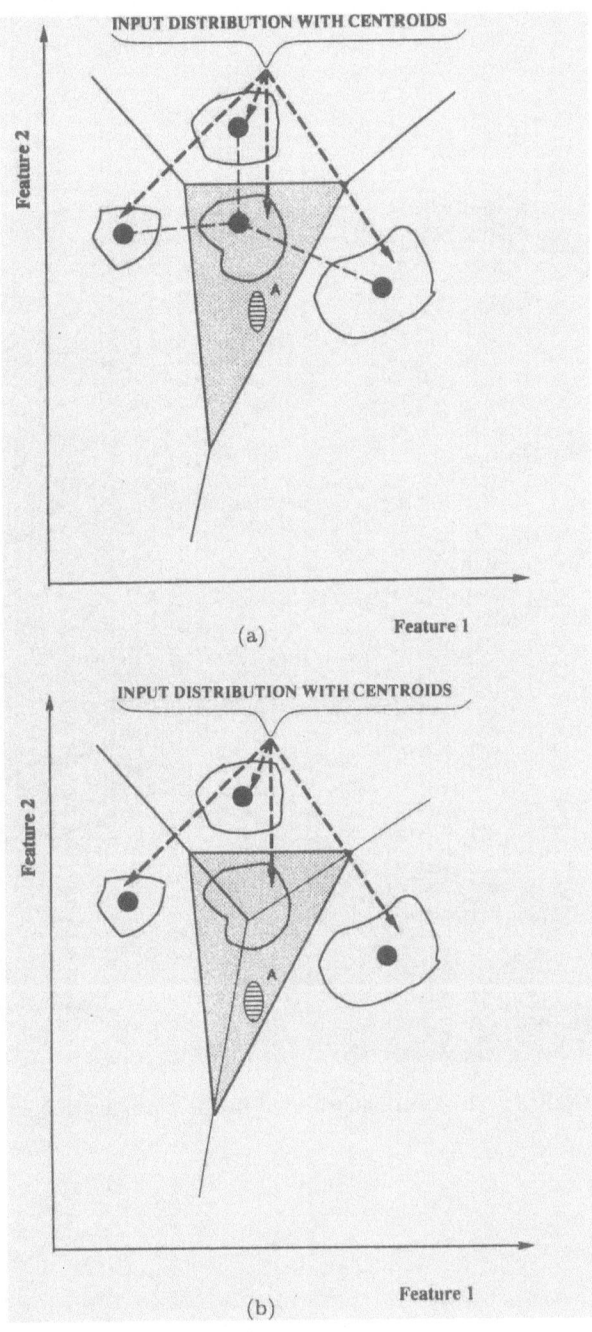

FIGURE 8.6. An Example of Input Pattern Space Partitioning By Selective Competition: (a) Partitioning when all neurons compete (say at frequency = 0); (b) Region capturing by neighbors when a neuron is *knocked out* of competition.

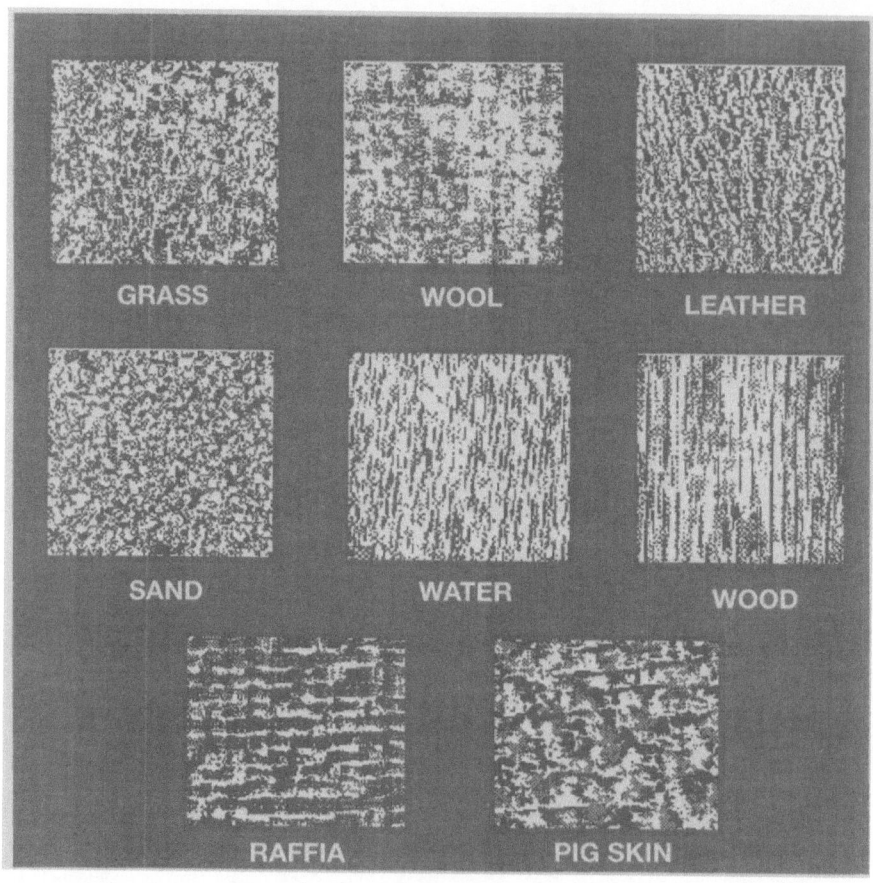

FIGURE 8.7. Texture images used in the classification experiment.

model into various parts having similar characteristics. In order to guarantee compatibility between model and test data parts, it is necessary to use a segmentation procedure which applies equally to both domains and uses features that are invariant to the parameterization of the surface. Fortunately Mean(H) and Gaussian(K) curvatures satisfy these conditions. We used zero-crossings of the determinant of the Hessian for segmentation, which determines convex, concave and planar regions [18]. Such a segmentation procedure applies to models and data, and is invariant to rigid motions.

Segmentation is to produce an efficient data structure for representing surface patches and their relational features. Such features need to optimize two somewhat contradictory goals: invariance and uniqueness. The former refers to the need to represent models which are invariant to rigid motions, pose etc., whereas the latter refers to the representations that uniquely define the model. Model surface features are typically of three generic forms: Morphological, Unary, and Binary features. The feature data set obtained from this 12-class synthetic object set is listed below:

1. Morphological features: Four features of the types perimeter, number of parts, genus-average shape over whole object and total area are ordered to become a single 4-dimensional feature vector. A set of 60 such feature vectors is obtained for simulations.

2. Unary features: Seven features of the types mean H, mean K, area, 3D spanning distance, part perimeter, mean curvature, mean torsion are combined into a single 7-dimensional feature vector and a total of 627 feature vectors are collected for use.

3. Binary features: Eight features of the types length of jumps, length of crease, bounding distance, centroid distance, maximum distance, differences in average normal angles across their common boundaries, average bounding angle between surfaces, average normal angle differences over the entire patches are combined to form a single feature vector and a set of 1728 feature vector of dimension 8 is obtained.

Furthermore, we normalized all the training feature data via the z-score transformation,

$$\hat{x} = \frac{x - E(x)}{\sigma_x}, \tag{8.15}$$

where $E(x)$ and σ_x correspond to the mean and standard deviation of x resulting in $\hat{x}$ having zero mean and unit variance. This guarantees an initial equal sensitivity to all features. For the purpose of FGNN setup, all these feature vectors are arranged as eight dimensional feature vectors by padding zeros and a total of 2415 vectors are available for training and testing.

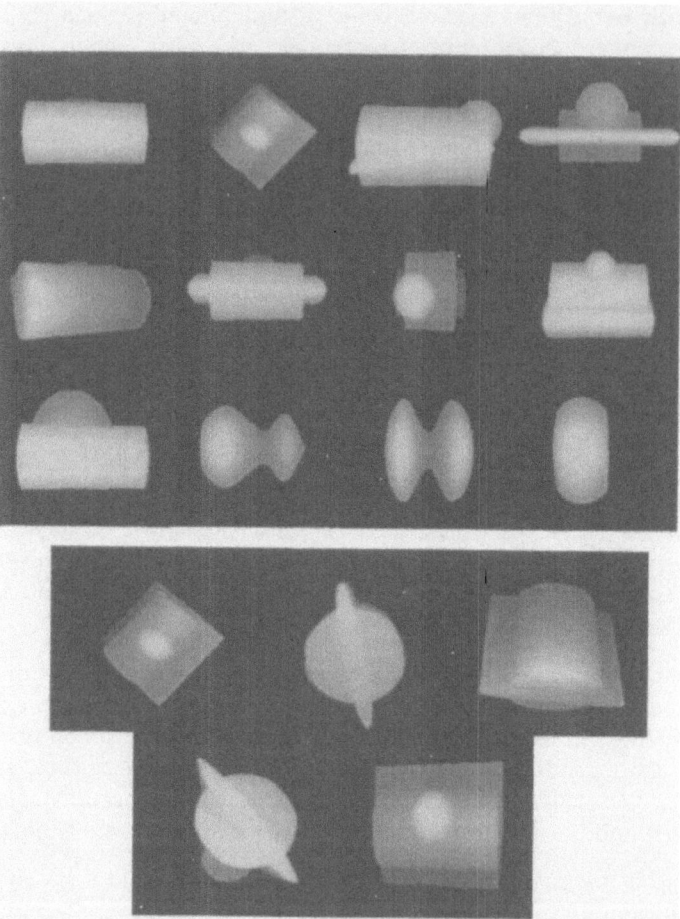

FIGURE 8.8. An example view of each object (top) and a full set of five views for a single object.

8.5.3   Real Range Images

A set of range images for 12 real-world objects was selected; each object
has five views. Fig. 8.9 shows a typical view of each object used in the test.

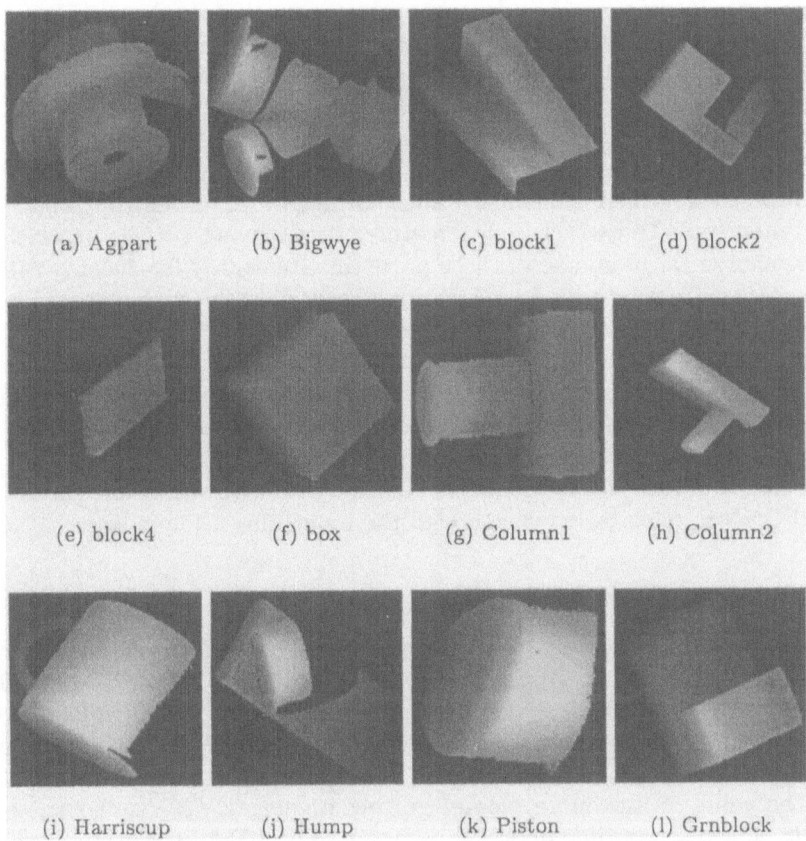

<table>
<tr><td>(a) Agpart</td><td>(b) Bigwye</td><td>(c) block1</td><td>(d) block2</td></tr>
<tr><td>(e) block4</td><td>(f) box</td><td>(g) Column1</td><td>(h) Column2</td></tr>
<tr><td>(i) Harriscup</td><td>(j) Hump</td><td>(k) Piston</td><td>(l) Grnblock</td></tr>
</table>

FIGURE 8.9. A typical view of the 3D Object set

These images were segmented using crease and jump edges of the objects. These edges are found by a dynamical neural network architecture
described in [19]. The dynamic neural network uses a class of fractional
derivative-based edge operators obtained by varying the fractional deriva-
tive index from a very low value to one in small steps. The network starts
with a set of real edge pixels and then links them with other real edge pixels
that are generated at subsequent stages of convolution with edge operators.
The edges that are considered noisy are deleted as the process continues.
Finally these real edges are linked using conventional image processing algorithms to obtain closed boundaries and well defined surface patches [19].

From the segmented patches, we extracted unary, binary, morphological and edge-based features. The feature extraction process is different from that in Section 8.5.2 for the 3D synthetic images. The process here is automated by extracting nine types of temporal features using sparse decomposition of the segmented images [16, 20]. Each subimage is decomposed into 31 images. The decomposition network consists of an array of GNs whose gates are controlled by a cosine function with a time-dependent frequency term and the image pixel value as its parameters. By changing the frequency from its maximum (say 100 Hz) to zero in 31 steps, the array provides a time-sequence of the decomposed images. These images are considered to be distributions satisfying a criterion whose characteristics can be specified statistically such as mean median, entropy, and central moments. To extract binary features between two surface patches, we have chosen range images of these parts simultaneously for decomposition. Whereas extracting edge-based features needs only the edge map, the morphological feature extraction requires the whole image for decomposition. The resulting feature dimension is 9×31 which can be treated as a single 279D vector. The total number of such features extracted from the 60 views of the 12 objects is 3191. All the feature vectors are normalized via the z-score transformation described in Section 8.5.2.

Using this data set, we demonstrated the ease with which one can set up an FGNN for this task, even though the input dimension is large.

8.5.4   Results and Discussions

FGNNs for the above classification tasks are organized as follows. The network size described by (Input Dimension × FGNs × Output Class Nodes) is set to be $(8 \times 100 \times 8)$ for the texture images, $(8 \times 400 \times 12)$ for the 3D synthetic range images, and $(279 \times 400 \times 12)$ for the 3D real range images.

The value of maximum spatial grating frequency f_{max} is chosen to be 30Hz. The discrete time period T is chosen to be 100 for the texture images and 3D synthetic range images. For the real range images, the value of T is chosen to be 60.

The feature data of all three types is split into two parts for training and testing purposes. 75% of the feature data set is picked randomly to become the training set and the rest is kept as validation set to verify the generalization capability.

From above training data set, representative vectors are picked directly for assigning weight connections to FGNs at i) regular intervals as well as ii) random for both texture images and 3D synthetic range images. Only random selection is done for the real range images.

The classification performance evaluation based on both the mean and Bayesian aggregation approaches is implemented for the texture images and 3D synthetic range images. For the real range images, we evaluated

TABLE 8.1. Classification Performance Results of FGNN over three different data sets: textures (TEXTURE), 3D synthetic objects (SYNTHETIC), and real objects (REAL), with FMAX = 30 in all cases.

DATA	TEXTURE T=100		SYNTHETIC T=100		REAL T=60
	Mean	Bayes	Mean	Bayes	Mean
Weight Space by Random Picking					
Train Data	83.74	97.66	85.60	86.70	71.55
Test Data	78.43	76.86	44.04	50.12	67.67
Weight Space at Regular Intervals					
Train Data	81.79	96.10	67.83	97.13	NA
Test Data	72.94	74.12	38.58	57.29	NA

the performance using only the mean aggregation method.

The results of the classification performance are shown in Table 8.1. From the results, we can see that FGNN has a satisfactory classification performance on a range of difficult pattern recognition tasks. If we used conventional networks such as MLFNN to deal with 279D patterns, it would require many hours of computation time. In comparison FGNN needs only one epoch presentation of the training patterns to create the class frequency table described earlier. Furthermore, the proposed system has many advantages, such as quick setup time, high classification performance, good generalization of the unseen data, and iterative learning modes: When a new pattern is to be learned, it is simply a matter of presenting and updating the class frequency table. Learning in FGNN has been extremely good, because it has demonstrated a high level of classification performance on "seen" data.

8.6   Improvements to FGNN

8.6.1   Performance under Noisy Data

Both Spatio-Temporal Feature Map (STFM) [1] and FGNN exhibit a high level of sensitivity to noise. The ability to predict accurately suffer when the patterns are allowed to migrate to neighboring regions under the influence of noise. It must be noted that the features extracted from images will vary over time depending on the degradation in the sensitivity and the nonlinearities associated with the imaging equipment. To simulate such an environment, the input features are perturbed with additive white Gaussian noise for different signal-to-noise ratios (SNRs) and a set of data with SNR values at 1000, 100, 50 and 10 is obtained using 3D synthetic objects data set. The noise is added based on the following definition of SNR,

$$\text{SNR} = \frac{\sigma_s^2}{\sigma_n^2}, \tag{8.16}$$

where σ_s^2 is the variance of signal and σ_n^2 is the variance of noise. The signal power σ_s^2 is assumed to be unity. The simulation results on STFM are tabled in [1].

In this section, we focus on FGNN only. The classification performance of FGNN with 400 nodes, $f_{\max} = 30$ Hz and $T = 100$ under the mean and Bayesian aggregation methods at the class nodes are shown in Table 8.2. In the test, all the training data were noiseless, but the test data were noisy.

It is evident that FGNN performs poorly under noise. Also, the type of knowledge aggregation method at the classification layer is an important factor. The product terms under Bayes rule results in useless predictions. In the following subsections, we will explore methods to improve the robustness of FGNN.

TABLE 8.2. Classification Performance Results of FGNN with noisy data for the 3D objects, where MA stands for Mean Aggregation, and BA for Bayes Aggregation.

SNR	0	1000	100	50	10
MA	77.35%	40.75%	29.57%	26.71%	18.01%
BA	90.27%	9.07%	5.18%	6.09%	8.03%

8.6.2 Noise Cover in FGNN

A "sufficient cover" for noise perturbations for patterns within the fragments of the feature space partitions can be accomplished by setting a thresholding distance from the decision planes. However, on account of complex boundaries of these fragments, provision of a such a cover at the boundaries may increase the computational complexity. The question is how to deal with the input patterns that fall within the cover. We address this issue below.

A noise cover is set at 3σ level with its mean centered at the decision boundary. When an input pattern P_r of dimension n is close to a decision boundary between the nodes i and j represented by weights W_i and W_j, its distance d_r to the boundary is computed as follows (see Fig. 8.10).

$$d_{ij} = \sqrt{\sum_{k=1}^{n}(w_{ik} - w_{jk})^2}, \tag{8.17}$$

$$V_{ij} \bullet V_{ir} = \sum_{k=1}^{n}(w_{ik} - w_{jk}) * (w_{ik} - p_{rk}), \tag{8.18}$$

$$V_{ji} \bullet V_{jr} = \sum_{k=1}^{n}(w_{jk} - w_{ik}) * (w_{jk} - p_{rk}), \tag{8.19}$$

$$d_r = abs(\frac{V_{ij} \bullet V_{ir}}{d_{ij}} - \frac{d_{ij}}{2}), \quad \text{or} \tag{8.20}$$

$$= abs(\frac{V_{ji} \bullet V_{jr}}{d_{ij}} - \frac{d_{ij}}{2}), \tag{8.21}$$

where d_{ij} is the Euclidean distance between W_i and W_j, V_{tu} is the n-dimensional vector from location t to u and $\bullet$ represents the dot product between two vectors. If the computed distance d_r is greater than $1.5\sigma_n$, then adequate noise cover/immunity for the specified SNR is deemed to be present. If one of the nodes in the set i, j is a "winner" according to minimum distance criterion and the pattern is close to the decision boundary between them, then both nodes are to be treated as winners.

Let the winning node set at any time instant t be $WS(t)$. Its identification process commences with the selection of node J having the minimum distance to the pattern among the competition set at time instant t. Then the presence of noise cover is verified at the decision planes between this minimum distance node J and all other nodes in the competition set. Based on the winning node criterion, the competition set is then reduced to a set of nodes that share the input pattern in the fuzzy boundary region. The above approach is intended to provide information in respect of class frequencies of patterns falling in crisp and fuzzy regions separately. With the possibility of multiple winning nodes at any time t, we need to devise suitable methods to acquire knowledge at these "winning" nodes for use in the recognition stage. This is addressed in the following subsection.

8.6.3 Knowledge Acquisition and Aggregation

KNOWLEDGE ACQUISITION
During the training phase, the class frequency counts of all such "crisp" nodes are updated by incrementing the associated class count by one. However, when the pattern falls in a fuzzy region, the class label of the input pattern is shared in inverse proportion to the distance to the nodes in the set $WS(t)$. Let $m(t)$ be the cardinality of the set $WS(t)$. When $m(t) = 1$, the

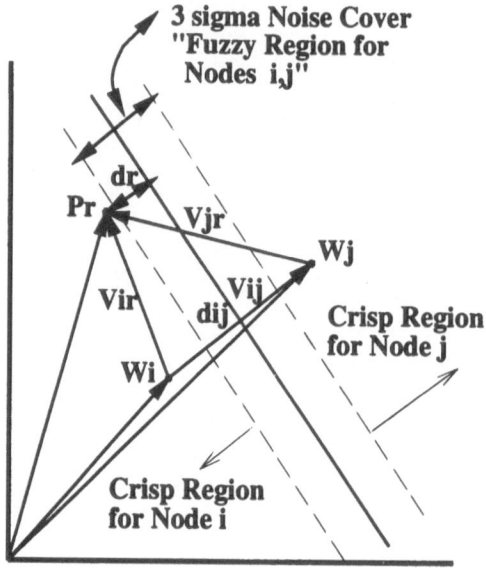

FIGURE 8.10. Noise cover near a boundary

winning node J confirms that the input pattern is in the crisp region. When $m(t) > 1$, the winning nodes $Q \in WS(t)$ jointly confirm the pattern being in a fuzzy region. Let the class label set be $\mathcal{L} = \{c_1, c_2, \cdots, c_q\}$. In the following sections, we explain the updating of frequency counts $n_c(C(p), J)(t)$ and $n_f(C(p), Q)(t)$ for crisp and fuzzy parts for an input pattern with class label $C(p)$.

The belief measures $B_c(c_i, M)(t)$ and $B_f(c_i, M)(t)$ associated with the crisp and fuzzy parts of a winning node M at time t are computed at the end of the training session. The symbol $+ =$ indicates incrementing the accumulator on the LHS by the value on RHS.

DURING TRAINING
a) For crisp winning node J:

$$n_c(C(p), J)(t) \quad + = \quad 1.0; \tag{8.22}$$

b) For fuzzy part of all "winner nodes" $Q \in WS(t)$:

$$sum(t) \quad = \quad \sum_{Q \in WS(t)} d_{Q,p}, \tag{8.23}$$

$$wt(Q, p) \quad = \quad \frac{sum(t) - d_{Q,p}}{sum(t) * (m - 1)}, \tag{8.24}$$

$$n_f(C(p), Q)(t) \quad + = \quad wt(Q, p); \tag{8.25}$$

c) At the end of training:

$$B_c(c_i, M)(t) = \frac{n_c(c_i, M)(t)}{\sum_{c_j \in \mathcal{L}} n_c(c_j, M)(t)}, \tag{8.26}$$

$$B_f(c_i, M)(t) = \frac{n_f(c_i, M)(t)}{\sum_{c_j \in \mathcal{L}} n_f(c_j, M)(t)}. \tag{8.27}$$

KNOWLEDGE AGGREGATION

As is evident, the presence of a noise cover fuzzifies the boundaries and results in multiple winning node representation as against a single node in a winner-take-all strategy. This modified process requires information fusion at every winning instant prior to performing any combination over the time-sequence of winning nodes. During the running (testing) mode, the belief measures obtained using class frequency counts of the crisp and fuzzy parts are used as appropriate to obtain a measure of "local" opinion. Let $L(c_i)(t)$ be the local opinion for class label c_i at time t.

$$L(c_i)(t) = \begin{cases} B_c(c_i, J)(t), & \text{for m=1} \\ \sum_{Q \in WS(t)} wt(Q, p) * B_f(c_i, Q)(t), & \text{otherwise.} \end{cases}$$

8.7   The Improved FGNN

The proposed improvements to FGNN are tested using the noisy data generated from the 3D synthetic range images under varying degrees of noise cover near the boundaries.

8.7.1   Mean and Bayesian Aggregation Methods

We conducted simulations with the mean aggregation and Bayesian approach. Table 8.3 shows some test results. We found that FGNN with evidence combination, the Bayesian approach performs poorly under noisy environment: The noise cover degrades the performance under the Bayes rule.

The mean aggregation has shown consistent improvement at noise cover set at SNR=1000. Thus, we conclude the space available near the boundaries to provide a reasonable cover is an important factor. Due to fragmentation in FGNN, we can implement only a small cover to improve the classification performance.

8.7.2   Alternative Aggregation Methods

As is evident, the choice of aggregation method is important. We have explored the aggregation of "local" opinions provided by the time-sequence of winning nodes using fuzzy integrals of Sugeno and Choquet. The performance results of the above trials are not discussed in this paper as they have not resulted in a better performance over mean aggregation. The definitions, testing and the findings are published in [22].

TABLE 8.3. Test data have different levels of noise: SNR = 1000, 100, 50, 10. For FGNA: FMAX = 30, T=100, NC-Noise Cover, M-Mean, B-Bayesian.

NC ⇒	NC = SNR 1000		NC = SNR 100		NC = SNR 50	
Test Data ⇓	M	B	M	B	M	B
SNR = 1000	43.27	7.58	36.69	17.89	32.22	21.49
SNR = 100	31.59	4.22	29.80	12.26	26.92	15.98
SNR = 50	27.95	4.60	27.33	11.27	25.13	13.95
SNR = 10	18.63	6.71	19.88	8.16	20.66	11.55

8.8   Conclusions

We have presented a new pattern recognition technique and demonstrated its performance on a variety of difficult pattern recognition tasks. The FGNN network has the following important properties:

- It is a topology-constraint fee feature map.
- It requires only one epoch for learning the distribution of the pattern space.
- It has an excellent classification performance on "seen" patterns.
- It can generalize well over "unseen" patterns from the same input distribution.
- It can perform real-time learning due to its topology-constraint fee property and quick setup times.

The elimination of the need for a topology preserving map has enabled drastic reduction in training time. The training time for FGNN is just one epoch presentation in contrast to many thousands of epochs in other popular neural networks.

This chapter has also dealt with the aspect of robustness to white Gaussian noise introduced by the feature extraction process. We conclude that the provision of a noise cover at the region boundaries improves the performance, if the maximum noise level present in the input data is close to the cover; any additional cover above the required level degrades the classification performance; therefore, it is necessary to estimate the anticipated noise level in the recognition system; the class label prediction under mean aggregation is demonstrated to be a robust operation.

8.9 REFERENCES

[1] V. Chandrasekaran, M. Palaniswami, and T.M. Caelli, "Spatio-temporal feature map using gated neuronal architecture," *IEEE Transactions on Neural Networks*, Vol.6, No.5, pp.1119-1131, September 1995.

[2] T. Kohonen, "The self-organizing map," *Proceedings of IEEE*, Vol.78, No.9, pp.1464-1480, September 1990.

[3] V. Chandrasekaran, M.Palaniswami, and T.M. Caelli, "Pattern recognition by topology free spatio-temporal feature map," *Proceedings of the International Conference on Systems, Man, and Cybernetics*, Vol.2, pp.1136-1149, October 1995.

[4] V. Chandrasekaran, Z.Q. Liu, and M. Palaniswami, "Fuzzy gated neuronal architecture for pattern recognition," *Proceedings of the International Conference on Neural Networks*, Vol.4, pp.1622-1627, November 1995.

[5] C. Koch and I. Segev, (Eds.), *Methods in Neuronal Modeling: From Synapses to Networks*, MIT Press, Cambridge, MA, 1989.

[6] W.S. McCulloch and W.H. Pitts, "A logical calculus of the ideas imminent in nervous activity," *Bulletin Math. Biophy.*, No.5, pp.115-133, 1943.

[7] F. Rosenblatt, "The perceptron: A probabilistic model for information storage and organization in the brain," *Psychol. Rev.*, Vol.65, pp.386-408, 1958.

[8] B. Widrow, "Adaline and madaline - 1963," *IEEE First International Conference on Neural Networks*, 1987, Vol.1, pp.143-157.

[9] M.N. Oguztoreli, G.M. Steil, and T.M. Caelli, "Underlying neural computations for some visual phenomena," *Biological Cybernetics*, No.60, pp.89-106, 1988.

[10] J.M. Zurada, *Introduction to Artificial Neural Systems*, West Publisching, 1992.

[11] S.C. Lee and E.T. Lee, "Fuzzy neural networks," *Mathematical Biosciences*, Vol.23, pp.151-177, 1975.

[12] Y.C. Lee, G. Doolen, H.H. Chen, G.Z. Sun, T. Maxwell, Y. Lee, and C.L. Giles, "Machine learning using higher order correlation network," *Physica D*, pp.276-306, 1986.

[13] V. Chandrasekaran, M. Palaniswami, and T.M. Caelli, "An extended self-organizing map with gated neurons," *Proceedings of the IEEE International Conference on Neural Networks*, Vol.III, pp.1474-1479, March 1993.

[14] V. Chandrasekaran, M. Palaniswami, and T.M. Caelli, "Performance evaluation of spatio-temporal feature maps with gated neuronal architecture," *Proceedings of the World Congress on Neural Networks*, Vol.IV pp.112-118, July 1993.

[15] H.K. Kwan and Y.L. Cai, "A fuzzy neural network and its application to pattern recognition," *IEEE Transactions on Fuzzy Systems*, Vol.2, No.3, pp.185-193, August 1994.

[16] V. Chandrasekaran, *Gated Neural Networks for Three Dimensional Object Recognition Systems*, PhD thesis, The University of Melbourne, Parkville, Vic-3052, Australia, 1995.

[17] K. Laws, *Textured Image Segmentation*, PhD thesis, University of Southern California, USA, 1980.

[18] N. Yokoya and M.D. Levine, "Range image segmentation based on differential geometry: A hybrid approach," *IEEE Transactions on Pattern Analysis and Machine Intelligence*, Vol.11, No.6, pp.643-649, June 1989.

[19] V.Chandrasekaran, M.Palaniswami, and T.M. Caelli, "Range image segmentation by dynamic neural network architecture," *Pattern Recognition*, Vol.29, No.2, pp.315-329, 1996.

[20] V. Chandrasekaran and Z.Q. Liu, "Robust face image retrieval by fuzzy gated neuronal architecture," *Proceedings of the International Conference on Neural Information Processing*, Vol.1, pp.432-437, September 1996.

[21] H. Sawai P. Haffner, A. Waibel and K. Shikano, "Fast backpropagation learning methods for large phonemic neural networks," *Proceedings of the European Conference on Speech Communication and Technology*, Paris, pp.553-556, 1989.

[22] V. Chandrasekaran, Z.Q. Liu, and T.M. Caelli, "On the use of fuzzy gated neural networks for pattern recognition in a noisy environment," *Proceedings of the International Conference on Control, Automation, Robotics and Vision*, Vol.1, pp.645-649, December 1995.

9

Soft Computing Technique in Kansei (Emotional) Information Processing

Takehisa Onisawa

9.1   Introduction

In human face-to-face communication, not only language but also voice pitch, facial expressions and a gesture are employed in order to have a smooth communication. The former is called verbal information and the latter is non-verbal information [1]. On the other hand, in human-computer interaction only a character, a numeric character and a symbolic character, which are a kind of verbal information, were used as the main conveyance way in the early days of a computer. Human-computer interaction was done by only verbal information because of poor technology, and it cannot help being recognized that a computer system was designed by a machine-oriented way in these days. As the recent development of multimedia technology, however, human-computer interaction can be performed by the use of sound information and image information as well as language information. And studies on human interface and human-computer interaction aiming at a human-friendly system have started [2]. These recent studies deal with non-verbal information aiming at having smooth communication between human and computer like human face-to-face communication. The non-verbal information is not confined to the above-mentioned information such as image information and sound information, and includes voice pitch, facial expressions, a gesture, so called human feelings information. Hereafter in this chapter these pieces of human feelings information are called *Kansei information* [3]-[7].

A computer system mainly processed numerical information in the early days of a computer system since a computer was literally a calculating machine. In the field of artificial intelligence in 1960's a computer played games or solved puzzles [8]-[9]. In these kinds of studies a computer generally processed numerical information. After that a computer was also used for data management, and processed verbal information such as a character and a symbolic character as well as a numeric character. From the second half of the 1970's to the first half of the 1980's at the second boom of artificial intelligence, a computer was recognized as a processing machine of intelligent information in studies on expert systems and pattern recognition [10]–[11]. In particular, in the field of pattern recognition, a computer processes pattern information which is given as sound or image. Intelligent information processed by a conventional computer system, however, is represented logically and is limited to have objectivity, uniqueness, universality and reappearance. These concepts are bases of natural science. On the other hand *Kansei information* which has been paid attention to recently, has subjectivity, ambiguity, vagueness and circumstances dependence. This piece of information has been eliminated from conventional information processing, i.e., natural science, as noise. A study on human feelings and its process, i.e., just the Kansei information processing becomes necessary for the construction of a human-friendly system.

This chapter explicates *Kansei information*, and mentions that a so-called soft computing technique is applicable to the Kansei information processing. This chapter also describes examples of studies on facial expressions which play an important role as one piece of *Kansei information*.

9.2   Concept of Kansei Information

This chapter explicates not feelings(Kansei) themselves but *Kansei information*. A study on human feelings(Kansei) themselves aims at making clear human feelings themselves, for example, how human feels *love, happiness* and so on. On the other hand a study on *Kansei information* aims at constructing a model that responds as human does. That is to say, the study pursues not feelings(Kansei) themselves in detail but the construction of the model that feels and manifests *beauty, surprise* and so on as if human does.

9.2.1   Difference Between Intelligent Information and Kansei Information

Conventional natural science has dealt with and deals with information which has objectivity, uniqueness, universality and reappearance. Information not belonging to this category has been dealt with as a noise and has not been a main stream in the information processing. Intelligent information dealt with in conventional fields of computer science and artificial intelligence has been limited to this category, and is always represented and processed logically. Even fuzzy set theory [12] that deals with subjective information almost always processes fuzzy sets logically since fuzzy theory asserts that a fuzzy set is an extension of an ordinal crisp set and that conventional set operations system can be extended as fuzzy set operations system. On the other hand *Kansei information* described in this chapter has subjectivity, ambiguity, vagueness and circumstances dependence, and is far different from information which conventional natural science deals with. Natural science has eliminated this piece of information from the information processing. The difference between information which has been dealt with in natural science and *Kansei information* which has been eliminated from natural science is explained in the following.

Objectivity and Subjectivity

Objectivity is based on the standpoint that it is reasonable for everyone. Natural science, which pursues objectivity, deals with only reasonable information for everyone. A mathematical theory and a physical phenomenon have objectivity. Is it possible, however, to assert that every physical phenomenon is objective? For example, Darwinism on organic evolution is not necessarily approved among biologists [13]–[14]. Of course, Darwinism seems to be reasonable. Some groups, however, protest against its theory. The justification of Darwinism cannot be verified easily because it takes years to perform experiments on organisms for the substantiation of Darwinism. It may be possible to substantiate Darwinism by a computer simulation. In a computer simulation, however, it is necessary to simplify the complicated structure of organisms. It sometimes becomes difficult to regard such a simplification as objectivity. Even the field of an optimization problem seems to have the problem of objectivity since a solution is obtained based on a given criterion. Is the criterion reasonable and objective? It is possible to obtain different solutions based on other criteria for the same optimization problem. Seeing above-mentioned points about objectivity, even if conventional natural science pursues objectivity, it seems to have some problems about objectivity.

On the other hand, subjectivity is dependent on an individual. Subjectivity is based on the consideration that it is reasonable for an individual or for a group. For example, let us consider *a 180cm tall person*. For Onisawa(the author of this chapter, 175cm tall), *a 180cm tall person* belongs to a set of tall persons to some extent. For a basketball player, who is, for example, 210cm tall, *a 180cm tall person* is not tall. Even if an object is the same person, the judgement is dependent on the individual who judges whether the person is tall or not. Furthermore, the judgement is not clear. For example, for the author it is not clear whether *a 180cm tall person* belongs to a set of tall persons completely or not. It is a matter of degree. The borderline between *tall* and *not tall* is not clear, either. Fuzzy set theory is proposed based on these concepts by Prof. Zadeh [12]. *Tall* as seen by the author of this chapter is defined by, for example, Figure 9.1. In the early days of fuzzy set theory, there had many objections [15]: How is the membership function, for example, as shown in Figure 9.1, is fixed? Judgements dependent on an individual should not be dealt with in the field of natural science. These objections are manifestations that natural science has attached great importance to objectivity and that natural science has excluded subjectivity from the study. Fuzzy set theory, however, has been acknowledged in many fields of science and engineering recently. The Kansei information processing becomes one of these fields.

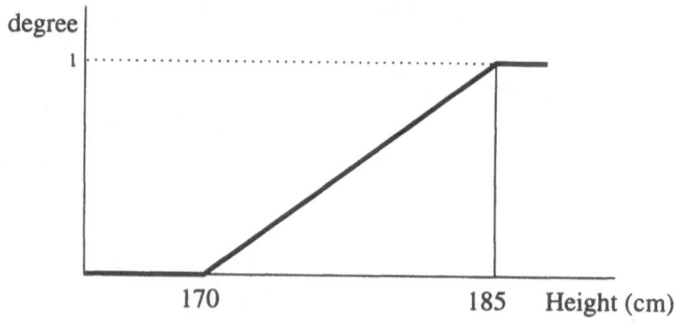

FIGURE 9.1. Fuzzy Set *Tall*

Uniqueness and Ambiguity

Objectivity based on the way of thinking that it is reasonable for everyone leads to the concept of uniqueness. If system behavior is represented by differential equations with initial conditions, everyone can obtain the same solution of the differential equations and the same simulation result about the system behavior. It is not too much to say that natural science has progressed based on this concept. In fact, if the interpretation of a mathematical theorem is dependent on a researcher, the progress of mathematics is prevented. The interpretation of a mathematical theorem is dependent on only premised axioms. If the interpretation of a mathematical theorem obtained from same axioms is different, mathematics does not hold good. Therefore, if premised axioms are different from one another, different theorems are, of course, obtained. This good example is Euclidean geometry [16] and non-Euclidean geometry [17]. All fields of natural science, however, have not uniqueness. For example, as a Chaotic phenomenon, even if system behavior is represented deterministically, the little difference of initial states often leads to quite different system behavior [18]. It is also true that if uniqueness is adhered to in all fields of natural science, some fields may not progress.

On the other hand, the concept of subjectivity leads to ambiguity. The ambiguity means that the interpretation of everything is dependent on an individual. The ambiguity arises from human subjectivity. The ambiguity must be considered so long as a research object is human in the field of such as *Kansei information*. Permission of ambiguity is to permit diversity of sense of values. The concept of uniqueness is obtained by consideration based on only one sense of values, and leads to an idea that only one correct solution exists. Permission of diversity of sense of values does not establish only one way of interpretation or judgement but turns our eyes to many ways. Diversity of sense of values has an idea that a correct solution may not necessarily exist, or that the existence of many solutions is recognized. Sense of values is dependent on the regional traits, the national traits, the period flavor, the age bracket, etc. From the viewpoint of the national traits, a study on *Kansei information* may be accessible to Japanese but may be difficult to be understood by Western.

As long as *Kansei information* is related to human, the ambiguity arising form subjectivity, i.e., from diversity of sense of values, should be dealt with.

Universality and Vagueness

The concept of universality arises from the law of cause and effect that has a clear relationship between cause and effect. Of course, this concept is also connected with objectivity and uniqueness. The law of cause and effect does not permit any exception at all or eliminates every exception as noise. When an inconsistency is recognized in the relation between cause and effect, it is usually dealt with as an exception or is eliminated from the relation as noise.

There is an analysis and synthesis methodology for the solution of a complex problem. When it is difficult to consider a complex problem as a whole, the complex problem is divided into detailed problems. And the given complex problem is solved by synthesizing analysis results of the divided problems. The analysis and synthesis methodology is useful to understand and solve the complex problem. This methodology is based on the law of cause and effect, after all, based on objectivity, uniqueness and universality. But is it right that diversity of sense of values is always eliminated from natural science and that an inconsistency is always disregarded as noise? Do concepts of objectivity, uniqueness and universality always hold good? Do they hold good under a particular condition, they don't? Against Euclidean geometry which arises from the axiom of parallels; *parallel lines do not cross each other*, non-Euclidean geometry arises from the axiom which denies the axiom of parallels. Therefore, Euclidean geometry is not recognized under the axiom denying the axiom of parallels. Gödel's incompleteness theorems show that there is no mathematical system without inconsistency [19]. In consideration of the above-mentioned points it seems to be doubtful that diversity of sense of values and inconsistency are disregarded. Still more, in the field of *Kansei information* in which human is the main object of a research, adherence of universality will become a big obstacle.

By the way, does human have clear cause-effect relationship? Does human always make a decision based on fixed cause-effect relationship? Such a decision may have inconsistencies. In fact, human sometimes has an incoherent feeling towards an object in logic and sometimes has different judgement from the previous one. This may be related to the circumstances dependence mentioned later. Thus, in the study with respect to human beings, the relationship between cause and effect cannot be expressed clearly, and all factors that have influences on the relationship cannot be mentioned. If *Kansei information* is expressed by the concept of the relationship between cause and effect, vagueness exists in the relationship certainly. Human *Kansei information*, with vagueness may not be recognized clearly even by the analysis and synthesis methodology. Although there may be no problem about universality when the object of a study is physical, the study in the field of human *Kansei information* has many problems about universality.

There are many and different phases among concepts of vagueness: Incompleteness which arises from the lack of knowledge, incorrectness, randomness which probability theory deals with, ambiguity which arises form many meanings, fuzziness which fuzzy theory deals with. Ambiguity, fuzziness and imprecision are considered in this chapter where imprecision is assumed to include mistake and inconsistency.

Reappearance and Circumstances Dependence

Reappearance means that the same result is obtained under the same condition at any time and anywhere by anyone. There is a consideration fundamentally that regarding to an objective, unique and universal fact, same results should be obtained at any time and anywhere by anybody in even repeated experiments. The concept of reappearance is a natural one in the field of the study which is physical or is not related to human itself. In the field of human *Kansei information*, however, it is difficult to guarantee reappearance. For example, even if a same person appreciates a same picture or same music, its feeling is dependent on the appreciating place, e.g., a museum, a concert hall, a living room in a house, or on the situation in which the person is relaxed after finishing a work, or on the situation in which the person is heavy-hearted because of work troubles. Of course even a music player cannot always play music under the same conditions since psychology of the player changes every time. Therefore, considering human psychology, it is unnatural to consider human *Kansei information* under the same condition all the time. When human is a study object, human mental sate has an influence on *Kansei information*. Circumstances in a wide sense including psychology have an influence on it. Evidently, *Kansei information* has circumstances dependence.

From the above considerations, it is found that it is difficult to deal with *Kansei information* by conventional approaches in natural science. What approach can deal with *Kansei information*? In the next section, the Kansei information processing is considered from the viewpoint of soft computing approaches.

9.2.2   Kansei Information from Soft Computing Approaches

We have a rating scale method [20], a paired comparison method [21], a multivariate analysis [22] which are used as approaches of the Kansei information processing. We also have a Semantic Differential method [23] using factor analysis [24]. As mentioned before, *Kansei information* includes many and different phases among concepts of vagueness. Vagueness dealt with in natural science is only randomness by probability theory. Randomness is uncertainty of random occurrence of an event. The approach using probability theory is not applicable to the processing of vagueness of human *Kansei information* since vagueness mentioned in this chapter is dependent on subjectivity rather than randomness.

We have fuzzy theory among theories dealing with vagueness. Which interval of ages does *middle-aged* mean? From 35 years old to 55 years or from 40 years old to 55 years old? Suppose that persons from 35 years old to 55 years are defined as *middle-aged*. Figure 9.2 shows this definition. According to this definition, a 34 years old person is not middle-aged. When this person is aged one year and 35 years old, this person joins in the party of middle-aged persons suddenly. This idea is based on the binary state, and is the origin of conventional set theory. Furthermore, this idea becomes the foundations of a computer. It is not too much to say that natural science has progressed based on this concept. This idea, however, is quite different from human understanding and is quite unnatural as the model of human understanding and recognition. This unnaturalness arises from the clear boundary line between *middle-aged* and *non-middle-aged.*

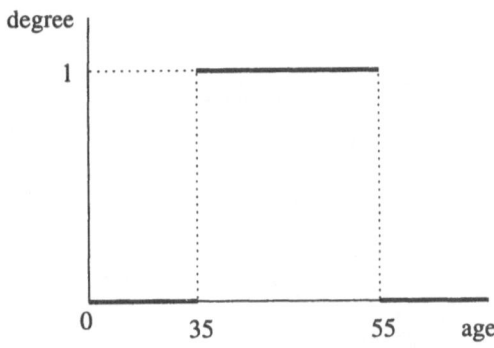

FIGURE 9.2. Crisp Definition of *Middle-aged*

If the boundary line becomes more blurred, the unnaturalness may become less. Figure 9.3 shows this idea. The degree of a 30 years old person in a middle-aged person party is assumed to be 0.3 since a 30 years old person does not join in the party completely. The degree of a 35 years old person in a middle-aged party, however, is higher than that of a 30 years old person

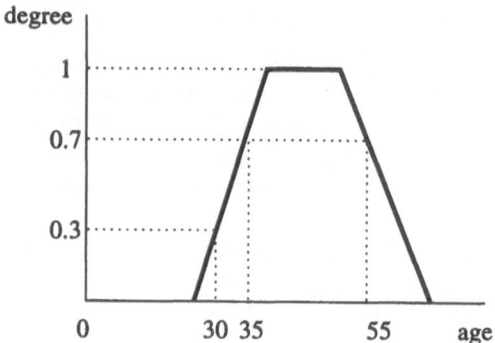

FIGURE 9.3. Fuzzy Definition of *Middle-aged*

in the party. This definition is called a fuzzy set [12]. If the boundary line becomes blurred, a seemingly inconsistent statement, for example, *He is young and middle-aged*, is not inconsistent truly. Let us define *young* by a fuzzy set as shown in Figure 9.4. *Young and middle-aged* corresponds to the shaded area in Figure 9.4, and *young and middle-aged* is not necessarily inconsistent, where *young and middle-aged* is defined by

$$\mu_{young \cap middle\text{-}aged}(x) = \mu_{young}(x) \wedge \mu_{middle\text{-}aged}(x) \tag{9.1}$$

where membership functions of fuzzy sets *young* and *middle-aged* are μ_{young} and $\mu_{middle\text{-}aged}$, respectively, and $\wedge$ stands for min.

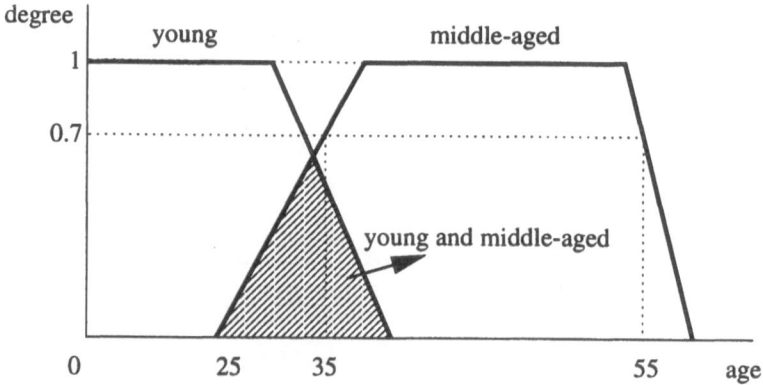

FIGURE 9.4. Definition of *Young* and *Middle-aged*

Next, is this definition of *middle-aged* fixed uniquely? No, not uniquely. It may be good in the field of fuzzy set theory that the definition, e.g., *middle-aged*, is fixed by an individual. It is not necessary for all people to recognize this definition. The definition reflects individual subjectivity. Therefore, fuzzy set theory is one of pioneers which drive a wedge into natural science based on objectivity, uniqueness, universality and reappearance.

In fuzzy set theory, however, these defined fuzzy sets, i.e., membership functions of fuzzy sets, are operated logically. For example, the fuzzy set *young and middle-aged* is obtained by Equation (9.1). The reason why logical operations are taken in fuzzy set theory is that fuzzy set theory is an extension of conventional set theory and that the conventional set operation system can be used as it is. That is, the framework in fuzzy set theory is objective and does not deviate from the standard framework in natural science. Let us consider a fuzzy system as shown in Figure 9.5. Although an input is a fuzzy set which is dependent on individuals and is not unique, an output fuzzy set is obtained by the approach which does not deviate from the standard framework in natural science largely. As mentioned before, fuzzy set theory is expected to be applicable to the Kansei information processing. But as long as fuzzy set theory approach adheres to the logical operating system, fuzzy set theory may not be expected so much in its application to the Kansei information processing. It is necessary to propose another framework which breaks down the conventional approach and deals with subjectivity, ambiguity, vagueness and circumstances dependence at the same times. As the extension of the conventional fuzzy set operations, several types of t-norms and t-conorms, e.g., parameterized t-norm and t-conorm, are considered [25]. These operations have possibility to reflect human subjectivity.

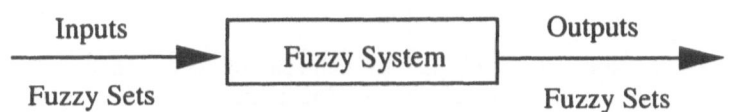

FIGURE 9.5. Fuzzy System

The field of human reliability also has difficulty of dealing with human as an object of the study. Probability theory has been the only approach to deal with uncertainty in the field of reliability. Probability theory has been also applied to the analysis of human reliability [26]. The application of probability theory to the analysis of human reliability has its limits since subjectivity, ambiguity, vagueness and circumstances dependence are also seen in human behavior as well as in *Kansei information*. Even if a basic error probability is given to each basic human task, the modification of the error probability is recommended, for example, as shown in Table 9.1, according to the stress level and the type of task [27]. And let us consider the case where human performs consecutive similar tasks. Dependence should

be considered in this case. Dependence means that if a human operator
fails or succeeds in one task, the operator is liable to fail or succeed in a
consecutive task. If we have a great deal of data about dependence, the
conditional error probability is used in order to deal with dependence. In
practice, however, we often have not so many data to estimate the conditional probability. Then the dependence model as shown in Table 9.2 is
proposed when we have not many data to estimate the conditional probability [28]. In these tables, however, there are many problems that borderlines of psychological levels and borderlines of dependence levels are not
clear. From various reasons applications of fuzzy set theory to reliability
engineering fields are also proposed [28]–[30].

TABLE 9.1. Comparison of Estimated Human Error Probability of Experienced
Personnel and Novices Under Different Stress Levels

Stress Level	Type of Task			
	Step-by-Step Procedures		Dynamic Interaction	
	Experienced	Novice	Experienced	Novice
Very Low	*BHEP* × *2*	*BHEP* × *2*	No Account	No Account
Optimum	*BHEP*	*BHEP*	*BHEP*	*BHEP*
Moderately High	*BHEP* × *2*	*BHEP* × *2*	*BHEP* × *5*	*BHEP* × *10*
Extremely High	0.25	0.25	0.25	0.25

BHEP : *Basic Human Error Probability*
Novice : a person with less than 6 months on the job in which he has
been licensed or other qualified.

TABLE 9.2. Equations for Conditional Probabilities of Failure on Task N, Given
Failure on Task N-1, for Different Levels of Dependence

Dependence Level	Failure Equations
Zero Dependence	$Pr(F_N|F_{N-1}) = Pr(F_N)$
Low Dependence	$Pr(F_N|F_{N-1}) = \dfrac{1 + 19 \times Pr(F_N)}{20}$
Moderate Dependence	$Pr(F_N|F_{N-1}) = \dfrac{1 + 6 \times Pr(F_N)}{7}$
High Dependence	$Pr(F_N|F_{N-1}) = \dfrac{1 + Pr(F_N)}{2}$
Complete Dependence	$Pr(F_N|F_{N-1}) = 1.0$

$Pr(F_N)$: Human Error Probability in Task N
$Pr(F_N|F_{N-1})$: Human Error Probability in Task N
on the condition of the error in Task N-1

Although fuzzy set theory is applicable to express fuzziness with respective to subjectivity, it is unsuitable for acquisition of fuzzy information and

knowledge. A neural network model [31] and genetic algorithms [32] are applicable to acquisitions of them. Unfortunately, however, these techniques are unsuitable for representations of acquired information and knowledge. That is, it is found that each technique has merits and demerits [33]. A soft computing technique is proposed as the complementary way of an advantage and a disadvantage of each technique [33].

As mentioned before, the law of cause and effect is almost unclear in *Kansei information*. It is necessary to express it somehow or other in order to implement the Kansei information processing in a computer. That is, it is necessary to express an input-output relationship from given physical quantity expressing degrees of stimuli to evaluations of feelings. The multivariate analysis with the statistical methodology is applied to evaluations of feelings because there are many statistics processing program libraries in the multivariate analysis. The relationship between cause and effect is processed linearly in the multivariate analysis. Since the input-output relationship from physical quantity to evaluations of feelings is considered to be non-linear, it cannot be dealt with by a linear map sufficiently.

The relationship between cause and effect is recently modeled by a hierarchical neural network model which is obtained by a supervised learning method, e.g., a back propagation method, with given physical quantity and evaluations of feelings as input-output data.

Genetic algorithms are known as a probabilistic algorithm. In genetic algorithms, an evaluation function is defined beforehand and an individual with a set of chromosomes is selected according to the evaluation value. Genetic operators such as a selection, a crossover and a mutation are considered. In *Kansei information*, the evaluation is almost always subjective and an evaluation function cannot be very often defined beforehand. A study on the relationship between cause and effect in *Kansei information* is performed evolutionally interactively by the use of genetic algorithms. This study is called the interactive evolutional computation method, i.e., the interactive GA [34]. *Kansei information* generalized by an individual which has a set of chromosomes obtained by a genetic algorithm is evaluated by users in stead of an evaluation function. By this interactive method, the relation between *Kansei information* and physical quantity fitting a user individual's evaluation is obtained.

The soft computing technique has more flexibility in processing *Kansei information* than the conventional clear and crisp computing techniques. In Japan many studies on *Kansei information* has been done recently [3]–[7]. On the other hand in Europe the concept of human machine symbiosis with a keyword *a human centered system* has been known [35]. Although this concept is not necessarily associated with *Kansei information* since European has not the concept of *Kansei* itself, the concept of human machine symbiosis permits diversity of sense of values, and considers a human factor a more important existence rather than a mere factor in a system. For example, in a human computer system, the concept of human machine symbiosis considers not only computational efficiency and speed but also a system design including human users. The time will come soon when everyone has a personal computer literally since the internet has come into so much wide use. Then the tendency requiring a private computer system will become stronger. A variety of users, i.e., the wide range from an expert user to a novice, will use a private computer system. Therefore, the stream is evidently from the age of mass production of small kinds of goods to the age of a little production of many kinds of goods. That is, a system design suiting user's personal taste is required in an age of diversity of sense of values, and surely designer's taste suiting user's one is required in these days. A designer is one of users and a user should be also a designer.

9.2.3 Facial Expressions as Kansei Information

Among many pieces of *Kansei information*, there are many and different pieces of information, for example, image information, sound information [3]–[7]. Image information gotten from human eyes includes a static image and a dynamic image. On the other hand there are music and environmental sound in sound information from human ears. Of course, a voice uttered by human is also included among sound information. It is possible to guess person's feelings by a voice pitch heard through telephone. In addition to a voice, there are facial expressions, a gesture and so on in information which arises from human. In human face to face communication besides verbal information, non-verbal information, e.g., facial expressions, a gesture, a voice pitch, is used and human has smooth communication by understanding others feelings. Especially non-verbal information plays an important role in a smooth communication. The importance of non-verbal information in human face to face communication has been known for a long time. For example, about the importance of facial expressions Descartes says as follows [1]. "Even a foolish manservant guesses his master's feelings by the look in his face whether the master is anger or not. Although the faculty of eyes is recognized easily and the meaning of the look is known easily, it is not easy to express the faculty and the meaning in words. The faculty of a face is rather learned consciously than acquired by nature." In this way Descartes says that it is difficult to express faculties of eyes and facial expressions in words. Although Descartes does not call them *Kansei information*, he points out difficulty of the Kansei information processing. By the way, it is said that there are six basic emotions, happiness, sadness, fear, disgust, anger and surprise, in facial expressions [36]. Basic emotion is expressed in a face according to the cause of feelings. It is difficult to express the general relation between the cause of feelings and facial expressions since the relation depends on culture, etc. For example, it is said that Japanese facial expressions are vague and that unpleasant emotions should not be expressed in public. On the other hand it is said that emotions should be expressed frankly in Western.

In the field of engineering the importance of non-verbal information is also recognized. Characters including numeric characters and symbols are inputted through a keyboard and these are outputted through a monitor TV and/or a printer in a main conventional human computer interaction. Recently, non-verbal information such as human facial expressions, a gesture, a sound, a voice, anthropomorphic emotions as well as verbal information can be processed through the highly developed multimedia technology. Human computer interaction modeling human face-to-face communication is tried to be implemented. If a computer system can deal with non-verbal information, we have a smooth and natural communication with a computer without consciousness that a computer is a machine.

Human facial expressions are applied to a study on a human interface [37]–[39]. Study examples are; an anthropomorphic agent by which a human computer interaction is implemented in the way that a human user is not conscious of a computer; monitoring for identification by the use of a face image in the field of security. Furthermore, if a computer can deal with impressions of a human face, this is the computer system which can process *Kansei information*, and this system will be expected to be one of human centered systems.

There are two main streams of studies on facial expressions: One is the study on face recognition for given facial expressions. The other is the study on drawing facial expressions for given emotions. With respect to the study on face recognition, many methods are proposed [1][37]–[38]. For example, feelings are estimated by movements of features points defined in a wooden face, or by learned neural network models. On the other hand with respect to the study on drawing facial expressions, there are also many methods [1][37]–[38]. There is the Cheanof's face graph which is changed according to the change of multi variable data on the premise that each feature in a face is assigned to each variable. There is an idea that a face is divided into finite elements and that facial expressions are animated according to the change of their parameter values. There is also the wire frame method by which a face is drawn in three dimensions. The wire frame method is applied to draw a smiling monkey face and a laughing Mona Lisa's face [40]. A monkey does not smile and the Mona Lisa cannot laugh. This study shows which facial expressions they make if they smile/laugh. There is also the study of the construction of a face robot that makes facial expressions [41].

In the study on facial expressions Ekman's Facial Action Coding System is well known [42]. The FACS defines 44 Action Units based on anatomical knowledge of facial muscles and expresses the movements of facial muscles by them globally and objectively. There are many studies applying the FACS to facial expressions, e. g., the wire frame method, a face robot.

Hereafter this chapter describes two examples of studies on facial expressions. The first one is a study of neural network models that recognize emotions of facial expressions according to situations [43]. The recognition model is applied to the interaction with the route decision system [44] that tries to go to a destination, given linguistic instructions about a route from a start point to the destination. Feelings of the route decision system are expressed through facial expressions. The other is a facial caricature drawing based on linguistic expressions of features in a face [45].

9.3   Study Examples of Facial Expressions

9.3.1   Recognition Model of Emotions Through Facial Expressions Considering Situations

An artificial neural network model is used as the recognition model of emotions through facial expressions. Given facial expressions and situations, the model recognizes emotions of facial expressions. With respect to facial expressions, eight parameters values, i.e., the size of eyes, the slant of eyes, the shape of eyes, the slant of eyebrows, the shape of eyebrows, the size of a mouth, the width of a mouth and the shape of a mouth, are given. And with respect to situations, many kinds of situations are given not directly but indirectly. That is, information on situations is given by the transformation into emotions, i.e., happiness, sadness, fear, surprise, anger and disgust. Figure 9.6 shows the structure of the recognition model.

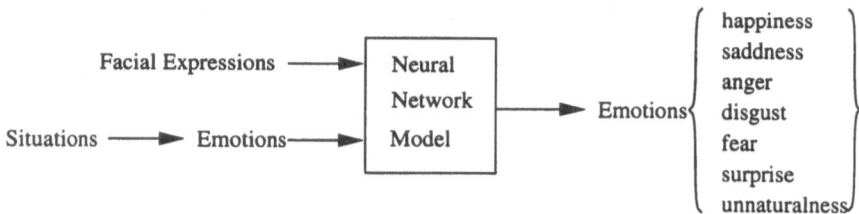

FIGURE 9.6. Structure of Recognition Model

As mentioned in Section9.2, the relationship among emotions, facial expressions and situations is not expressed clearly. Given data about the relation among them, the neural network model is applicable to the relation model, where it is not necessary to represent the relation linguistically.

Questionnaire about Situations

The relationship between situations and emotions is considered in order to transform information on situations into emotions. To this end, the questionnaire about situations is performed in subject experiments. Subjects mark the questionnaire sheet as shown in Figure 9.7 with one of three scale evaluations, a weak feeling, a moderate feeling, or a strong feeling of happiness, surprise, anger, disgust, sadness and fear according to the degree of each emotion under the presented situations. For example, as in Figure 9.7 let us consider the situation in which a subject is forced the work on. If the subject has a moderate feeling of anger and weak feelings of disgust and fear for the situation, the subject marks the sheet as shown in Figure 9.7. Subjects are undergraduate or graduate students, and 8 students are males and 2 females. As shown in Table 9.3, 18 kinds of situations are presented in the questionnaire.

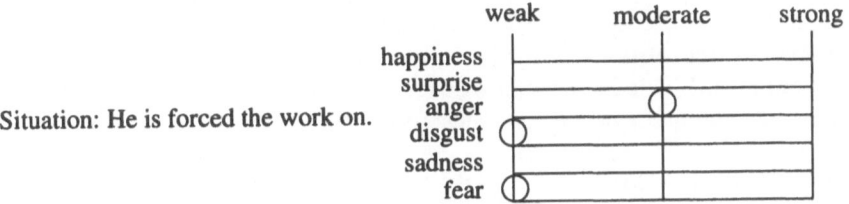

FIGURE 9.7. Questionnaire Sheet for Situations

TABLE 9.3. 18 Kinds of Situations Presented in the Questionnaire

Number	Situation
1	He passes an exam unexpectedly.
2	He is going to see a pass announcement.
3	He is scolded.
4	He is scolded unfairly.
5	He is praised.
6	He is parted from familiar friend.
7	He gets bad marks in an exam.
8	He passes an exam.
9	He gets bad marks in an exam unexpectedly.
10	He loses a wallet.
11	He is barked at by a big dog.
12	He is going to have a picnic.
13	He is barked at suddenly by a big dog.
14	He is preached.
15	He is forced the work on.
16	He is going to give a talk in large company.
17	He is forced to choose one between the two.
18	He is ill spoken of.

Questionnaire about Facial Expressions and Situations

In the questionnaire about facial expressions and situations, faces by line drawing are used since they can be controlled more easily than real expressions such as face photographs. Subjects compare the facial expression at the right side of the arrow with that at the left side as shown in Figure 9.8, and mark the questionnaire sheet with one of three scale evaluations, a weak feeling, a moderate feeling or a strong feeling of happiness, surprise, anger, disgust, sadness and fear according to the degree of each emotion which subjects feel under the presented situation. The left side of the expression is a standard expression that is assumed to be an expressionless face. If a subject feels that the combination of a situation and facial expressions is unnatural, the subject marks unnatural in the sheet. In this questionnaire, 38 kinds of facial expressions and 12 kinds of situations are combined, where situations are selected out of 18 situations in 9.3.1.

Neural Network Model

Eight parameters values about facial expressions, i.e., the size of eyes, the slant of eyes, the shape of eyes, the slant of eyebrows, the shape of eyebrows, the size of a mouth, the width of a mouth and the shape of a mouth, belong to the [0, 1] interval. Figure 9.9 shows the parameter value range according to the location of each feature in a face. Each parameter value of the standard facial expression is assigned to 0.5 except for the size of a mouth. And the natural language expressions with respect to the feeling of emotions in both questionnaires for situations and for the combination of facial expressions and situations correspond to numerical values in the interval [0, 1] as follows:

$$\begin{cases} No \ mark: & 0.00 \\ Weak: & 0.33 \\ Moderate: & 0.67 \\ Strong: & 1.00 \end{cases} \tag{9.2}$$

If unnatural in the questionnaire sheet is marked; the numerical value about the unnaturalness is assigned 1. Otherwise, the value is 0.

Seven neural network models are obtained by the back propagation algorithm. Each model has an input layer, an output layer and two hidden layers. The model has 8 input nodes about facial expressions and 6 input nodes about emotions transformed from situations. The model has an output node about one of 6 kinds of emotions or unnaturalness. The model has 20 nodes in each hidden layer. The recognition model is composed of seven neural networks. Outputs of the neural network models are expressed by natural language expressions as shown in Figure 9.10. If three or more unpleasant emotions such as anger, disgust, sadness and fear are felt even a little, the natural language expression *a feeling of unpleasant*

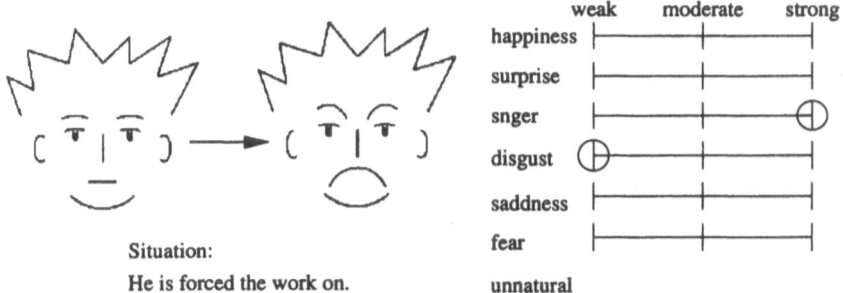

Situation:
He is forced the work on.

	weak	moderate	strong
happiness	├─────────┼─────────┤
surprise	├─────────┼─────────┤
snger	├─────────┼──────⊕
disgust	⊕─────────┼─────────┤
saddness	├─────────┼─────────┤
fear	├─────────┼─────────┤

unnatural

FIGURE 9.8. Questionnaire Sheet for Combination of Facial Expressions and Situations

Value Range	the slant of eyes	the shape of eyes	the size of eyes	the shape of eyebrows	the slant of eyebrows	the shape of a mouth	the size of a mouth	the width of a mouth
1 ° 0								

FIGURE 9.9. Parameter Value Range for Location of Each Feature in a Face

is used in order to express those unpleasant feelings together, where the degree of unpleasantness is the largest output degree in those unpleasant feelings. Natural language expressions about unnaturalness are shown in Figure 9.11. If the output value of unnaturalness is in the interval [0, 0.33], *natural* is not expressed, and natural language expressions of emotions are used instead.

Model Performance

First of all, let us show how obtained models respond when the same facial expressions are given under different situations. Figure 9.12 shows some examples. In these examples, three kinds of situations are considered: He is forced the work on; he gets bad marks in an exam; he is ill spoken of. How do the models respond for the same facial expressions in these situations? Figure 9.12 shows the results: He has a weak feeling of disgust and rather a weak feeling of sadness; he has a weak feeling of sadness and fear; he has a feeling of unpleasant. It is found that the models infer emotions depending on situations.

Next, let us show the verification of model performances. In order to verify the model performance, two kinds of questionnaires are performed. The one is subjects evaluations about model recognition results. The other is the comparison evaluations between model recognition results and subjects

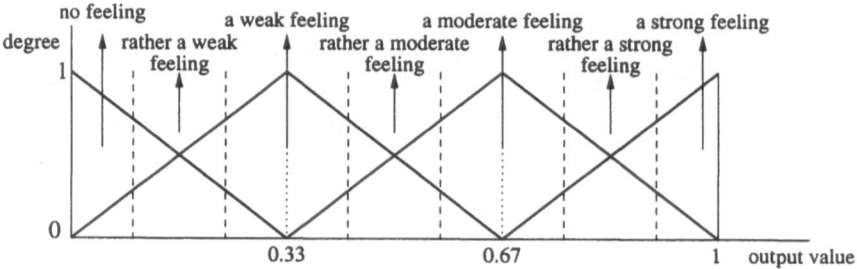

FIGURE 9.10. Natural Language Expressions about Feelings of Emotions and Their Fuzzy Sets

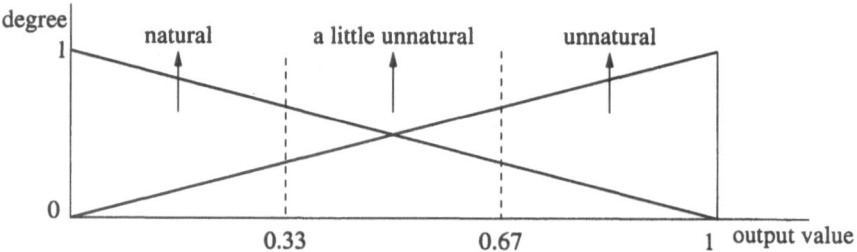

FIGURE 9.11. Natural Language Expressions about Unnaturalness and Their Fuzzy Sets

ones. In the first evaluation, 10 subjects evaluate model recognition results by the use of the questionnaire sheet as shown in Figure 9.13. In this questionnaire, 57 combinations of facial expressions and situations are used. Some of them are not used in the questionnaire described in 3.1.2. Each result gets the point according to subject's evaluation as shown Figure 9.4. The average point of the evaluation of recognition results in 57 combinations among 10 subjects is 0.90. The best average evaluation point among 10 subjects is 2.00 and the worst one is -0.43 as shown in Figure 9.14.

In the second evaluation, 10 subjects mark the same questionnaire sheet as shown in Figure 9.8. In this questionnaire, 22 kinds of pairs of facial expressions and situations are selected, where some of them are not used in the questionnaire described in 3.1.2. The degree of model recognition result is no feeling, a weak feeling, a moderate feeling or a strong feeling. The comparison between the model recognition result and the questionnaire result is performed where let the difference of the degree between, for example, a weak feeling and a moderate feeling be one scale. Figure 9.15 shows each average difference scale among 10 subjects for each kind of emotion in one facial expression. The average difference scale of the degree per kind of emotion in one facial expression is 0.63. In order to evaluate the value 0.63,

facial expressions			
situations	He is forced the work on.	He gets bad marks in a exam.	He is ill spoken of.
results	He has a weak feeling of disgust and rather a weak feeling of sadness.	He has a weak feeling of sadness and fear.	He has a feeling of unpleasant.

FIGURE 9.12. Recognition Results Under Different Situations

TABLE 9.4. Evaluation Points

Evaluations	Points
A subject does not agree with the model recognition result at all.	-2
A subject does not agree with the model recognition result.	-1
A subject does not conclude whether he/she agrees with the model recognition result or not.	0
A subject agrees with the model recognition result.	+1
A subject agrees with the model recognition result completely.	+2

the numbers 0, 1, 2 and 3 are generated random. It is assumed that the numbers 0, 1, 2 and 3 correspond to no feeling, a weak feeling, a moderate feeling, and a strong feeling of emotions, respectively. The average of the difference between the generated random number and model recognition results is 1.3 per kind of emotion in one facial expression. Therefore it is found that the obtained models gets a high evaluation point.

The models can recognize emotions very well when facial expressions and situations are given.

9.3.2 Application of Recognition Model of Emotions Through Facial Expressions Considering Situations

The combination of the recognition model of emotions and the route decision model [46]–[47] is considered as an application example of the recog-

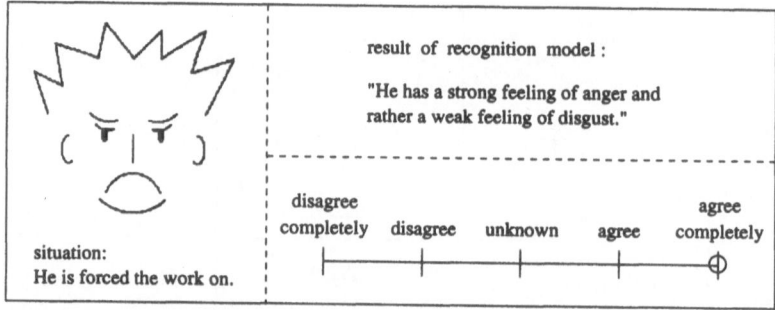

FIGURE 9.13. Questionnaire Sheet for Evaluation of Model Recognition Results

facial expressions	 (face)	 (face)	
situations	He is praised.	He gets bad marks in an exam.	
evaluation points	2.00 (best point)	-0.43 (worst point)	0.90 (average point)

FIGURE 9.14. Average Evaluation Point of Model Recognition Results

nition model of emotions through facial expressions considering situations. The route decision system searches a route to an instructed destination and represents a system state with line drawing facial expressions. The recognition model has a communication with the route decision system by knowing its situations and recognizing its anthropomorphic emotions. It is possible to recognize the state of the route decision system by anthropomorphic emotions through facial expressions and the recognition of emotions is useful for the human computer interaction. The Kansei information process is necessary for a anthropomorphic system which expresses emotions through facial expressions [48]–[50].

9.3.3   Facial Caricature Drawing

We can recognize a friend's face or a famous person's face even in a crowd. Especially, even looking a facial caricature with a simple line, we can recognize whose face it is because a facial caricature is drawn with exaggeration and catches his/her features well. There have been some studies on a facial

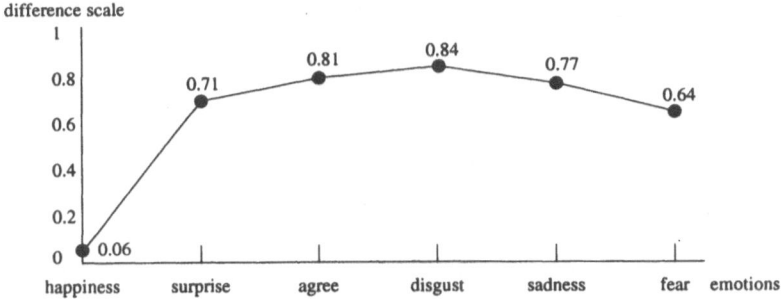

FIGURE 9.15. Average Difference Scale For Each Kind of Emotion in One Facial Expressions

caricature drawing [51]–[52]. These studies are based on the image processing of a face picture. That is, these studies draw a facial caricature based on feature points which are extracted from the face image by the image processing. The image processing is usually used for the analysis of face images rather than the facial caricature drawing. Although the facial caricature may be drawn exactly by the image processing, it is difficult for this processing to reflect the impression of a face in a caricature which is dependent on the person who looks at the face. We usually have the face image as its impression and express it with natural language subjectively. For example, words play an important role to make up a montage picture based on a witness. It is possible to express the feature of a face, e.g., the image of a human face, with linguistic terms [53]. Therefore, although facial caricatures of the same person's face give different impressions one another individually, the drawn facial caricature may be evaluated to reflect impressions of the face well by the person who draws it.

This section tries facial caricature drawing with linguistic information on a human face. This kind of linguistic information is certainly *Kansei information* dependent on the person who looks at the face. This study is also considered a kind of computation with words [54] as well as the Kansei information processing since human and the facial caricature drawing system interaction is performed with words.

Facial Caricature Drawing Process

Input Section

The input section deals with a face shape and features of a face by the use of natural language. With respect to the face shape, the shape is chosen from five kinds of shapes, a moon face, an egg-shaped face, a triangular-shaped face, a rectangular-shaped face and a home base shaped face. Eyes, a nose, a mouth, eyebrows and ears are considered features of a face. The feature of a face has some elements as shown in Table 9.5. The image of

the feature's element is expressed in the form of the linguistic term, called
a feature term here, and an adverb. The adverb is chosen from five kinds of
adverbs, *completely, very, fairly, slightly* and *only a little*. Features terms
are classified according to the feature's elements as shown in Table 9.6.
Table 9.6 shows features terms about eyes as an example of features terms.
In Table 9.6, *normal* means the image of the feature's element in the average
face which is drawn referring to [55]. The feature term and an adverb are
inputted and the sex is also inputted.

Output Section

The feature term has several parameters of which values are expressed by
fuzzy sets and adjusted by adverbs as mentioned afterward. In this study
a facial caricature is drawn by changing the average face according to the
parameters values. When each feature of a face is inputted as *normal*, the
average face is drawn.

TABLE 9.5. Elements of Features and Positional Relation

	Items	Elements
F	Eyes	size, slant, others
e	Nose	size, height, size of wings of a nose, direction
a	Mouth	size, thickness of lips
t	Eyebrows	thickness, depth of color, length, slant
u	Ears	size
r	Profile	width of face, length of chin, cheek
e	Hair	length, volume
position		eye and eye, eyes and nose, nose and mouth, eyes and eyebrows

TABLE 9.6. Feature Terms about Eyes' Elements

Size	Slant	Others
Big	Lower	Big Pupils
Small	Upper	Small Pupils
Thin	Normal	Sleepy-looking in Eyes
Normal		Normal

Modification Section

When a drawn facial caricature does not suit the impression and the image
of the face in parts or as a whole, it is necessary to modify the caricature.
In this section, the feature of the caricature which should be modified is

also expressed by the use of the linguistic term and the adverb. In the modification section the adverb is chosen from three kinds of adverbs, *slightly, more*, and *much more*.

Parameters of Each Feature of a Face

Parameters

The feature of a face has several elements which are represented by parameters, e.g., width, height. Figure 9.16 shows parameters of each feature of a face. Each parameter has a fuzzy value which represents fuzziness of natural language in feature terms and adverbs. The fuzzy value is a fuzzy set defined on the interval [-1, 1] with a triangular typed membership function $N(r - 0.2, r, r + 0.2)(-1 \leq r \leq 1)$, where the sign of r shows the increase or the decrease of the parameter value and its absolute value is the change of the parameter value from an average face's one. When $r - 0.2 \leq -1$, then $r - 0.2 = -1$, and when $r + 0.2 \geq 1$, then $r + 0.2 = 1$. The increase and the decrease of the parameter value are defined beforehand according to the feature term. The $N(-0.2, 0, 0.2)$ is a special fuzzy set which shows a parameter value of an average face. The change of the parameter value is adjusted by the adverb. That is, the fuzzy set representing the parameter value is transformed in parallel according to the adverb. The move distance is set up beforehand as shown in Table 9.7. The larger the absolute value of the parameter value is, the more strongly the feature of a face is drawn with exaggeration. The center value of each parameter value concerning to elements of eyes, a nose, a mouth, eyebrows and ears has the value range. Table 9.8 shows the value range concerning to elements of a nose as an example. Values of parameters height and width are varied between 1.0 and -1.0 according to a large nose or a small nose as the size of a nose. And values of parameters height and width are varied between 1.0 and -0.5, or -0.5 and 1.0 according to a long nose or a short nose. Parameters values, which are not given in Table 9.8, are assumed to be 0.

TABLE 9.7. Parallel Transformation Distance by Each Adverb

Adverb	Distance
Completely	0.8
Very	0.7
Fairly	0.5
Slightly	0.2
Only a little	0.1

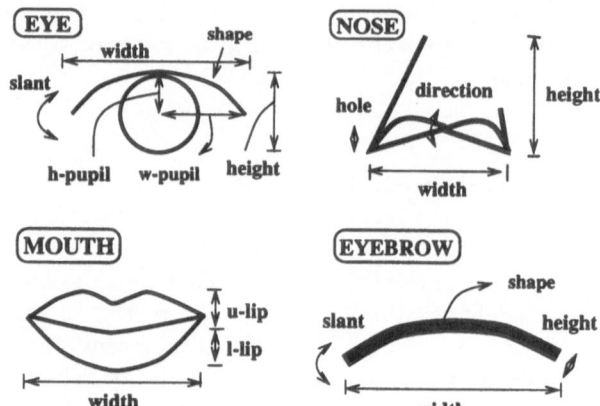

FIGURE 9.16. Parameters of Each Feature of a Face

Calculation of Parameters Values

The center value of the membership function is related to the element i and the parameter j. The fuzzy set, which represents the value of the parameter j in a feature of a face, is obtained by the following process.

Process 1: $(i = 1, 2, \cdots, m)$
The case where the parameter value is increased;

$$r'(i, j) = f \times r_{max}(i, j) \tag{9.3}$$

The case where the parameter value is decreased;

$$r'(i, j) = f \times r_{min}(i, j) \tag{9.4}$$

The case where the feature term is normal;

$$r'(i, j) = 0 \tag{9.5}$$

Process 2:
$i = 1$

$$r(1, j) = r'(1, j) \tag{9.6}$$

$i = 2, 3, \cdots, m$

$$r(i, j) = (1 - |r(i - 1, j)|) \times r'(i, j) + r(i - 1, j) \tag{9.7}$$
$$r(final, j) = r(m, j) \tag{9.8}$$

where $j = 1, 2, \cdots, n$, $r'(i, j)$ is the center value of the membership function of the j-th parameter in the i-th element of a feature of a face, f is the

TABLE 9.8. Maximum and Minimum Values of Parameters Concerning to Elements of a Nose

(1)Element : size

Parameter	the Largest	the Smallest	the Longest	the Shortest
Height	1.0	-1.0	1.0	-0.5
Width	1.0	-1.0	-0.5	1.0

(2)Element : size of wings of a nose

Parameter	the Largest	the Smallest
Width	1.0	-1.0
Hole	1.0	-1.0

(3)Element : direction

Parameter	the Highest	the Lowest
Direction	1.0	-1.0
Hole	1.0	-1.0

(4)Element : height

Parameter	the Highest	the Lowest
Height	1.0	-1.0
Width	1.0	-1.0
Direction	-1.0	1.0

value that is defined according to the adverb as shown in Table 9.7, and $r_{max}(i,j)$ and $r_{min}(i,j)$ are the maximum value and the minimum value of the j-th parameter in the i-th element of the feature of a face variable in an increasing and in a decresing directions, respectively, and n and m are the number of the parameters in the i-th element and the number of elements in each feature of a face, respectively. The feature term, for example, *big* or *small*, plays the role to determine which of Equation (9.3) and Equation (9.4) is used.

Modification of Drawn Facial Caricature

If a drawn caricature does not suit the impression and/or the image of a face in parts or as a whole, the caricature is modified. The modification is performed by the use of a linguistic term in the form of the feature term and an adverb. Three kinds of adverbs are used such as *a little*, *more* and *much more*. The meanings of these adverbs for the modification are represented by fuzzy sets as shown in Figure 9.17. The same feature terms as the ones used in the input section are prepared in the modification. The width of a face, the length of a chin and the feature of a cheek about a face shape as well as the feature about eyes, a nose, a mouth, eyebrows and ears can be modified. Moreover, distances between eyes, between eyes and a nose, between eyes and a mouth, and between eyebrows and eyes can be also modified, which are considered as the positional relation between features in a face.

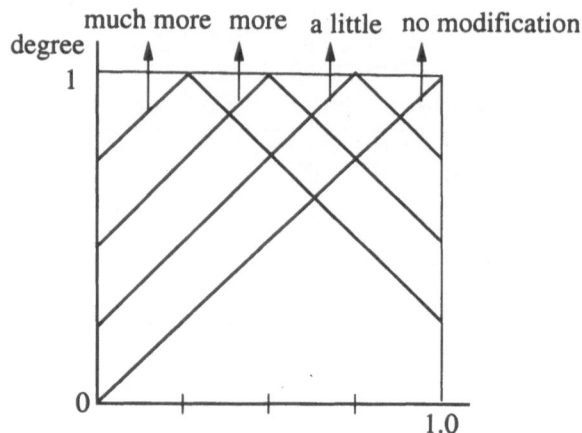

FIGURE 9.17. Fuzzy Sets Representing Meanings of Adverbs for Modification

The truth qualification is applied to the fuzzy set(FT) representing the meaning of the feature term and the fuzzy set(T) representing the meaning of the adverb in the modification process.

$$FT' = FT \text{ is } T, \qquad (9.9)$$

where FT' is a fuzzy set obtained by the truth qualification. The membership function of FT' is obtained by

$$
\begin{aligned}
\mu_{FT'} &= \mu_T \circ \mu_{FT} \\
&= \mu_T(\mu_{FT}), \qquad (9.10)
\end{aligned}
$$

where $\mu_{FT'}$, μ_{FT} and μ_T are membership functions of FT', FT and T, respectively. The FT' has usually an M typed shape membership function as shown in Figure 9.18. The FT' shows both parallel transformations to the positive direction and to the negative direction. So the M typed shape membership function is separated into two fuzzy sets, which are shown by the dotted line in Figure 9.18. The right hand membership function A^+ is used when the adverb and the feature term mean the parallel transformation to the positive direction. On the other hand the left side one A^- is used when they mean the transformation to the negative direction.

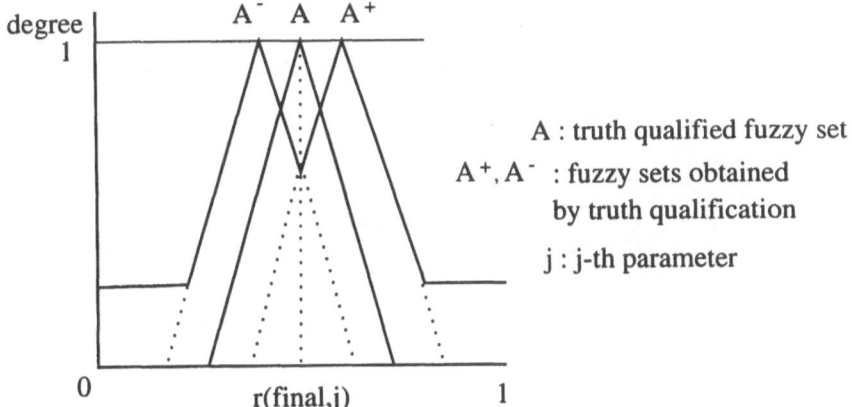

FIGURE 9.18. Fuzzy Sets Obtained by Truth Qualification

The modification can be performed many times. The transformation value at the k-th time is dependent on the transformation directions in the $(k-1)$-th time and the $(k-2)$-th time. There are four combinations according to the transformed directions in the $(k-1)$-th time and in the $(k-2)$-th time as shown in Figure 9.19, where A is the k-th fuzzy set, B is the $(k-1)$-th one and C is the $(k-2)$-th one. Let $r(k)$ be the center value of the membership function $N(r(k)-0.2, r(k), r(k)+0.2)$ representing the meaning of the feature term at the k-th modification, and p be the transformed value that is obtained by the truth qualification, respectively, where $r(i,j)$ at the k-th modification is rewritten as $r(k)$ for simplicity.

(1) In the case of the positive direction at the $(k-2)$-th and at the $(k-1)$-th times.

$$r(k) = r(k-1) + p \times (r_{max} - r(k-1)), \qquad (9.11)$$

where r_{max} is the maximum value which can be transformed by the modification.

(2) In the case of the positive direction at the $(k-2)$-th time and the negative direction at the $(k-1)$-th time.

$$r(k) = r(k-1) - p \times (r(k-1) - r(k-2)) \qquad (9.12)$$

(3) In the case of the negative direction at the $(k-2)$-th time and the positive direction at the $(k-1)$-th time.

$$r(k) = r(k-1) + p \times (r(k-2) - r(k-1)) \qquad (9.13)$$

(4) In the case of the negative direction at the $(k-2)$-th and at the $(k-1)$-th

times.

$$r(k) = r(k-1) - p \times (r(k-1) - r_{min}), \qquad (9.14)$$

where r_{min} is the minimum value which can be transformed by the modification.

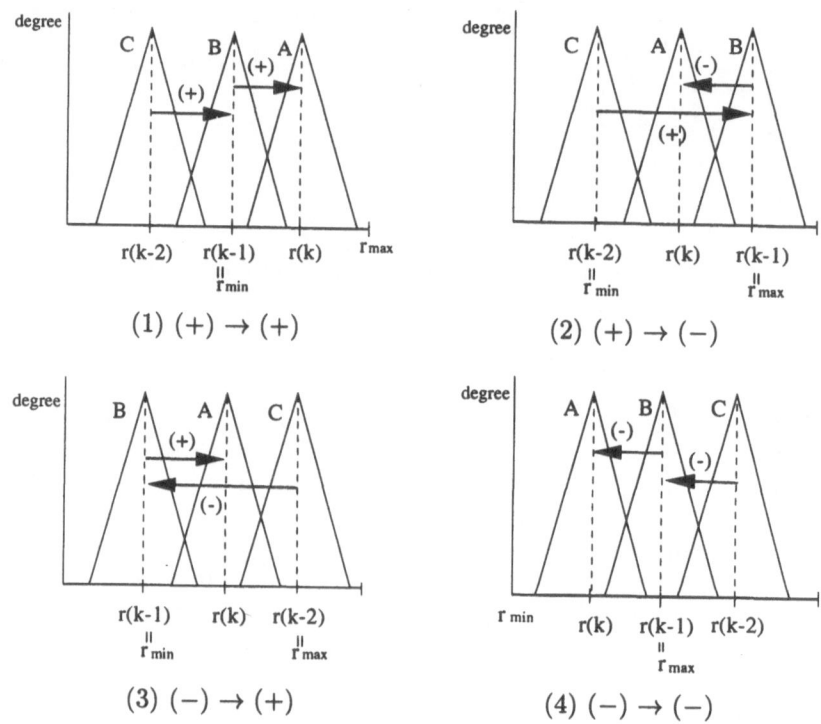

FIGURE 9.19. Change of Parameter Value in Modification

Simulation Example

The facial caricature of Prof. Zadeh is drawn as an example.
1. Terms of initial input
1-1. A face shape is home base shaped face.
1-2. Eyes are fairly big and pupils are slightly big.
1-3. The size of a nose is fairly big. The size of wings of a nose is fairly big. The height of a nose is very high.
1-4. A mouth is slightly small. The thickness of lips is fairly thin.
1-5. The thickness of eyebrows is very thick.
1-6. Ears are very big.
2. Terms for modification

2-1. Eyes are a little small. Pupils are a little small.
2-2. The direction of a nose is more low.
2-3. A mouth is a little small.
2-4. The distance between eyes and eyebrows is more short.
2-5. The distance between eyes and eyebrows is more short.
2-6. The distance between a nose and a mouth is much more long.

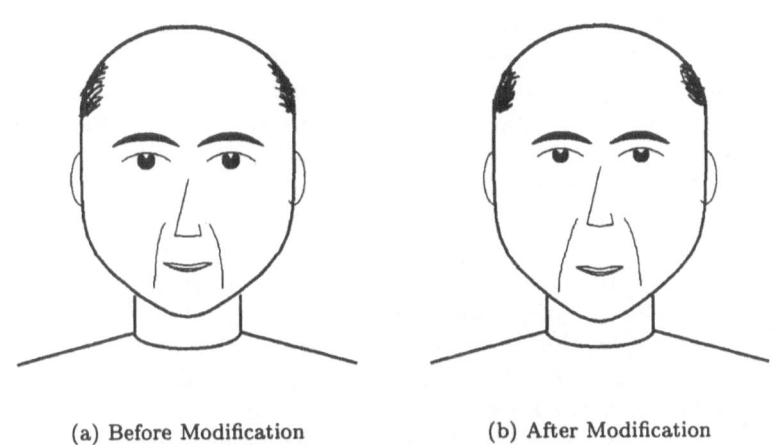

(a) Before Modification   (b) After Modification

FIGURE 9.20. Facial Caricature of Prof. Zadeh

Facial caricatures before the modification and after the modification are shown in Figure 9.20. The facial caricatures seem younger than present Prof. Zadeh since the caricature is drawn based on his photograph taken more than ten years ago. And there are other reasons. The presented facial caricature drawing system has not necessarily a sufficient ability to draw a caricature yet. For example, the system cannot draw a face with exaggeration, wrinkles, light and shade in a face, and a thick and a thin lines, and does not reflect the usefulness of a fuzzy set completely. There are many other open problems to solve. But Figure 9.20 reflects the impression of Prof. Zadeh in the photograph to some extent for the person who draws his caricature.

9.4   Conclusions

This chapter mentions the concept of *Kansei information*. *Kansei information* is understood easily by Japanese, but it is difficult to understand it in Western. As mentioned in this chapter, however, in Europe, the concept of human machine symbiosis in a human centered system is known. Although this concept is not necessarily associated with *Kansei information* directly, the concept of human machine symbiosis is necessary to be considered from the viewpoint of *Kansei information*.

This chapter describes the difference between the Kansei information processing and the conventional natural science approach. The conventional natural science is based on objectivity, uniqueness, universality and reappearance. On the other hand *Kansei information* is necessary to be processed by the approach based on subjectivity, ambiguity, vagueness and circumstances dependence. These concepts are related to diversity of sense of values. Therefore, it is difficult to deal with *Kansei information* by the conventional natural science approach.

This chapter also describes *Kansei information* from the soft computing technique including fuzzy set theory, a neural network model, genetic algorithms, and so on. Soft computing approaches seem to have much possibility to process *Kansei information*. Finally this chapter mentions two studies examples of facial expressions as *Kansei information*, using the soft computing technique, recognition model of emotions through facial expressions considering situations and facial caricature drawing. In the former study a neural network model and in the latter fuzzy set theory are mainly applied, respectively. But in these studies, the soft computing technique is merely employed as a methodology. Although the soft computing technique is applicable to the Kansei information processing, a large framework that deals with subjectivity, ambiguity, vagueness and circumstances dependence at the same time is not proposed from soft computing techniques at the present time. It is hoped that the large framework will be proposed in a near future.

The author is grateful to Dr. S.Kitazaki and a candidates for the Ph.D., Ms. S.Iwashita for their preparing figures.

9.5   REFERENCES

[1] T. Kurokawa, *Non-verbal Interface*, Ohmusha, Tokyo, 1994.

[2] Laboratory for International Fuzzy Engineering Research (Ed.) *Intelligent Information Processing by Fuzzy Thinking*, Computer Age, Tokyo, 1995.

[3] Japan Interdisciplinary Council (Ed.) *Kansei and Information Processing –New Possibility to Computer Science*, Kyoritsu, Tokyo, 1993.

[4] S. Iguchi,*et.al. Kansei information Processing*, Ohmusha, Tokyo, 1994.

[5] S. Tsuji, "Science of Kansei –an approach to Kansei information processing," Saiensu-sha Co., Ltd. Publishers, Tokyo, 1997.

[6] M. Nagamachi, "Kansei engineering and comfort–preface," *International Journal of Industrial Ergonomics*, Vol.19, No.1, pp.79-80, 1997.

[7] Japan Society for Fuzzy Theory and Systems (Ed.), "Panel discussion: Kansei engineering," *Journal of Japan Society for Fuzzy Theory and Systems*, Vol.10, No.3, pp.426-444. 1998.

[8] N.J. Nilsson, *Problem-Solving Methods in Artificial Intelligence*, McGraw-Hill Book Company, New York, 1971.

[9] P.H. Winston, *Artificial Intelligence*, Addison-Wesley Publishing Company, Inc., Mass., 1977.

[10] Nikkei Computer, *An Extra Issue, A.I. –Dawn of Its Practicality*, Nikkei McGraw-Hill, Inc., Tokyo, 1985.

[11] P.S. Shell, *Expert Systems: A Practical Introduction*, Macmillan, Hampshire, 1985.

[12] L.A. Zadeh, "Fuzzy Sets," *Information and Control*, Vol.8, pp.338-353, 1965.

[13] B. Halstead, "Anti-darwinian Theory in Japan," *Nature*, Vol.317, 587, 1985.

[14] H. Nakahara and S. Sakawa, "Theory of Evolution Is Changing – Molecular Biology Shakes Darwinism", Blue Backs, B-852, Kodansha, Tokyo, 1995.

[15] H. Tanaka, "The past, present and future states of the art of fuzzy systems theory," *Journal of Japan Society for Fuzzy Theory and Systems*, Vol.1, No.1, pp.48-62, 1989.

[16] C.W. Dodge, *Euclidean Geometry and Transformation*, Addison-Wesely Pub. Co., MASS., 1972.

[17] D. Gans, *An Introduction to Non-Euclidean Geometry*, Academic Press, New York, 1973

[18] J.L. McCauley, *Chaos, Dynamics and Fractals, An Algorithmic Approach to Deterministic Chaos*, Cambridge University Press, Cambridge, 1993.

[19] R.M. Smullyan, *Gödel's Incompleteness Theorems*, Oxford University Press, New York, 1992.

[20] S.S. Stevens (Ed.), *Handbook of Experimental Psychology*, p.43, John Wiley & Sons, Inc., New York, 1951.

[21] H.A.David, *The Method of Paired Comparison*, Charles Griffin, London, 1963.

[22] C. Chatfield and A.J.Collins, *Introduction to Multivariate Analysis*, Chapman and Hall, London, 1980.

[23] J.G. Snider and C.E. Osgood (Eds.), *Semantic Differential Technique*, A Sourcebook, Aldine, Chicago, 1969.

[24] R.P. McDonald, *Factor Analysis and Related Methods*, Lawrence Erlbaum Associates, Hillsdak, N.J., 1985.

[25] M. Mizumoto, "Pictorial Representations of Fuzzy Connectives, Part I: Cases of t-norms, t-conorms and Averaging Operators," *Fuzzy Sets and Systems*, Vol.31, pp.217-242, 1989.

[26] E. Hollnagel, *Human Reliability Analysis, Context and Control*, Academic Press, London, 1993.

[27] A.D. Swain and H.E. Guttmann, *Handbook of Human Reliability Analysis with Emphasis on Nuclear Power Plant Applications*, NUREG/CR-1278, 1983.

[28] Special Issue on Fuzzy Methodology in System Failure Engineering, *Fuzzy Sets and Systems*, Vol. 83, No.2, 1996.

[29] T. Onisawa and J. Kacprzyk (Eds.), *Reliability and Safety Analyses Under Fuzziness*, Physica-Verlag, A Springer-Verlag Company, Heidelberg, 1995.

[30] E. Kerre, T. Onisawa, B. Cappele, and I. Gazdik, "Reliability," pp.391-420, in *Fuzzy Sets in Decision Analysis, Operations Research and Statics*, R.Slowinski (Ed.) The Handbooks of Fuzzy Sets Series, D.Dubois and H.Prade (Eds.), Kluwer Academic Publishers, London, 1998.

[31] L. Fausett, "Fundamentals of Neural Networks: Architectures, Algorithms and Applications," Prentice Hall, N.J., 1994.

[32] D.E. Goldberg, emphGenetic Algorithm in Serach, Optimization, and Machine Learning, Addison-Wesley, 1989.

[33] X. Li, D. Ruan, and A.J. van der Wal, "Discussion on Soft Computing at FLINS'96," *International Journal of Intelligent Systems*, Vol.13, pp.287-300, 1998.

[34] Special Issue on Technical Papers, Interactive Evolutionary Computing, *Journal of Japanese Society for Artificial Intelligence*, Vol.13, No.5, pp.691-745, 1998.

[35] K.S.Gill (Ed.), *Human Machine Symbiosis −The Foundations of Human-Centred Systems Design*, Springer-Verlag, London, 1996.

[36] M.Ito, H.Ushida, A.Yamadori, T.Ono, A.Tokisumi and K.Ikeda, *Cognitive Science 6, Jodo (Emotions)*, Iwanami, Tokyo, 1994.

[37] Special Issue on Human Face, *The Transactions of The Institute of Electronics*, Information and Communication Engineers, Vol.J80-A, No.8 and Vol.J80-D-II, No.8, 1997.

[38] O.A. Uwechue and A.S. Pandya, *Human Face Recognition Using Third-Order Synthetic Neural Networks*, Kluwer Academic Publishers, Mass., 1997.

[39] H. Doi and M. Ishizuka, "A visual software agent connected to WWW/Mosaic," *The Transactions of The Institute of Electronics*, Information and Communication Engineers D-II, Vol.J79-D-II, No.4, pp.585-591, 1996.

[40] H. Harashima, "Face, expression, and emotion –an approach from video-robot technology," *Journal of the Japan Society of Mechanical Engineering*, Vol.95, No.883, pp.503-507, 1992.

[41] F. Hara, "Artificial emotion: generation model and expression," *Mathematical Sciences*, Saiensu-sha, Tokyo, No.373, pp.52-59, 1994.

[42] P. Ekman and W.V. Friesen, *Facial Action Coding System (FACS): A Technique for the Measurement of Facial Action*, Consulting Psychology Press, California, 1978.

[43] T. Onisawa ans S. Kitazaki, "Recognition of facial expressions and its application to human computer interaction," in *Intelligent Biometric Techniques in Fingerprint and Face Recognition*, L.C.Jain et al. (Eds.) pp.423-458, CRC Press LLC, Florida, 1999.

[44] S. Kitazaki and T. Onisawa, "Communication model considering facial expressions and situations," *Proc. of 1998 IEEE International Conference on Fuzzy Systems*, Vol. 1, pp.171-176, Anchorage, 1998.

[45] S. Iwashita and T. Onisawa, "A study on facial caricature drawing by fuzzy theory," *Proc. of Sixth IEEE International Conference on Fuzzy Systems*, Vol.2, pp.933-938, Barcelona, 1997.

[46] T. Onisawa and M. Iwata, "A model of route decision and facial expression," *Proc. Of the Sixth International Fuzzy Systems Association World Congress*, pp.629-632, San Pãulo, July, 1995.

[47] T. Onisawa, Y. Masuda and M. Iwata, "A route decision system with human interaction and facial expression model," *Proc. of the 1996 International Fuzzy Systems and Intelligent Control Conference*, pp.137-146, Hawaii, 1996.

[48] T. Onisawa and D. Date, "Personified Game Playing Support System," *Japanese Journal of Ergonomics*, Vol.35, No.3, pp.157-167, 1999.

[49] T. Naito and A. Takeuchi, "Interaction with vision-based synthesized facial display," *Computer Vision and Image Media*, May 24, pp.39-46, 1996.

[50] O. Hasegawa, K. Sakaue, K. Ito, T. Karita, S. Hayamizu, K. Tanaka, and N. Otsu, "A multimodal anthropomorphic agent which learns visual information through interactions," *Computer Vision and Image Media*, May 24, pp.33-38, 1996.

[51] J. Hayashi, K. Murakami and H. Koshimizu, "A method for automatic generation of caricature profile in PICASO system," *The Transactions of The Institute of Electronics*, Information and Communication Engineers D-II, Vol.J80-D-II, No.8, pp.2102-2109, 1997.

[52] H. Shimada and M. Shiono, "A method for caricature generation based on 3D facial model," *The Transactions of The Institute of Electronics*, Information and Communication Engineers D-II, Vol.J76-D-II, No.12, pp.2513-2521, 1993.

[53] J. Nishino, T. Kameyama, H. Shirai, T. Odaka, and H. Ogura, "Facial caricature drawing and modification using linguistic expressions," *Proc. of the 5th International Conference on Soft Computing and Information/Intelligent Systems*, Vol.1, pp.205-208. Iizuka, 1998.

[54] L.A. Zadeh, "Fuzzy logic = computing with words," *IEEE Transactions on Fuzzy Systems*, Vol.4, No.2, pp.103-111, 1996.

[55] *History Reader, Extra Number ⟨ Japanese Series ⟩: Japanese Faces*, Shin-Jinbutsu Ohrai-sha, Tokyo, 1995.

10

Vagueness in Human Judgment and Decision Making

Kazuhisa Takemura

10.1 Introduction

In many situations we use vague or fuzzy judgment in forming our own opinions or making decisions. The vagueness or fuzziness is inherent in people's perception and judgment. Traditionally, psychological and philosophical theories implicitly had assumed the vagueness of thought processes [25, 26]. For example, Wittgenstein [43] pointed out that lay categories were better characterized by a "family resemblance" model which assumed vague boundaries of concepts rather than a classical set-theoretic model. Rosch [18] and Rosch and Mervice [19] also suggested vagueness of lay categories in her prototype model and reinterpreted the family resemblance model. Moreover, the social judgment theory [24] and the information integration theory [2] for describing judgment and decision making assumed that people evaluate the objects using natural languages which were inherently vague.

However, psychological theories did not explicitly treat the vagueness in judgment and decision making with the exception of using random error in judgment and decision making. For the measurement and representation of vagueness in judgment and decision making, the vagueness was treated as a random error which is represented by the probability theory. On the contrary, the vagueness is different from uncertainty due to randomness inherent in the probability system [26].

Recently, the vagueness of judgment and decision making have been conceptualized by fuzzy set theory. Fuzzy set theory provides a formal framework for the presentation of the vagueness. Fuzzy sets were defined by Zadeh [45] who also outlined how they could be used to characterize complex systems and decision processes [46]. Zadeh [45, 46] argues that the capacity of humans to manipulate fuzzy concepts should be viewed as a major asset, not a liability. The complexities in the real world often defy precise measurement and fuzzy logic defines concepts and its techniques provide a mathematical method able to deal with thought processes which are often too imprecise and ambiguous to be deal with by classical mathematical techniques.

In this chapter, the vagueness in human judgment and decision making is discussed through evaluating the behavioral decision theory, fuzzy measure theory, and fuzzy set theory in section 10.2. In section 10.2, it was concluded that the representation of the vagueness for judgment and decision making by the fuzzy set theory is more appropriate. In section 10.3, the methods of measurement and representation of vagueness for judgment and decision making are introduced. The method of the fuzzy rating to measure a certain type of vagueness in evaluation is introduced, and the representation of the rating data by the fuzzy set theory is described using some examples of psychological experiments. In section 10.4, the method of analysis for the fuzzy rating data is described through a comparison between the possibilistic linear regression analysis and the fuzzy regression analysis using the least square method. In section 10.5, as a conclusion, properties of human judgment and decision making which are suggested by the findings of the psychological studies are stated.

10.2   Theoretical Representation of Vagueness in Judgment and Decision Making

There are in general three types of theories regarding mathematical aspects of the model for explaining vagueness in judgment and decision making. The first type is a cluster of theories which adopt a probabilistic approach such as the expected utility theory [42] and the subjectively expected utility theory [21]. The second type is a cluster of theories which adopt a fuzzy-measure approach such as the non-linear utility theory [9, 22], the prospect theory [11, 40], and the fuzzy integral theory [15, 27]. Although the origin of fuzzy measure is different from the mathematical concept of capacity, the fuzzy measure and the capacity are mathematically equivalent. Therefore, those theories can be categorized as the same type of theories regarding mathematical aspects of the model. The third type is a cluster of theories which adopts fuzzy set approach such as fuzzy utility theories [17, 23] and fuzzy multi-attribute attitude model [31].

In the first type of approach, von Neuman and Morgenstern [32] axiomatized the expected utility theory in which the expected utility of an alternative is the weighted average of the utilities of the possible outcomes where the weights are the objective probabilities of each outcome. Savage [21] also axiomatized the subjective expected utility theory which allows the derivation of the decision maker's own subjective probabilities for events.

However, many experimental studies indicated that people systematically make judgments and decisions that violate properties required by the expected utility theory and the subjective expected utility theory [1, 8, 11, 13, 32]. The two most famous violations are Allais [1] and Ellsberg [8] paradoxes. The Allais paradox is known as a violation of the sure-thing property [21]. Suppose a lottery D, formed by reducing the compound lottery $PA + (1 - P)Z$ by having a P chance (probability) of getting lottery A and a $(1 - P)$ chance of getting lottery Z, is preferred over E, the reduced compound lottery corresponding to $PB + (1 - P)Z$. The sure-thing property requires that whenever some lottery D is preferred over E, then D' must be preferred over E', where D' and E' are formed by replacing the common consequences Z with a new "sure-thing" consequence Z' [13]. On the other hand, the Ellsberg paradox is known as a violation of ambiguity indifference [13]. The property of ambiguity indifference requires that indifference must hold between two options that are identical except that one portion has a non-vague and crisp subjective probability for an event, and the other has a vague or unknown probability for a corresponding event.

The Allais paradox and the Ellsberg paradox stimulated new approaches of explaining judgment and decision-making. The most-widely accepted approach in the field of judgment and decision making is the so called generalized utility theory which includes various types of non-linear utility theories [9, 22] and the prospect theory [11, 40]. This type of theory can be viewed as the fuzzy measure approach in which a fuzzy measure (or capacity) holds monotonicity but not always holds additivity. The fuzzy measure approach has succeeded in explaining the Allais and the Ellsberg paradoxes. For example, in the prospect theory, subjectively transformed probabilities called decision weights(which holds properties of fuzzy measure) multiply subjectively transformed outcomes, called a value function, is defined on losses and gains relative to a psychological reference point. The decision weights in the prospect theory holds a sub-additivity in which the weights function is concave below and convex above a certain point of probability. The weights are dependent on the rank of the associated outcome in the distribution of possible outcomes. The rank dependent property of the decision weights can be represented by the Choquet integral [4], which is a kind of fuzzy t-conorm integral with respect to the fuzzy measure [15].

On the contrary, there is a certain problem of the fuzzy measure approach for describing judgment and decision making. The reason is a finding that the property of monotonicity for fuzzy measure is violated in the probability judgment of conjunctive events [39]. If the property of monotonicity holds, the probability of conjunction for event A and event B, $P(A \cap B)$ must not exceed a probability of A, $P(A)$ or a probability B, $P(B)$ because $A \cap B$ is a subset of A or B. Although there are a few exceptions which assume the property of non-monotonicity in the the generalized theory [16], the proper fuzzy measure approach can not explain the violation of monotonicity in judgment and decision making such as the conjunction fallacy [39].

Moreover, the recent research on judgment and decision making suggests that people make judgments and decisions in vague ways. Firstly, people tend to use verbal phrases which have inherently vague ranges for the extension of objects [3, 33, 45, 48]. Secondly, people often use such vague information in order to make judgment and decision making through communicating with each other [3, 33]. Thirdly, people tend to adopt expressions indicating ranges of values such as "from 100 to 110 cm" and vague meanings such as "almost zero" [33]. It is almost impossible for describing such vague phenomenon in judgment and decision making using the probabilistic approaches and the fuzzy measure approaches, because those approaches can hardly represent the vagueness of natural languages and thought processes in judgment and decision making.

In this chapter, the fuzzy set approach is adopted because this approach can avoid above problems. Of course, there are some methodological problems in this approach. Firstly, the fuzzy set approach has not been well developed and is not sophisticated in mathematical axiomatization, comparatively, than other approaches. Secondly, the methods of measurement and representation of fuzzy data for judgment and decision making have not been well developed, whereas many traditional methods for non-fuzzy data exist and have been well developed in the field of psychology. Thirdly, the method of data analysis for fuzzy data has not been well developed and not sufficiently examined by empirical research.

For the descriptive perspective of judgment and decision making, it would be more important to describe the vagueness than to make a very rigorous mathematical model. Therefore, there is no reason that the fuzzy set approach should be rejected for describing judgment and decision making although the first problem still exists. Moreover, to cope with the second and third problems, some researches have started. In the following sections, such methodological issues and empirical examples will be introduced.

10.3   Measurement and Fuzzy-Set Representation of Vagueness in Judgment and Decision Making

Fuzzy Rating Method

Traditional approaches to the measurement of social judgment have involved methods such as the semantic differential, the Likert scale, or the Thurstone scale. Although insights into the ambiguous nature of social judgment were identified early in the development of measurement of social judgment, the subsequent methods used failed to capture this ambiguity, no doubt because traditional mathematics was not well developed for dealing with vagueness of judgment [10].

In order to measure the vagueness of human judgment, the fuzzy rating method has recently been proposed and developed [10, 33]. In the fuzzy rating method, respondents select a representative rating point on a scale and indicate lower or upper rating points if they wish depending upon the relative vagueness of their judgment. For example, the fuzzy rating method would be useful for measuring perceived temperature indicating the representative value and the lower or upper values. This rating scale allows for asymmetric, and overcomes the problem, identified by Smithson [25], of researchers arbitrarily deciding most representative value from a range of scores. By making certain simplifying assumptions (not uncommon within fuzzy set theory), the rating can be viewed as a triangular fuzzy number, hence making possible the use of fuzzy set theoretic operations [10]. An example of the fuzzy rating scale and of the representation of the rating data using the triangular fuzzy number are shown in Fig.1.

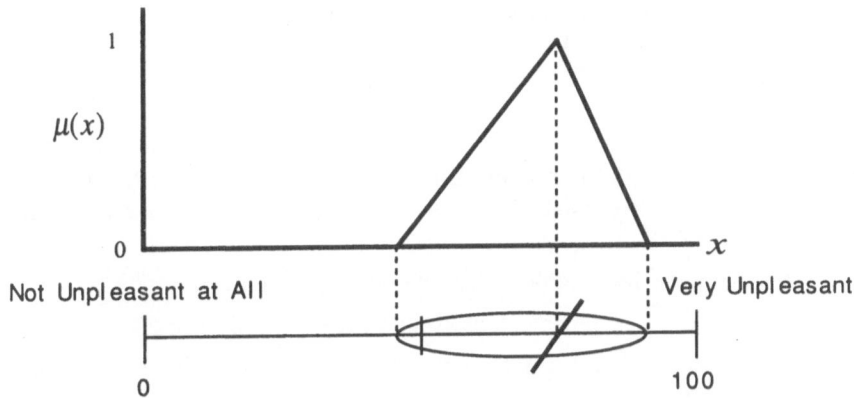

FIGURE 10.1. An Example of Fuzzy Rating Data and Its Representation by the Triangular Fuzzy Number

Fuzzy Set and Rating Data

Fuzzy set

A fuzzy set A is defined as follows. Let X denote a universal set, such as $X = \{x_1, x_2, \dots, x_n\}$. Then, the membership function $\mu_A \subseteq X$ by which a fuzzy set A is defined has the form

$$\mu_A : X \to [0, 1],$$

where $[0, 1]$ denotes the interval of real numbers from 0 to 1, inclusive.
    The concept of a fuzzy set is the foundation for analysis where fuzziness

exists [45]. a fuzzy set may be expressed as:

$$A = \mu_A(x_1)/x_1 + \mu_A(x_2)/x_2 + \cdots + \mu_A(x_n)/x_n$$
$$= \sum_{i=1}^{n} \mu_A(x_i)/x_i,$$

where $\mu_A(x_i)$ represents the "grade of membership" of x_i in A, or the degree to which x_i satisfies the properties of the set A. It should be noted that here the symbol "+" does not refer to the ordinary addition.

μ_A is called a membership function, or a possibility function. The x_i values are drawn from a global set of all possible values, X. The grade of membership take values between 0 and 1. The membership function has a value of 0 when the properties of the fuzzy set are not at all satisfied, and 1 when the properties of fuzzy set are completely satisfied.

Fuzzy number

Hesketh *et al.* [10] pointed out that fuzzy rating data can be represented as fuzzy sets by making certain simplifying assumptions, which are not uncommon within fuzzy set theory. According to Hesketh *et al.* [10], those assumptions are:

1. The fuzzy set has a convex membership function.
2. The global set X is represented along the horizontal axis.
3. The fuzzy membership function takes its maximum value, one, at the point on the support represented by the representative point.
4. The extent of the support is represented by a horizontal segment.
5. The membership function tapers uniformly from its value of one at the representative point to a value of zero beyond the support or the left and right extensions. The membership value of the lower point and the upper point is 0.

Making those assumptions, fuzzy rating data in this study can be expressed as a fuzzy number which is a kind of fuzzy sets . The concept of the fuzzy number can be defined from the concept of the fuzzy subset [12]. The properties of fuzzy numbers are the convexity and the normality of a fuzzy subset.

Firstly, the convexity of the fuzzy subset is defined as follows: A fuzzy subset $A \subseteq R$ is convex if and only if every ordinary

$$A_\alpha = \{x \mid \mu_A(x) \geq \alpha\}, \alpha \in [0,1],$$

subset is convex(that is a closed interval of the set of real numbers).

Secondly, the normality of the fuzzy subset is defined as follows: A fuzzy subset $A \in R$ is normal if and only if

$$\max_{x \in R} \mu_A(x) = 1.$$

One of the best known fuzzy numbers is the $L - R$ fuzzy number [7]. The $L - R$ fuzzy number is defined as follows:

$$\mu_A(x) = \begin{cases} L\left(\frac{(x-m)}{u}\right) & (-\infty < x < m), \\ 1 & (x = m), \\ R\left(\frac{(x-m)}{v}\right) & (m < x < \infty) \end{cases}$$

where $L\left(\frac{(x-m)}{u}\right)$ is a increasing monotonic function, $R\left(\frac{(x-m)}{v}\right)$ is a decreasing monotonic function, $u > 0$, and $v > 0$.

For the simplicity, $L - R$ fuzzy number can be approximated to the triangular fuzzy number. The triangular fuzzy number is a particular case of $L - R$ fuzzy number. The triangular fuzzy number [12] is defined as:

$$\mu_A(x) = \begin{cases} 0 & (x \leq a_1) \\[2mm] \dfrac{x - a_1}{a_2 - a_1} & (a_1 < x \leq a_2) \\[3mm] \dfrac{a_3 - x}{a_3 - a_2} & (a_2 < x \leq a_3) \\[2mm] 0 & (x > a_3) \end{cases}$$

where a_1, a_2, and a_3 are finite.

A triangular fuzzy number is represented by a triplet such as:

$$A = (a_1, a_2, a_3).$$

It should be noted that the representation of the triplet used in the definition is different from the that of the L-R fuzzy number.

At the level α the interval of confidence (α-level set) is given by:

$$\begin{aligned} A\alpha &= [a_1(\alpha), a_3(\alpha)] \\ &= [a_1 + \alpha(a_2 - a_1), a_3 - \alpha(a_3 - a_2)]. \end{aligned}$$

Information Integration and Extension Principle

In order to examine the information integration rule of subjects, the predictive integrated values from the data concerning evaluation of attributes can be calculated using Zadeh's [47] extension principle, which is one of the most basic ideas of fuzzy set theory. It provides a general method for extending non-fuzzy mathematical concepts in order to deal with fuzzy quantities.

Let X and Y be two universal sets and $f : X \to Y$ be a given function. For a non-fuzzy set (a crisp set), $C \subseteq X, f(C)$ is a crisp set of Y. As a generalization of the image of a crisp set, the extension principle was proposed by Zadeh [47]. Let A be a fuzzy set of X. According to the extension principle, an image of $f(A)$ is defined as follows:

$$\mu_{f(A)}(y) = \begin{cases} \sup_{y=f(x)} \mu_A(x), & y \in f(X), \\ 0 & y \notin f(X), \end{cases}$$

where the supremum is taken for all $x \in X$ such that $y = f(x)$.

More generally, the extension principle for the Cartesian product of fuzzy sets is given as follows. Let $S_1, S_2, \ldots, S_n$ be universal sets, $f : S_1 \times S_2 \times \ldots \times S_n \to Y$ be a given function, and $A_i \subseteq S_i, i = 1, \ldots, n$ be fuzzy sets.

$$\mu_{f(A_1, A_2, \ldots, A_n)}(y) = \sup_{y=f(x_1, x_2, \ldots, x_n)} \mu_{f(A_1 \times A_2 \times \ldots \times A_n)}(x_1, x_2, \ldots, x_n)$$

$$= \sup_{y=f(x_1, x_2, \ldots, x_n)} \min\left[\mu_{A_1}(x_1), \mu_{A_2}(x_2), \ldots, \mu_{A_n}(x_n)\right].$$

According to Dubois and Prade [7], binary operation $*$ can be extended into R to combine two fuzzy numbers (i.e., convex and normalized fuzzy sets) M and N. Moreover, μ_M and μ_N are assumed to be continuous functions on R:

$$\mu_{M \circledast N}(z) = \sup_{z=x*y} \min\left\{\mu_M(x), \mu_N(y)\right\}.$$

A weighted fuzzy average model postulates that the the total evaluation E is given directly by the strategy as:

$$E = \sum_{i=1}^{n} w_i \otimes X_i = (w_1 \otimes X_1) \oplus (w_2 \otimes X_2) \oplus \cdots \oplus (w_n \otimes X_n) \quad (10.1)$$

where $\sum_{i=1}^{n} w_i = 1, w_i \geq 0$.

X_i is an evaluation of the i th attribute (fuzzy number) and w_i is a weight associated with the i th attribute(nonfuzzy number), '$\otimes$' is the product operator based on the extension principle, and '$\sum$' is the summation operator based on the extension principle [47]. This type of model for the crisp data is very common in social psychological research, particularly as used in the information integration model literature [2]. The above representation of the fuzzy weighted average model in the Equation (10.1) is a natural extension of the ordinary weighted average model of the crisp data.

10.4   Experimental Studies of Vagueness of Judgment and Decision Making Using the Fuzzy Rating Method

Overview of Two Studies

In the following empirical studies, the fuzzy rating method which can directly measure the vagueness in social judgment is used. In the fuzzy rating method, a respondent rates (1) the lower point, (2) the upper point,(3) the representative point for each item. A width between the value of the lower point and the value of the upper point is considered to be an indicator of vagueness of judgment. In study 1, the width between the lower point and the upper point is compared with the traditional indicator of vagueness of judgment, that is, the standard deviation of rating value. In study 2, the information integration rule is examined in several social judgment tasks. Predicted values of final integration were calculated using the extension principle and the predicted values are compared with the final integration values.

Study 1: Comparison of the Fuzzy Rating Method with Traditional Treatment of Vagueness

Purpose of the study

In many studies on social judgment, a value of standard deviation has been widely used as an indicator of vagueness of judgment [29]. If a value of standard deviation is an indicator of vagueness of judgment, a width of judgment is positively proportional to a value of the standard deviation of a representative point (It is also expected that the width of judgment and the standard deviation of the representative point are positively correlated). The present study aims to test this hypothesis. If the hypothesis is rejected, the result suggests that the traditional interpretation of the standard deviation as the indicator of the vagueness in judgment and decision making is not adequate, and implies that the fuzzy rating method is more effective for measuring a vagueness in the judgment and decision making than the traditional rating methods. If the hypothesis is not rejected, it would be concluded that the fuzzy rating method is not necessary for measuring vagueness in judgment and decision making.

Method

Subjects. Subjects were 396 college students, officers, and teachers.
Procedure. Subjects were asked to complete a questionnaire about their impressions of an actor who had given a lecture at the college. Subjects

rated their impressions on 20 adjective scales. They rated (1) the lower point,(2) the upper point, and (3) the representative point for each item. Before the fuzzy rating, various possible manipulations were demonstrated, including ratings with no spread on either side, one with spread only on the one side, and other variations.

Results and discussion

Data from each rating comprised three numbers $[0, 100]$ reflecting position on the x-axis (the lower point, the upper point , and the representative point). In order to examine the width of judgment, the values of upper point minus the values of lower point were calculated. In order to examine a relationship between the standard deviation of the representative point and the width of judgment, a regression analysis was performed. The independent variable was the standard deviation of the representative point (SD). The dependent variable was the mean value of width for judgment (WIDTH). The result of analysis indicated the following relationship: WIDTH = -0.19 SD + 17.74. Moreover, SD and WITH were negatively correlated -.564, and the proportion of explained variance was 31.8%.

These results indicate that the width of judgment is not directly proportional to the standard deviation. Because the hypothesis was rejected, it can be concluded that the traditional interpretation of the standard deviation as the indicator of the vagueness in judgment and decision making is not adequate and that the fuzzy rating method could be more effective for measuring the vagueness than the traditional rating methods.

Study 2

Purpose of the study

In this study, the information integration rule is examined in several social judgment tasks. Predicted values of final integration such as Min, Max, and Average rules are calculated using the extension principle, and then compared with the final overall values of judgment using the measure of possibility.

Method

Subjects. One hundred and eighty-six male and female undergraduates who were enrolled in the Introductory Psychology course participated in the experiment.

Procedure. Each subject was assigned into one of four groups. Each subject was asked to complete one of four types of questionnaires, which were related to (1) evaluative judgment of marital partner, (2) evaluative judgment of personality, (3) evaluative judgment of romantic partner, (4) evaluative judgment of occupation . Each questionnaire consisted of two parts: (1)evaluative judgment of partial information for two attributes, (2) evaluative judgment of combined information.

For evaluative judgment of partial information, subjects rated desirability of three levels of each attribute. For evaluative judgment of combined information, subjects rated desirability of 9 possible combinations (3 × 3) of partial information.

For all rating , the fuzzy rating method was utilized. The fuzzy rating method was explained to subjects using an example not related to the stimuli. They rated (1) the lower point, (2) the upper point, (3) the representative point for each item. Before the fuzzy rating, various possible manipulations were demonstrated, including ratings with no spread on either side, one with spread only on the one side, and other variations.

All experimental sessions took place in groups of about 10 persons and took about a half hour.

Results and discussion

Data from each rating comprised three numbers [0, 100] reflecting position on the x-axis (the lower point, the upper point , and the representative point). In order to examine the width of judgment, the values of the upper point minus the values of the lower point were calculated.

Tables 10.1 to 10.4 contain the means and standard deviations (the lower point, the upper point, the representative point, and the width for the ratings). To summarize, very desirable or very undesirable items have smaller standard deviations than moderately desirable or undesirable items. The widths of judgment are varied among items. In general, very desirable or very undesirable items have a smaller width of judgment than moderately desirable or undesirable items. The results also indicated that the width of the integrated information was amplified especially when the most ambiguous information was presented.

We checked the validity of the fuzzy weighted average model which contains the minimum, the maximum, and the equally weighted average models. Much literature on human judgment shows that a weighted average model is valid for data of information integration in social judgment [2, 5, 44].

Valid conditions for the fuzzy weighted average model are as follows:

$$\min(x_{11},\dots,x_{i1},\dots,x_{n1}) \le e_1 \le \max(x_{11},\dots,x_{i1},\dots,x_{n1})$$
$$\min(x_{12},\dots,x_{i2},\dots,x_{n2}) \le e_2 \le \max(x_{12},\dots,x_{i2},\dots,x_{n2})$$
$$\min(x_{13},\dots,x_{i3},\dots,x_{n3}) \le e_3 \le \max(x_{13},\dots,x_{i3},\dots,x_{n3})$$

where x_{i1} is a value of the lower point for attribute i, x_{i2} is a value of the representative point for attribute i, x_{i3} is a value of the upper point for

TABLE 10.1. Mean value of judgment for marital partner

Item		Lower Point	Upper Point
Physical Attraction			
	Above average(A)	56.72 (19.23)	84.24 (16.69)
	Below average (B)	22.54 (14.17)	47.46 (17.40)
	Average (C)	62.01 (19.44)	82.68 (16.40)
Consideration			
	Above average(A)	81.96 (12.99)	97.23 (6.83)
	Below average (B)	3.17 (5.71)	17.57 (13.21)
	Average (C)	44.80 (20.01)	66.00 (20.90)
Physical (A)	Consideration(A)	79.29 (17.58)	96.41 (7.89)
Attraction(A)	(B)	8.44 (12.53)	24.08 (17.50)
(A)	(C)	48.01 (20.66)	71.03 (21.15)
(B)	(A)	48.01 (18.11)	74.58 (18.85)
(B)	(B)	0.96 (2.43)	10.16 (9.31)
(B)	(C)	24.87 (15.81)	44.08 (20.16)
(C)	(A)	71.56 (17.55)	91.63 (11.73)
(C)	(B)	9.73 (12.49)	26.61 (17.52)
		46.50 (18.06)	65.47 (19.03)

Item		Representative Point	Width
Physical Attraction			
	Above average(A)	71.56 (18.07)	27.52 (14.31)
	Below average (B)	33.46 (14.79)	24.91 (13.03)
	Average (C)	73.17 (16.90)	20.67 (13.20)
Consideration			
	Above average(A)	90.87 (9.29)	15.27 (10.82)
	Below average (B)	8.15 (7.87)	14.10 (11.72)
	Average (C)	56.18 (20.42)	21.21 (14.38)
Physical (A)	Consideration(A)	89.53 (11.44)	17.12 (14.50)
Attraction(A)	(B)	13.97 (14.08)	15.65 (10.99)
(A)	(C)	60.76 (20.45)	23.01 (13.12)
(B)	(A)	61.85 (18.37)	26.56 (15.23)
(B)	(B)	4.26 (5.12)	9.20 (8.78)
(B)	(C)	34.89 (17.86)	19.22 (11.65)
(C)	(A)	83.62 (13.68)	20.07 (12.69)
(C)	(B)	16.36 (13.87)	16.88 (11.55)
		56.07 (18.13)	18.97 (12.96)

TABLE 10.2. Mean value of judgment for personality

Item		Lower Point	Upper Point
Kindness			
	Above average(A)	74.42 (14.97)	91.60 (14.14)
	Below average (B)	14.62 (16.40)	31.83 (19.24)
	Average (C)	51.38 (15.11)	71.06 (14.82)
Responsibility			
	Above average(A)	75.74 (16.95)	92.98 (13.06)
	Below average (B)	11.96 (17.95)	28.24 (21.23)
	Average (C)	48.56 (18.51)	67.34 (18.22)
Kindness(A)	Responsibility (A)	81.86 (13.89)	95.38 (11.35)
(A)	(B)	35.06 (21.14)	52.66 (22.53)
(A)	(C)	60.22 (16.50)	79.23 (15.77)
(B)	(A)	40.93 (21.31)	59.26 (21.95)
(B)	(B)	6.76 (17.02)	19.13 (20.77)
(B)	(C)	25.83 (17.56)	42.37 (17.99)
(C)	(A)	57.85 (15.68)	76.06 (15.08)
(C)	(B)	29.55 (17.87)	47.12 (16.81)
(C)	(C)	45.13 (17.46)	65.48 (15.83)

Item		Representative Point	Width
Kindness			
	Above average(A)	83.56 (15.56)	17.18 (10.06)
	Below average (B)	21.73 (17.16)	17.21 (11.97)
	Average (C)	60.35 (15.01)	19.68 (11.01)
Responsibility			
	Above average(A)	85.13 (15.46)	17.24 (10.16)
	Below average (B)	18.91 (19.76)	16.28 (10.29)
	Average (C)	58.43 (18.80)	18.78 (9.34)
Kindness(A)	Responsibility (A)	89.94 (12.55)	13.53 (7.96)
(A)	(B)	43.27 (21.87)	17.60 (9.14)
(A)	(C)	69.17 (15.72)	19.01 (7.72)
(B)	(A)	50.77 (20.97)	18.33 (10.90)
(B)	(B)	11.67 (19.12)	12.37 (8.35)
(B)	(C)	33.59 (17.50)	16.54 (7.61)
(C)	(A)	66.70 (15.32)	18.21 (8.43)
(C)	(B)	37.85 (16.84)	17.56 (7.86)
(C)	(C)	55.38 (15.83)	20.35 (13.21)

TABLE 10.3. Mean value of judgment for romantic partner

Item		Lower Point	Upper Point
Romantic			
	Above average(A)	57.38 (16.62)	78.78 (15.58)
	Below average (B)	35.39 (16.73)	56.82 (17.70)
	Average (C)	56.55 (19.89)	75.98 (16.90)
Realistic			
	Above average(A)	45.80 (15.98)	68.15 (17.07)
	Below average (B)	27.26 (14.28)	47.53 (14.26)
	Average (C)	50.00 (17.27)	71.10 (15.93)
Romantic(A)	Realistic(A)	61.04 (21.19)	81.25 (17.09)
(A)	(B)	30.12 (17.02)	50.33 (18.91)
(A)	(C)	49.37 (12.89)	69.40 (15.71)
(B)	(A)	28.51 (18.05)	48.33 (20.85)
(B)	(B)	15.54 (10.85)	33.87 (13.53)
(B)	(C)	29.08 (14.60)	47.77 (14.49)
(C)	(A)	55.45 (13.22)	75.63 (12.49)
(C)	(B)	34.49 (17.20)	54.35 (16.89)
(C)	(C)	50.00 (18.23)	71.61 (18.44)

Item		Representative Point	Width
Romantic			
	Above average(A)	68.75 (15.60)	21.40 (9.79)
	Below average (B)	45.15 (17.47)	21.43 (10.92)
	Average (C)	66.70 (17.68)	19.43 (10.91)
Realistic			
	Above average(A)	57.17 (15.74)	22.35 (12.92)
	Below average (B)	36.58 (14.15)	20.27 (9.47)
	Average (C)	59.26 (15.61)	21.10 (15.90)
Romantic(A)	Realistic(A)	72.56 (19.01)	20.21 (9.78)
(A)	(B)	39.17 (18.89)	20.21 (9.78)
(A)	(C)	60.54 (13.66)	20.03 (10.03)
(B)	(A)	37.59 (20.22)	19.82 (10.60)
(B)	(B)	24.11 (12.51)	18.33 (8.49)
(B)	(C)	38.60 (14.38)	18.69 (8.26)
(C)	(A)	66.07 (12.51)	20.18 (10.23)
(C)	(B)	44.52 (16.69)	19.85 (8.40)
(C)	(C)	61.04 (16.61)	21.61 (15.16)

TABLE 10.4. Mean value of judgment for occupation

Item		Lower Point	Upper Point
Hardness			
	Above average(A)	54.18 (20.19)	74.82 (19.95)
	Below average(B)	30.28 (17.78)	51.17 (18.73)
	Average (C)	60.74 (18.17)	82.55 (16.68)
Salary			
	Above average(A)	77.83 (14.78)	93.57 (9,72)
	Below average(B)	12.22 (13.22)	31.15 (19.15)
	Average (C)	46.02 (15.68)	65.92 (17.13)
Hardness(A)	Salary(A)	72.37 (20.23)	90.00 (15.49)
(A)	(B)	24.80 (17.57)	43.78 (21.58)
(A)	(C)	48.60 (18.16)	68.93 (18.84)
(B)	(A)	55.94 (15.97)	78.49 (15.95)
(B)	(B)	6.05 (10.27)	20.66 (18.14)
(B)	(C)	25.48 (15.48)	43.90 (17.33)
(C)	(A)	70.82 (18.12)	91.81 (9.32)
(C)	(B)	25.48 (16.11)	43.90 (17.00)
(C)	(C)	46.20 (17.14)	65.89 (17.31)
Item		Representative Point	Width
Hardness			
	Above average(A)	66.12 (20.09)	20.64 (13.22)
	Below average(B)	38.52 (17.90)	20.89 (12.10)
	Average (C)	72.60 (17.19)	21.81 (13.53)
Salary			
	Above average(A)	87.96 (10.89)	15.74 (11.61)
	Below average(B)	20.48 (14.57)	18.93 (13.98)
	Average (C)	55.94 (16.65)	19.90 (9.23)
Hardness(A)	Salary(A)	82.40 (16.88)	17.63 (12.95)
(A)	(B)	32.63 (18.59)	18.98 (10.87)
(A)	(C)	59.31 (18.56)	20.33 (10.52)
(B)	(A)	68.14 (15.36)	22.55 (11.72)
(B)	(B)	13.20 (14.09)	14.62 (11.82)
(B)	(C)	34.36 (17.08)	18.42 (9.69)
(C)	(A)	83.34 (12.75)	20.99 (14.76)
(C)	(B)	33.29 (15.57)	18.42 (8.75)
(C)	(C)	56.12 (16.74)	19.69 (12.94)

TABLE 10.5. Mean value of validity for each model in judgment of marital partner, where PA: physical attraction, C: considerate.

Item		Model			Validity
PA	C	Min	Max	Average	
(A)	(A)	0.35(10.7%)	0.72(69.6%)	0.47(14.3%)	51.8%
(A)	(B)	0.47(55.4%)	0.03(0.0%)	0.15(17.9%)	75.0%
(A)	(C)	0.12(23.2%)	0.01(0.0%)	0.03(0.0%)	3.6%
(B)	(A)	0.25(44.6%)	0.01(0.0%)	0.03(1.8%)	14.3%
(B)	(B)	0.47(46.4%)	0.25(16.1%)	0.31(16.1%)	60.7%
(B)	(C)	0.29(44.6%)	0.06(1.8%)	0.14(7.1%)	14.3%
(C)	(A)	0.04(10.7%)	0.01(0.0%)	0.01(0.0%)	1.8%
(C)	(B)	0.47(57.1%)	0.04(1.8%)	0.15(16.1%)	73.2%
(C)	(C)	0.11(21.4%)	0.03(0.0%)	0.05(0.0%)	1.8%
	Total	0.28(34.9%)	0.13(9.9%)	0.15(8.1%)	32.9%

attribute i, e_1, e_2, e_3 are the lower, the representative, and the upper points for integrated information values, respectively.

If all of the above conditions are satisfied, the data are valid for the weighted average model. Proportions of valid cases for the model are shown in Tables 10.5 to 10.8.

In order to examine the validity for the Max, Min, Average (equal weighted average) models, the measure of possibility for two fuzzy numbers [7] was used as a validity indicator according to Sugeno [28]. The measure is as follows:

$$Pos(M = D) \quad = \quad \sup_{x \in R} \min \{\mu_M(x), \mu_D(x)\}$$

where ω is the validity measure, $\mu_M(x)$ is a membership function of the predictive model, M, and $\mu_D(x)$ is a membership function of the data, D. The value of the measure ranges from 0 to 1.

Tables 10.5 to 10.8 indicated mean values of $Pos(M = D)$ for fuzzy information integration models. Values in parenthesis for Tables 10.5 to 10.8 indicated the number of cases as being the most consistent with each model.

As shown in Tables 10.5-10.8, the mean values of validity for a max rule tended to be highest among three rules when positive information(level-A) and positive information(level-A) were integrated. The mean values of validity for a min rule tended to be highest among three rules when negative information(level-C) and negative information(level-C) were integrated or when positive information(level-A) and negative information(level-C) were integrated.

Thus, it can be concluded as follows:

TABLE 10.6. Mean value of validity for each model in judgment of personality, where K: kindness, RS: responsibility

Item		Model			Validity
K	RS	Min	Max	Average	
(A)	(A)	0.44(5.1%)	0.66(64.1%)	0.57(15.4%)	38.5%
(A)	(B)	0.14(25.6%)	0.10(12.8%)	0.29(35.9%)	84.6%
(A)	(C)	0.31(46.2%)	0.08(2.6%)	0.18(10.3%)	30.8%
(B)	(A)	0.16(28.2%)	0.10(12.8%)	0.23(25.6%)	82.1%
(B)	(B)	0.15(12.8%)	0.17(30.8%)	0.16(7.7%)	10.3%
(B)	(C)	0.14(12.8%)	0.32(43.6%)	0.31(28.2%)	56.4%
(C)	(A)	0.27(33.3%)	0.08(2.6%)	0.18(12.8%)	30.8%
(C)	(B)	0.14(17.9%)	0.27(35.9%)	0.29(30.8%)	61.5%
(C)	(C)	0.27(20.5%)	0.29(28.2%)	0.29(5.1%)	15.4%
	Total	0.22(22.5%)	0.23(25.9%)	0.28(19.1%)	45.6%

TABLE 10.7. Mean value of validity for each model in judgment of romantic partner, where RE: realistic, RO: romantic

Item		Model			Validity
RE	RO	Min	Max	Average	
(A)	(A)	0.36(21.4%)	0.41(42.9%)	0.41(14.3%)	31.0%
(A)	(B)	0.42(52.4%)	0.16(9.5%)	0.28(19.0%)	45.2%
(A)	(C)	0.24(38.1%)	0.13(7.1%)	0.17(9.5%)	14.3%
(B)	(A)	0.52(64.3%)	0.23(9.5%)	0.32(14.3%)	33.3%
(B)	(B)	0.50(54.8%)	0.41(26.2%)	0.42(9.5%)	33.3%
(B)	(C)	0.46(59.5%)	0.22(14.3%)	0.31(7.1%)	26.2%
(C)	(A)	0.27(45.2%)	0.12(9.5%)	0.15(2.4%)	4.8%
(C)	(B)	0.44(59.5%)	0.12(9.5%)	0.21(11.9%)	38.1%
(C)	(C)	0.19(33.3%)	0.15(14.3%)	0.13(0.0%)	2.4%
	Total	0.38(47.6%)	0.22(15.9%)	0.27(9.8%)	25.4%

TABLE 10.8. Mean value of validity for each model in judgment of occupation, where HA: hardness, SA: salary

Item		Model			Validity
HA	SA	Min	Max	Average	
(A)	(A)	0.35(14.3%)	0.61(49.0%)	0.46(18.4%)	49.0%
(A)	(B)	0.24(26.5%)	0.18(14.3%)	0.38(40.8%)	71.4%
(A)	(C)	0.31(36.7%)	0.15(2.0%)	0.23(8.2%)	12.2%
(B)	(A)	0.28(42.9%)	0.02(2.0%)	0.17(20.4%)	38.8%
(B)	(B)	0.26(26.5%)	0.27(34.7%)	0.26(10.2%)	36.7%
(B)	(C)	0.30(30.6%)	0.25(16.3%)	0.31(18.4%)	26.5%
(C)	(A)	0.16(28.6%)	0.02(0.0%)	0.04(0.0%)	6.1%
(C)	(B)	0.25(26.5%)	0.16(12.2%)	0.38(42.9%)	73.5%
(C)	(C)	0.31(36.7%)	0.13(2.0%)	0.21(8.2%)	12.2%
	Total	0.27(29.9%)	0.20(14.7%)	0.27(18.6%)	36.3%

1. A max rule tended to be adopted when positive information and positive information were integrated.

2. A min rule tended to be adopted when negative information and negative information were integrated.

3. A min rule tended to be adopted when positive information and negative information were integrated.

10.5 Regression Analyses for Fuzzy Rating Data

Fuzzy Rating Data and Fuzzy Regression Analysis

To analyze fuzzy rating data, it would be useful to apply fuzzy linear regression analysis, in which observed values and estimated values are assumed to have fuzziness. The original version of the possibilistic linear regression analysis proposed by Tanaka *et al.* [37]assumed that while output data and parameters are fuzzy numbers, input data are not fuzzy numbers. More recently, Sakawa and Yano [20] formulated and developed a possibilistic linear regression analysis, where both input data, output data, and parameters are fuzzy numbers. This method could be very effective for human-related sciences such as psychology, sociology, and ergonomics, because most of the input data and output data for such sciences, are considered to be fuzzy.

Unfortunately, some application studies using fuzzy rating data have indicated that the predicted variable for possibilistic linear analysis for fuzzy input-output data had too large a spread of fuzzy number for meaningful interpretation [32, 34]. Moreover, practically, it takes a lot of resources to compute the solution for the possibilitistic linear regression analysis by computer.

On the other hand, Diamond [6] proposed a fuzzy least square regression, in which both input data and output data were represented by the triangular fuzzy number. In his model of the analysis, the regression coefficients were assumed to be crisp numbers because a product by a multiplication of triangular fuzzy numbers is not a triangular fuzzy number. However, firstly, the assumption that the regression coefficients have fuzziness is rather natural for many human related sciences such as psychology. Secondly, a true product by a multiplication of triangular fuzzy number, which is a fuzzy number, can approximate a triangular fuzzy number [12, 14]. Thus, we adopted the assumption that both input-output data and regression coefficients had fuzziness which was represented by the triangular fuzzy number. However, this approach is more heuristic than Diamond's [6] approach because the present approach is considered to be the natural extension of the ordinary least square method for the crisp data rather than the extension of fuzzy set theory.

Fuzzy linear regression model (where both input and output data are fuzzy numbers) is represented as follows,

$$
\begin{aligned}
Y(X_i) &= A \otimes X_i, i = 1, \ldots, k \\
&= (A_0 \otimes 1) \oplus (A_1 \otimes X_{i1}) \oplus \cdots \oplus (A_m \otimes X_{im}) \quad (10.2)
\end{aligned}
$$

where

$$
A = (A_0, \ldots, A_n), X_i = (1, X_{i1}, \ldots, X_{in})^T,
$$

the predicted value $Y(X_i)$, the parameter, A_j, and the observed input data, X_{ij} are fuzzy numbers, '$\otimes$' is the product operator based on the extension principle, and '$\otimes$' is the summation operator based on the extension principle [47].

In the following example, for simplicity, we assume that the fuzzy input data X_{ij}, fuzzy output data Y_i, and fuzzy parameter a_j are given as triangular fuzzy numbers defined by

$$
\mu_{X_{ij}}(x_{ij}) =
\begin{cases}
0 & (x_{ij} \leq x_{ij1}) \\[2mm]
\dfrac{x_{ij} - x_{ij1}}{x_{ij2} - x_{ij1}} & (x_{ij1} < x_{ij} \leq x_{ij2}) \\[2mm]
\dfrac{x_{ij3} - x_{ij}}{x_{ij3} - x_{ij2}} & (x_{ij2} < x_{ij} \leq x_{ij3}) \\[2mm]
0 & (x_{ij} > x_{ij3})
\end{cases}
\quad (10.3)
$$

$$\mu_{Y_i}(y_i) = \begin{cases} 0 & (y_{ij} \leq y_{ij1}) \\[2ex] \dfrac{y_{ij} - y_{ij1}}{y_{ij2} - y_{ij1}} & (y_{ij1} < y_{ij} \leq y_{ij2}) \\[2ex] \dfrac{y_{ij3} - y_{ij}}{y_{ij3} - y_{ij2}} & (y_{ij2} < y_{ij} \leq y_{ij3}) \\[2ex] 0 & (y_{ij} > y_{ij3}) \end{cases} \quad (10.4)$$

$$\mu_{A_j}(a_j) = \begin{cases} 0 & (a_{ij} \leq a_{ij1}) \\[2ex] \dfrac{a_{ij} - a_{ij1}}{a_{ij2} - a_{ij1}} & (a_{ij1} < a_{ij} \leq a_{ij2}) \\[2ex] \dfrac{a_{ij3} - a_{ij}}{a_{ij3} - a_{ij2}} & (a_{ij2} < a_{ij} \leq a_{ij3}) \\[2ex] 0 & (a_{ij} > a_{ij3}) \end{cases} \quad (10.5)$$

where x_{ij1}, y_{i1}, and a_{j1} are lower values, x_{ij2}, y_{j2}, and a_{j2} are representative values, and x_{ij3}, y_{i3}, and a_{j3} are upper values on the set of real numbers. Clearly, the grades of memberships are zero for the lower and upper values for the fuzzy variables $(X_{ij}, Y_i, and\ a_j)$, and the grades of memberships are one for the representative values for the fuzzy variables $(X_{ij}, Y_i, and\ a_j)$.

Triangular fuzzy numbers such as X_{ij}, Y_i, and a_j are often symbolically represented by

$$\begin{aligned} X_{ij} &= (x_{ij1}, x_{ij2}, x_{ij3}), \\ Y_i &= (y_{i1}, y_{i2}, y_{i3}), \\ A_j &= (a_{j1}, a_{j2}, a_{j3}) \end{aligned} \quad (10.6)$$

For simplicity, we assume that X_{ij} and Y_i are positive triangular fuzzy numbers such that:

$$x_{ij1} > 0, y_{i1} > 0. \quad (10.7)$$

In the following section, firstly, the possibilistic linear regression analysis for both input and output data [20] is reviewed. Then, to explore an alternative way to examine fuzzy input-output data, a fuzzy linear regression analysis using the least square method under linear constraints, in which the fuzzy input-output data and the regression coefficients were represented by the triangular fuzzy numbers, was proposed. The proposed analysis was compared to the possibilistic linear regression analysis proposed by Sakawa and Yano [20] using fuzzy rating data in psychological studies.

Possibilistic Linear Regression Analysis for Fuzzy Input-output Data

In the Sakawa and Yano's [20] original formulation of the possibilistic linear regression analysis for fuzzy input-output data, it was assumed that input - output data and parameters were the symmetric L-L fuzzy numbers. In the following, the formulation of the possibilistic linear regression analysis for fuzzy-input data is slightly generalized by using the asymmetric triangular fuzzy numbers, because the fuzzy rating data can be often represented by asymmetric fuzzy numbers.

In the possibilistic linear regression model, the α-level set of $A \otimes X_i$ in Equation (10.2) can be obtained as follows:

$$(A \otimes X_i)\,\alpha = \left[Z_{i\alpha}^L, Z_{i\alpha}^R \right] \qquad (10.8)$$

where

$$Z_{i\alpha}^L = \sum_{j=0}^{n} \{ \min\{ (a_{j1} + \alpha\,(a_{j2} - a_{j1}))\,(x_{ij1} + \alpha\,(x_{ij2} - x_{ij1})) ,$$

$$(a_{j1} + \alpha\,(a_{j2} - a_{j1}))\,(x_{ij3} - \alpha\,(x_{ij3} - x_{ij2})) \} \},$$

$$Z_{i\alpha}^R = \sum_{j=0}^{n} \{ \max\{ (a_{j3} - \alpha\,(a_{j3} - a_{j2}))\,(X_{ij1} + \alpha\,(X_{ij2} - X_{ij1})) ,$$

$$(a_{j3} - \alpha\,(a_{j3} - a_{j2}))\,(X_{ij3} - \alpha\,(X_{ij3} - X_{ij2})) \} \}.$$

It should be noted that $Z_{i\alpha}^L$ and $Z_{i\alpha}^R, i = 1, \ldots, k$, involve the minimization and maximization operators. In order to deal with these operators in $Z_{i\alpha}^L$ and $Z_{i\alpha}^R, i = 1, \ldots, k$, assume that the following relation hold for any fixed degree α:

$$a_{j1} + \alpha(a_{j2} - a_{j1}) \;\geq\; 0, \;\; j \in J_1,$$
$$a_{j1} + \alpha(a_{j2} - a_{j1}) \;\leq\; 0, \;\; a_{j3} - \alpha(a_{j3} - a_{j2}) \geq 0, \;\; j \in J_2,$$
$$a_{j3} - \alpha(a_{j3} - a_{j2}) \;\leq\; 0, \;\; j \in J_3,$$

where $J = \{0, \ldots, n\} = J_1 \cup J_2 \cup J_3, J_i \cap J_j = \phi, i(\neq j) = 1, 2, 3$.

On the basis of the above assumptions, the following set which depends on the index sets J_1, J_2, and J_3 is defined .

$$L\,(J_1, J_2, J_3) = \Big\{ (\bar{a}_1, \bar{a}_2, \bar{a}_3) \in R^{3(n+1)} \mid a_{j1} + \alpha\,(a_{j2} - a_{j1}) \geq 0,$$

$$a_{j1} \leq a_{j2} \leq a_{j3}, j \in J_1,$$
$$a_{j1} + \alpha\,(a_{j2} - a_{j1}) \leq 0, a_{j3} - \alpha\,(a_{j3} - a_{j2}) \geq 0,$$
$$a_{j1} \leq a_{j2} \leq a_{j3}, j \in J_2,$$
$$a_{j3} - \alpha\,(a_{j3} - a_{j2}) \leq 0,$$
$$a_{j1} \leq a_{j2} \leq a_{j3}, j \in J_3 \Big\} \qquad (10.9)$$

where

$$\bar{a}_1 = (a_{01}, \dots, a_{j1}, \dots, a_{n1})$$
$$\bar{a}_2 = (a_{02}, \dots, a_{j2}, \dots, a_{n2})$$
$$\bar{a}_3 = (a_{03}, \dots, a_{j3}, \dots, a_{n3})$$

Then, $Z_{i\alpha}^L$ and $Z_{i\alpha}^R, i = 1, \dots, k$, can be expressed as

$$
\begin{aligned}
Z_{i\alpha}^L &= \sum_{j \in J_1} (a_{j1} + \alpha \, (a_{j2} - a_{j1})) \, (x_{ij1} + \alpha \, (x_{ij2} - x_{ij1})) \\
&\quad + \sum_{j \in J_2 \cup J_3} (a_{j1} + \alpha \, (a_{j2} - a_{j1})) \, (x_{ij3} - \alpha \, (x_{ij3} - x_{ij2})) , \\
Z_{i\alpha}^R &= \sum_{j \in J_1 \cup J_2} (a_{j3} - \alpha \, (a_{j3} - a_{j2})) \, (x_{ij3} - \alpha \, (x_{ij3} - x_{ij2})) \\
&\quad + \sum_{j \in J_3} (a_{j3} - \alpha \, (a_{j3} - a_{j2})) \, (x_{ij1} + \alpha \, (x_{ij2} - x_{ij1})) ,
\end{aligned}
$$

The possibilistic linear model where both input and output data are fuzzy numbers can be constructed as follows,

$$P(\alpha) : \min J(\bar{a}_1, \bar{a}_2, \bar{a}_3) = \sum_{i=1}^{k} \left\{ (Z_{i1}^L - Z_{i0}^L) + (Z_{i0}^R - Z_{i1}^R) \right\}$$

subject to

$$
\begin{aligned}
&Pos \, (Y_i = a \otimes X_i) \geq \alpha, i = 1, \dots, k, \\
&(\bar{a}_1, \bar{a}_2, \bar{a}_3) \in L \, (J_1, J_2, J_3)
\end{aligned}
\tag{10.10}
$$

where the value α $(0 \leq \alpha \leq 1)$ represents the degree of conformity between the fuzzy out put data, Y_i, and the fuzzy linear regression model, $a \otimes X_i$.

The inequality of problem $P(\alpha)$ can be transformed to the following inequality,

$$Pos \, (Y_i = a \otimes X_i) \geq \alpha, \quad i = 1, \dots, k$$

if and only if

$$
\begin{aligned}
y_{i3} - \alpha \, (y_{i3} - y_{i2}) &\geq Z_{i\alpha}^L, \\
y_{i1} + \alpha \, (y_{i2} - y_{i1}) &\leq Z_{i\alpha}^R
\end{aligned}
\tag{10.11}
$$

Therefore, the problem $P(\alpha)$ can be written as follows,

$$P(\alpha : J_1, J_2, J_3) : \min J(\bar{a}_1, \bar{a}_2, \bar{a}_3) = \sum_{i=1}^{k} \left\{ (Z_{i1}^L - Z_{i0}^L) + (Z_{i0}^R - Z_{i1}^R) \right\}$$

subject to

$$
\begin{aligned}
y_{i3} - \alpha \left(y_{i3} - y_{i2}\right) &\geq Z_{i\alpha}^{L}, \quad i = 1, \ldots, k, \\
y_{i1} + \alpha \left(y_{i2} - y_{i1}\right) &\leq Z_{i\alpha}^{R}, \quad i = 1, \ldots, k \\
a_{j1} + \alpha \left(a_{j2} - a_{j1}\right) &\geq 0, \quad j \in J_1, \\
a_{j1} + \alpha \left(a_{j2} - a_{j1}\right) &\leq 0, \quad a_{j3} - \alpha \left(a_{j3} - a_{j2}\right) \geq 0, \quad j \in J_2, \\
a_{j3} - \alpha \left(a_{j3} - a_{j2}\right) &\leq 0, \quad j \in J_3, \\
a_{j1} \leq a_{j2} &\leq a_{j3}, \quad j = 0, 1, \ldots, n
\end{aligned}
\tag{10.12}
$$

The above problem can be reduced to the linear programming problem with respect to (a_1, a_2, a_3). Therefore, the problem can be solved by the linear programming method.

A Fuzzy Linear Regression Analysis Using the Least Square Method under Linear Constraints

In the proposed analysis of a fuzzy linear regression analysis using the least square under linear constraints, it is assumed that both input-output data and regression coefficients have fuzziness which is represented by the triangular fuzzy number as shown in Equations (10.4), (10.5), and (10.6). This method is more heuristic than Diamond's [6] approach because the present approach is considered to be the natural extension of the ordinary least square method for the crisp data rather than the extension of fuzzy set theory.

In order to estimate fuzzy parameters, we adopt the least square method under linear constraints, i.e., $a_{j1} \leq a_{j2} \leq a_{j3}$ for all j. That is, the problem is to obtain fuzzy parameters $a_j = (a_{j1}, a_{j2}, a_{j3}), j = 1, \ldots, k$, that minimize $F(a_1, a_2, a_3)$ subject to the above constraints. The problem $FP(w_1, w_2, w_3)$ is defined as follows:

$$
\begin{aligned}
FP\left(w_1, w_2, w_3\right) \quad : \quad & \min F\left(\bar{a}_1, \bar{a}_2, \bar{a}_3\right) \\
= \sum_{i=1}^{n} \Bigg\{ & w_1 \left(y_{i1} - \left(a_{01} + \sum_{j=1}^{k} a_{j1} x_{ij1} \right) \right)^2 \\
+ \; & w_2 \left(y_{i2} - \left(a_{02} + \sum_{j=1}^{k} a_{j2} x_{ij2} \right) \right)^2 \\
+ \; & w_3 \left(y_{i3} - \left(a_{03} + \sum_{j=1}^{k} a_{j3} x_{ij3} \right) \right)^2 \Bigg\}
\end{aligned}
\tag{10.13}
$$

subject to

$$
a_{j1} \leq a_{j2} \leq a_{j3}, \quad j = 1, \ldots, k
\tag{10.14}
$$

where w_1, w_2, or w_3 is relative importance of lower value, representative value, and upper value, respectively. In ordinary cases, w_1, w_2, and w_3 should be equal. This minimizing formula as shown in Equation (10.15) can be reduced to the quadratic programming problem with respect to $(\bar{a}_1, \bar{a}_2, \bar{a}_3)$. Therefore, it is easy to solve the problem by the quadratic programming method.

As described the above, the proposed analysis can be interpreted as the natural extension of the ordinary least square regression analysis. Assume that input and output data are crisp numbers.

The problem $FP(w_1, w_2, w_3)$ in Equation (10.13) can be reduced to the following ordinary least square problem.

$$\min F(a_2) = \sum_{i=1}^{n} \left(y_{i2} - \left(a_{02} + \sum_{j=1}^{k} a_{j2} x_{ij2} \right) \right)^2 \qquad (10.15)$$

Application to Psychological Study and Comparison between the Proposed Analysis and the Possibilistic Linear Regression Analysis

Study 3: On the effect of perceived temperature and humidity on an unpleasantness rating

Overview

This study was conducted to examine the effect of perceived temperature and perceived humidity on unpleasantness using the fuzzy rating method to measure vagueness in judgment.

Method

Subjects: Eight adults (Age: 21 - 34 years old).

Procedure: The experiment was conducted in hot midsummer (2th August 1995). Subjects were instructed to rate representative values, lower values, and upper values from 0 to 100 for perceived humidity(0: not humid at all; 100: very humid), perceived temperature(0: not hot at all; 100: very hot), and unpleasantness(0: not unpleasant at all; 100: very unpleasant). For example, a rating of the perceived humidity is 70, 85, 95 for the lower value, the representative value, and the upper value, respectively. In the same way, for example, a rating of the perceived temperature is 30, 40, 55 , and a rating of the unpleasantness is 60, 80, 85. The fuzzy rating data were represented by the triangular fuzzy numbers as shown in Fig.1.

Result of the fuzzy linear regression analysis using the least square method under linear constraints

For evaluating the fitness of the identified fuzzy linear function, there are the following types of measures for the fitness [35, 36, 38].

$$Pos\,(Y_i = a \otimes X_i) = \sup_{x \in R} \min\{\mu_{Y_i}(x), \mu_{A \otimes X_i}(x)\},$$

$$Nes\,(Y_i \supseteq a \otimes X_i) = \inf_{x \in R} \max\{\mu_{Y_i}(x), 1 - \mu_{A \otimes X_i}(x)\},$$

$$Nes\,(Y_i \subseteq a \otimes X_i) = \inf_{x \in R} \max\{1 - \mu_{Y_i}(x), \mu_{A \otimes X_i}(x)\},$$

where $Pos\,(Y_i = A \otimes X_i)$ is the possibility measure concerning the equality, $Y_i = A \otimes X_i$, $Nes\,(Y_i \supseteq A \otimes X_i)$ is the necessity measure concerning the relation, $Y_i \supseteq A \otimes X_i$, $Nes\,(Y_i \subseteq A \otimes X_i)$ the necessity measure concerning the relation, $Y_i \subseteq A \otimes X_i$, $\mu_{Y_i}(x)$ is a membership function of the observed dependent data, Y_i, and $\mu_{A \otimes X_i}(x)$ is a membership function of the predicted model $A \otimes X_i$. The values of the all measure ranges from 0 to 1.

In this study, the possibility measure concerning the predicted fuzzy variable and dependent fuzzy variable for each observation was used in order to compare with the results of the possibilistic linear regression analysis which maximize the possibility measure under the certain constraints. The possibility measure is considered to be useful for contrasting with the possibilistic linear regression analysis because this value of possibility is the same as the criterion for the possibilistic linear regression analysis.

The mean values of the possibility and the regression coefficients for the proposed analysis were shown in Table 10.9. The mean value of $Pos\,(Y_i = a \otimes X_i)$ was .65 for the inside room, and .74 for the outside. The estimated parameter was (.37, .46, .53) for the perceived temperature, and (.36, .43, .53) for perceived humidity in the room, respectively. Outside the room, the parameter was (.70 ,.78, .86.) for the perceived temperature, and (.51, .60, .62) for perceived humidity. The mean value of the predicted variable was (29.39, 37,73, 69.60) for the inside room, and (59.90, .73.54, .83.88) for the outside , respectively, whereas the mean value of the dependent variable was (30.63, 37.75, 46.38) for the inside room, and (63.87, .73.38, .80.62) for the outside. As shown in the result, the fitness defined by the possibility measure tended to be reasonably high.

Comparison to the possibilistic linear regression analysis

We analyzed the data by the possibilistic linear regression analysis method, which is the most popular version of fuzzy linear regression analysis. For the analysis, we manipulated α level as a threshold for minimum possibility from 0.5 to 0.1 (That is, 0.5, 0.4, 0.3, 0.2, 0.1). As a result, when the α

TABLE 10.9. Mean Values of the Regression Coefficients and the Possibilities for the Proposed Analysis and the Possibilistic Linear Regression Analysis in Study 1.

Model of Analysis	Inside the Room	
	The Proposed Analysis	Possibilistic Linear Regression Analysis
Regression Coefficients		
Perceived Temperature	(.37, .46, .53)	− (−.02, .08, .38)
Perceived Humidity	(.36, .43, .53)	(.85, .85, .85)
Possibility Measure	.65	.57
Model of Analysis	Outside the Room	
	The Proposed Analysis	Possibilistic Linear Regression Analysis
Regression Coefficients		
Perceived Temperature	(.70, .78, .86)	(.95, .95, .95)
Perceived Humidity	(.51, .60, .62)	(.00, .00, .00)
Possibility Measure	.74	.52

level was less than 0.2, meaningful parameters were obtained. when the α level was greater than 0.2, no meaningful parameters were obtained. Thus, we adopted the parameters in the case where α level equals 0.2.

The mean values of the possibility and the regression coefficients for the possibilistic linear regression analysis in comparison with the proposed analysis were shown in Table 10.9. The mean value of $Pos\,(Y_i = a \otimes X_i)$ was .57 for the inside room, and .52 for the outside. The estimated parameter was -(-.02, .08, .38) for the perceived temperature, and (.85, .85, .85) for perceived humidity in the room, respectively. Outside the room, the parameter was (.95 ,.96, .96.) for the perceived temperature, and (.00, .00, .00) for perceived humidity. The mean value of the predicted variable was (11.12, 28.30, 42.57) for the inside room, and (55.68, 64.06, 70.53) for the outside , respectively, whereas the mean value of the dependent variable was (30.63, 37.75, 46.38) for the inside room, and (63.87, .73.38, .80.62) for the outside.

The fitness of the model tended to be reasonably high. However, as shown in the results of Table 10.9, the width between the upper and lower values of the predicted model by possibilistic linear regression analysis was not nearer to the width of the dependent variable than that of the proposed fuzzy linear regression analysis. It was also found that the representative value of the predicted value for the proposed analysis was also nearer to that of the dependent variable compared with that of the possibilistic linear regression analysis.

For the both analyses, the spreads of the estimated fuzzy parameters were very small, and the paramerter were near to crisp numbers. The results indicate that the weight of attributes can approximate to crisp numbers.

Study 4 Consumer attitudes research

Overview

This study is conducted to examine the effect of multi-attribute attitudes on behavioral intention using fuzzy rating method to measure vagueness in judgment and decision making.

Method

Subjects: One hundred and seventy-seven male and female adults who were living in Taiwan.

Procedure: Subjects were instructed to rate representative values, lower values, and upper values within 0 and 100 for attitude scales and behavioral intention of shopping behavior for fashion goods by indicating the points on the scales in the questionnaire. The scales were as follows: (1) traffic access to the store, (2) price of goods, (3)variety of goods in the store, (4) customer service, (5) quality of the goods, (6)atmosphere of the store, (7)environment of the store, (8) perceived value of buying goods in the store. The former seven items were independent variables, and the last item was a dependent variable. These scales were extracted from a pilot study for forty-two men and women. The rating data were represented by the triangular fuzzy numbers as same as in the study 3.

Result of the fuzzy linear regression analysis using the least square method under linear constraints

We computed a value of possibility for intersection between the predicted fuzzy variable and the fuzzy dependent variable as fitness of the model for each observation. That is, we defined the possibility measure $Pos(Y_i = A \otimes X_i)$ as the same manner in the study 3.

The mean values of the possibility and the regression coefficients for the proposed analysis were shown in Table 10.10. The mean value of $Pos(Y_i = A \otimes X_i)$ was .91. The estimated parameter was (1) (.24, .40, .44) for the traffic access, (2) (.27, .27, .30) for the price of goods, (3) (.44, .44, .48) for the variety of goods in the store, (4)-(.28, .43, .43) for customer service, (5)-(.02, .02, .02) for the quality of the goods, (6) (.09, .55, .55) for the atmosphere of the store, (7)-(.20, .39, .39) for the environment of the store respectively. The mean value of the predicted variable was (23.02, 60.10, 88.04), whereas the mean value of the dependent variable was (53.79, 60.38, 66.76). As shown in the results of Table 10.10, the fitness defined by the possibility measure for the proposed analysis tended to be reasonably high.

TABLE 10.10. Mean Values of the Regression Coefficients and the Possibilities for the Proposed Analysis and the Possibilistic Linear Regression Analysis in Study 2.

Model of Analysis		The Proposed Analysis	Possibilistic Linear Regression Analysis
Regression Coefficients	Traffic Access	(.24, .40, .44)	(.08, .08, .08)
	Price	(.27, .27, .30)	−(.08, .44, .44)
	Variety	(.44, .44, .48)	(.06, .39, .39)
	Service	−(.28, .43, .43)	(.10, .10, .10)
	Quality	−(.02, .02, .02)	(.00, .00, .00)
	Atmosphere	(.09, .55, .55)	(.60, .60, .60)
	Environment	−(.20, .39, .39)	(.00, .00, .00)
Possibility Measure		.91	.84

Comparison to the possibilistic linear regression analysis

The data were analyzed by using the possibilistic linear regression analysis, which is the most popular version of fuzzy linear regression analysis. For the analysis, we manipulated α level as a threshold for minimum possibility from 0.5 to 0.1 (That is, 0.5, 0.4, 0.3, 0.2, 0.1). As a result, when the α level was less than 0.5, meaningful parameters were obtained. Thus, we adopted the parameters in the case where α level equals 0.5.

The mean values of the possibility and the regression coefficients for the possibilistic linear regression analysis in comparison with the proposed analysis were shown in Table 10.10. The mean value of $Pos\,(Y_i = a \otimes X_i)$ was .84. The estimated parameter was (1) (.08,.08,.08) for the traffic access, (2) -(-.08, .44, .44) for the price of goods, (3) (.06, .39, .39) for the variety of goods, (4)(.10, .10, .10) for the customer service (5) (.00, .00, .00) for the quality of the goods, (6) (.60, .60, .60) for the atmosphere of the store, and (7) (.00, .00, .00) for the environment of the store, respectively. The mean value of the predicted variable was (-42.00, 27.24, 132.35), whereas the mean value of the dependent variable was (53.79, 60.38, 66.76).

The fitness of the model tended to be reasonably high. Moreover, in the possibilistic linear regression model, we may have larger possibility grades by assigning larger values (e.g., 0.8) to α. Of course, this may have larger widths of the fuzzy parameters. As shown in the results of Table 10.10, the width between the upper and lower values of the predicted model by possibilistic linear regression analysis was not nearer to the width of the dependent variable than that of the proposed fuzzy linear regression analysis. It was also found that the representative value of the predicted value by the proposed analysis was also nearer to that of the dependent variable than that of the possibilistic linear regression analysis.

The results of the both analyses indicated that the spreads of the estimated fuzzy parameters were very small, and the parameters were near to crisp numbers. This property is the same as the finding in study 3, which suggests that the weights of attributes can approximate to crisp numbers.

10.6   Conclusion

In this chapter, the concepts and the methods to investigate the vagueness in human judgment and decision making were introduced and discussed. Firstly, several theoretical approaches based on probability theory, fuzzy measure theory and fuzzy set theory were evaluated in order to represent the vagueness of judgment and decision making. It was concluded that the concept of the fuzzy set is more appropriate for representing the vagueness in judgment and decision making.

In order to measure the vagueness of judgment and decision making by using fuzzy set theory, the fuzzy rating method [10] was introduced. By using the fuzzy rating method, psychological experimental studies of the vague integration were undertaken. The result of the present study indicates that the width of judgment is not positively proportional to the standard deviation. It can be concluded that the traditional interpretation of the standard deviation as the indicator of vagueness in judgment and decision making is not adequate, and that the fuzzy rating method is more effective for measuring vagueness in judgment and decision making. The results also indicated that the width of the integrated information was amplified especially when the most ambiguous information was presented.

Lastly, the method of analysis for the fuzzy rating data is described through a comparison between the possibilistic linear regression analysis and the fuzzy regression analysis using the least square method. The fuzzy linear regression analysis using the least square method under linear constraints, where input data, output data, and the regression coefficients were represented by triangular fuzzy numbers the shapes of which are basically asymmetric, was introduced. The solutions of estimated parameters can be easily obtained by the ordinary quadratic programming method.

The psychological study was undertaken in order to examine the effects of fuzzy independent variables on fuzzy dependent variable using fuzzy rating method. The results of these applications to psychological study indicated that the parameter estimation method for the proposed analysis effectively explains the fuzzy dependent variable.

The fuzzy regression analysis using the least square was compared to the possibilistic linear regression analysis using fuzzy rating data in the psychological studies. The major finding of the comparison were as follows: (1) Under the fuzzy linear regression analysis using the least square method, the width between the upper and lower values of the predicted model was nearer to the width of the dependent variable than that of the possibilistic linear regression analysis, (2) The representative value of the predicted variable by the fuzzy linear regression analysis using the least square method was also nearer to that of the dependent variable compared with that of the possibilistic linear regression analysis, (3) For the both analyses, the estimated weights of the attributes (fuzzy parameters) were nearer to crisp numbers rather than fuzzy numbers with larger spreads. However, these findings might be restricted on the rating data of the present studies. Further research which examines the validity of the several methods for analyzing the fuzzy rating data will be needed.

Acknowledgment

The author would like to thank Prof.S. Miyamoto, Prof.T.O nisawa, Mr.K. Umayahara, and Mr.D.Wakayama for their helpful comments, and, Mr.T. Ho, Mr.J. Wo, and Mr. T. Oshiyama for their assistance. This study was supported by projects of TARA (Tsukuba Advanced Research Alliance) in the University of Tsukuba, Yoshida Memorial Foundation, and the Ministry of Education, Science, Sports, and Culture, Japanese Government.

10.7 REFERENCES

[1] M. Allais, "Le comportement de l'homme rationnel devant le risque, critique des postulates et axiomes de I'école américaine," *Econometrica*, Vol.21, pp.503-546, 1953.

[2] N.H. Anderson, "A functional approach to person cognition," In T.K. Srull and R.S. Wyer (Eds.), *Advances in social cognition*, vol.1, Hiisdale, New Jersey: Lawrence Erlbaum Associates, pp.37-51, 1988.

[3] R. Beyth-Marom, "How probable is probable?: Numerical translation of verbal probability expressions," *Journal of Forecasting*, Vol.1, pp.267-269, 1982.

[4] G. Choquet, "Theory of capacities," *Annales de L'Institut Fourier*, Vol.5, pp.131-295, 1953.

[5] R.M. Dawes, "The robust beauty of improper linear models in decision making," Kahneman, P.Slovic, and A. Tversky (Eds.), *Judgment under uncertainty: Heuristics and biases*, Cambridge: Cambridge University Press, pp.391-407, 1982.

[6] P. Diamond, "Fuzzy least squares," *Information Sciences*, Vol.46, pp.141-157, 1988.

[7] D. Dubois and H. Prade, *Fuzzy sets and systems: Theory and applications*, New York: Academic Press, 1980.

[8] D. Ellsberg, "Risk, ambiguity, and the Savage axiom," *Quaterly Journal of Economics*, Vol.75, pp.643-669, 1961.

[9] P.C. Fishburn, *Nonlinear preference and utility theory*, Baltimore, MD: The Johns Hopkins University Press, 1988.

[10] B. Hesketh, R. Pryor, M. Gleitman, and T. Hesketh, "Practical applications and psychometric evaluation of a computerised fuzzy graphic rating scale," in T.Zetenyi, (Ed.), *Fuzzy sets in psychology*, New York: North-Holland, pp.425-454, 1988.

[11] D. Kahneman and A.Tversky, "Prospect theory: An analysis of decision under risk," *Econometrica*, 47, pp.263-291, 1979.

[12] A. Kaufman and M.M.Gupta, *Introduction to fuzzy arithmetic: Theory and applications*, New York: Van Nostrand Reinhold, 1985.

[13] L.R. Keller, "Properties of utility theories and related empirical phenomena," in W. Edwards(Ed.), *Utility theories: Measurements and applications*, Boston: Kluwer Academic Publishers, pp.3-23, 1992.

[14] M. Mizumoto, "Fuzzy number," in Japan Society for Fuzzy Theory and Systems, (Ed.), *Fuzzy sets*, Nikkan Kohgyo Shinbunsha, (in Japanese) pp.157-185, 1992.

[15] T. Murofushi and M. Sugeno, "Fuzzy t-conorm integral with respect to fuzzy measures: Generalization of Sugeno integral and the Choquet integral," *Fuzzy set and Systems*, Vol.42, pp.57-71, 1991.

[16] T. Murofushi, M. Sugeno, and M. Machida, "Nonmonotonic fuzzy measures and Choquet integral," *Fuzzy Set and Systems*, Vol.64, pp.73-86, 1994.

[17] K. Nakamura, "On the nature of intransitivity in human preferential judgments," in V. Novak, J.Ramik, M.Mares, M.Cherny, and J.Nekola (Eds.), *Fuzzy Approach to Reasoning and Decision-Making*, Dordrecht: Kluwer Academic Publishers, pp.147-162, 1992.

[18] E. Rosch, "Cognitive representation of semantic categories," *Journal of Experimental Psychology: General*, Vol.104, pp.192-233, 1975.

[19] E. Rosch, and C.B. Mervis, "Family resemblances: Studies in the internal structure of categories," *Cognitive Psychology*, Vol.7, pp.573-603, 1975.

[20] M. Sakawa and H.Yano, "Multiobjective fuzzy linear regression analysis for fuzzy input-output data," *Fuzzy Sets and Systems*, Vol.47, pp.173-181, 1992.

[21] I.R. Savage, *The foundations of statistics*, New York: Wiley, 1954.

[22] D. Schmeidler, "Subjective probability and expected utility without additivity," *Econometrica*, Vol.57, pp.571-587, 1989.

[23] F. Seo and I. Nishizaki, "Fuzzy multiattribute utility functions," *Discussion Paper, No.357, Kyoto Institute of Economic Research*, Kyoto University, 1992.

[24] M. Sherif and C.I. Hovland, *Social judgment: Assimilation and contrast effects in communication and attitude change*, New Haven: Yale University Press, 1961.

[25] M. Smithson, *Fuzzy set analysis for behavioral sciences*, New York: Springer-Verlag, 1987.

[26] M.Smithson, *Ignorance and uncertainty*, New York: Springer-Verlag, 1989.

[27] M.Sugeno, "Theory of fuzzy integrals and its applications," *Unpublished Ph.D. Thesis*, Tokyo: Tokyo Institute of Technology, 1974.

[28] M. Sugeno, *Fuzzy control*, (in Japanese) Nikkan Kogyo, 1988.

[29] K. Takemura, "Reconsideration on concept of the attitude," *Bulletin of Koka Woman's Junior College*, (in Japanese) Vol.28, pp.119-132, 1990.

[30] K. Takemura, "A vague information integration model: An application of fuzzy set theory to social psychology," *Bulletin of Koka Woman's Junior College*, Vol.29, pp.91-107, 1991 (in Japanese).

[31] K. Takemura, "An analysis of shopping choice behavior using fuzzy multiattribute attitude model: A proposal of a new psychological method for the area marketing," *Japanese Studies in Regional Science*, (in Japanese with English abstracts) Vol.22, pp.119-131, 1992.

[32] K. Takemura, "Prediction of behavioral intention by fuzzy multiattribute attitude model: An application of possibilistic linear regression analysis to fuzzy rating data," *Procedings of the 22th Annual Conference of the Behaviormetric Society of Japan*, (in Japanese) Tsukuba, Japan, pp.122-125, 1994.

[33] K. Takemura, *Psychology of Decision Making*, (in Japansese) Tokyo:Fukumura syuppan, 1996.

[34] K. Takemura, T. Hou, and J. Wo, "Study on consumer's multiattribute attitude in Taiwan: using possibilistic linear regression analysis for fuzzy rating data," *Proceedings of the 11 th Fuzzy System Symposium* (in Japanese), 1995, Okinawa, Japan, pp.413-416.

[35] H. Tanaka, "Fuzzy data analysis by possibilistic linear model," *Fuzzy Sets and Systems*, Vol.24, pp.363-375, 1987.

[36] H. Tanaka and J. Watada, "Possibilistic linear systems and their application to the linear regression model," *Fuzzy Sets and Systems*, Vol.27, pp.275-289, 1988.

[37] H.Tanaka, S.Uejima, and K.Asai, "Linear regression analysis with fuzzy model," *IEEE Transactions on Systems Man Cybernetics*, Vol.12, pp.903-907, 1982.

[38] H. Tanaka, I.Hayashi, and K.Nagasaka, "Interval regression analysis by possibilistic measures," *The Japanese Journal of Behaviormetrics*, (in Japanese) Vol.16, No.1, pp.1-7, 1988 .

[39] A. Tversky and D. Kahneman, "Extensional versus intuitive reasoning: The conjunction fallacy in probability judgment," *Psychological Review*, Vol.90, pp.293-315, 1983.

[40] A. Tversky and D. Kahneman, "Advances in prospect theory: Cumulative representation of uncertainty," *Journal of Risk and Uncertainty*, Vol.5, pp.297-323, 1992.

[41] A. Tversky, S. Sattath, and P. Slovic, "Contingent weighting in judgment and choice," *Psychological Review*, Vol.95, pp.371-384, 1988.

[42] J. von Neumann and O. Morgenstern, *Theory of Games and Economic Eehavior*, Princeton, NJ: Princeton University Press, 2nd ed, 1947.

[43] L. Wittgenstein, *Philosophical Investigations*, New York:MacMillan, 1953.

[44] J.F. Yates, *Judgment and Decision Making*, Englewood Cliffs, New Jersey: Prentice-Hall, 1989.

[45] A. Zadeh, "Fuzzy sets," *Information and Control*, Vol.8, pp.338-353, 1965.

[46] A. Zadeh, "Outline of a new approach to the analysis of complex systems and decison processes," *IEEE Transactions on Systems, Man and Cybernetics*, Vol.3, No.1, pp.28-44, 1973.

[47] A. Zadeh, "The concept of a linguistic variable and its application to approximate reasoning," *Information Sciences*, Vol.8, pp.199-249, 1975.

[48] R. Zwick, D.V. Budescu, and T.S. Wallsten, "An empirical study of the integration of linguistic probabilities," in T. Zetenyi (Ed.), *Fuzzy sets in psychology*, Amsterdam: North-Holland, pp.91-125, 1988.

11

Chaos and Time Series Analysis

Tohru Ikeguchi
Tadashi Iokibe
Kazuyuki Aihara

11.1   Introduction

Researches on deterministic chaos have been rapidly progressing during the last two decades and our understanding on low-dimensional chaos has been considerably deepened. Theoretical and numerical analyses have shown that a simple deterministic nonlinear system with a few degrees of freedom can naturally produce very complicated chaotic behavior. In addition, it has been reported that there have been discovered many experimental data that imply the presence of low dimensional chaos in various real-world systems.

As characteristics of deterministic chaos have been progressively revealed, these characteristics have affected on traditional time series analysis and modeling.

Namely, the theory of deterministic chaos motivates a new type of time series analysis based on possibility that even if an observed time series appears irregular, it may be actually produced from a nonlinear dynamical system with a few degrees of freedom, namely the phenomenon is actually governed by a deterministic rule. This concept is far different from the old paradigm that a system that produces a complicated response has many degrees of freedom, and conversely, a system with few degrees of freedom cannot produce a complicated response. Thus, we have been reexamining many time series data that had been discarded conventionally for the reasons that an experimental equipment did not work well, or phenomenon under study are really stochastic, irregular variations, even though a chaotic response, as it would now be called, was observed.

It is an intriguing question to ask whether or not given time series data are produced from deterministic chaos. To answer the question, one of the best methods is to observe bifurcation phenomena associated with a change of the system parameters, that is, to confirm the existence of a route to chaos. However, it is almost impossible to change parameter values in many interesting systems, such as biological, social and economic

Tohru Ikeguchi, Tadashi Iokibe, Kazuyuki Aihara

systems. Moreover it might actually be quite difficult to retry phenomena in these systems because of their nonstationarity. In addition, there are various kinds of noise superimposed into signals as well as finite length of observation and accuracy of measurements.

Then, our goal here is how to characterize chaos and how to produce appropriate nonlinear models based solely on observed time series data obtained under the limitations and conditions described above.

11.2   Embedding Time Series Data

When deterministic chaos is observed as time series data, its temporal behavior is so complicated that it appears impossible to discriminate chaos from random data. However, transforming the time series to a reconstructed state space with a sufficiently large dimension guarantees at least theoretically reconstructing of dynamical systems by the embedding theorems [41, 39].

When irregular time series data are to be analyzed from the viewpoint of deterministic chaos, the first step is this reconstruction of an attractor of an underlying dynamical system. One of the most popular methods of reconstructing attractors is to transform observed time series data to a time delayed coordinate system. The embedding theorems by Takens [41] and Sauer *et al.* [39] show that the transformation is embedding, under some assumptions.

For example, let us consider a k-dimensional discrete time dynamical system described as follows:

$$x_{t+1} = f_\mu(x_t) \qquad (11.1)$$

where x_t is a k-dimensional system state at time t, f_μ is a k-dimensional nonlinear function, and μ is a parameter vector of the system.

An attractor of a dissipative dynamical system is defined by its asymptotic behavior in the k-dimensional state space after the transient behavior has disappeared. Attractors of a dynamical system can be classified into (i) equilibrium points, (ii) limit cycles, (iii) tori, and (iv) strange attractors (chaos).

Usually, all the state variables x cannot be observed. Instead, it is often that only a single variable time series y can be observed through a measuring equipment. In the real world, perturbations called observation noise and dynamic noise are also included in measurements, then the relationship between a dynamical system (unknown) and an observed time series (known) in real environment is described as follows:

$$x_{t+1} = f_\mu(x_t) + \eta_t, \qquad (11.2)$$
$$y_t = g(x_t) + \xi_t, \qquad (11.3)$$

where η_t and ξ_t are additive dynamical noise and observation noise, respectively; the function g represents a measuring function.

Let us assume that time series data are observed at least for a single variable in the above situation. Then, the important and essential question is: can we really recover information concerning the hidden dynamical system in the original state space only from the observed time series y_t? The answer is yes and guaranteed by reconstruction of attractors. For example, the following m-dimensional vectors can be constructed using a time delay coordinate system based on time differences with a constant time delay:

$$v_t = (y_t, y_{t+\tau}, \ldots, y_{t+(m-1)\tau}). \tag{11.4}$$

When the dimension m of the reconstructed state space is sufficiently large, the transformation from an original state space to the reconstructed state space using Equation (11.4) with the observed time series is guaranteed to be embedding. In other words, even if the original dynamical system cannot be directly known, its topological and differential structure can be reconstructed [41, 39]. In the following, an extended theorem of the Takens theorem [41] proved by Sauer *et al.* [39] is described.

Theorem 11.2.1 (Embedding theorem) *[39] Let f be a diffeomorphism in an open subset U of $\mathbf{R}^k$, A which is assumed to be an attractor be a compact subset of U, D_0 be the box count dimension of A. Let assume that $m > 2D_0$, that for all positive integers $p \leq m$, the box count dimension of the set A_p of periodic points with the period p is smaller than $p/2$, and that the Jacobian matrix Df^p for each of these periodic orbits has different eigenvalues.*

Then, for almost every C^1 class function g on U, the transformation of Equation (11.4) is:

(I) one–to–one on A,

(II) an immersion on each compact subset of a smooth manifold contained in A.

Here, we also assume that there are no noise terms η_t and ξ_t in Equations (11.2) and (11.3).

In Figure 11.1, we show the schematic representation of the relation between an original manifold, a measuring function (readout map), the delay coordinate, and a reconstructed manifold.

It is very important that the transformation v be embedding, since we can make apply nonlinear prediction algorithms to estimate the future behavior of the time series. Moreover, we can estimate Lyapunov exponents, since a differential structure is preserved through the transformation of Equation (11.4).

In these embedding theorems [41, 39], the value of τ is arbitrary except for very special cases. However, when treating practical data, it is also

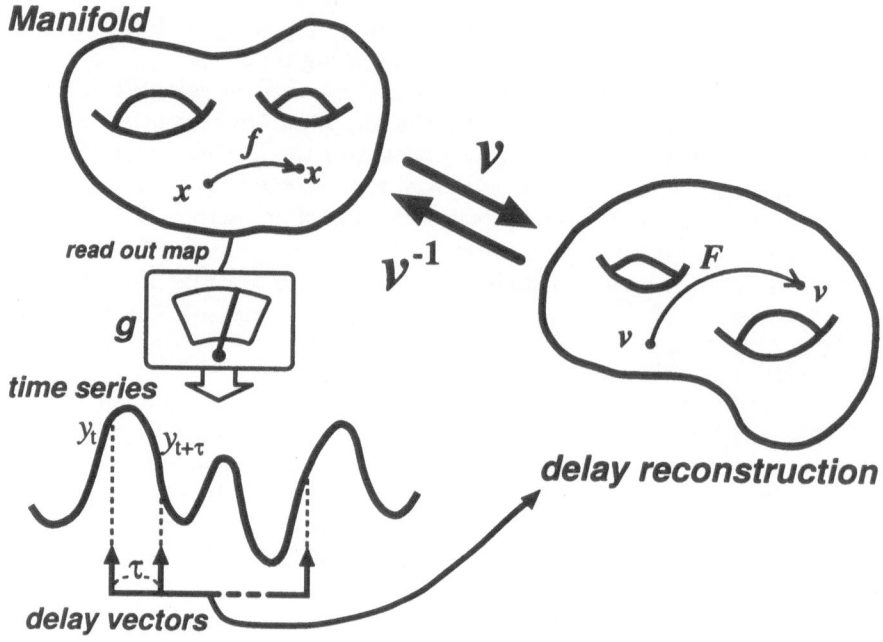

FIGURE 11.1. Schematic representation of the embedding theory.

necessary to set an appropriate time delay value. Many criteria have been proposed to determine appropriate time delays [20]. Recently, Judd and Mees showed that the method called variable embedding is effective where the dimensions of the state space and time delays are adaptively changed depending on the local structure of the attractors [26].

11.3   Deterministic Nonlinear Prediction

One of the important characteristics of chaos is a sensitive dependence on initial conditions and orbital instability and long-term unpredictability. In fact, it is essentially impossible to predict the future of a chaotic system for a long term. This characteristic of chaos is called long-term unpredictability. However, since there is a deterministic rule for any chaos, its complicated phenomenon could be predicted satisfactorily for a short term with a good nonlinear model. Moreover, if we can extract both short-term predictability and long-term unpredictability from the time series, it would be a good signature for deterministic chaos [40]. Modeling and predicting methods using these concepts are based on the assumption that the systems are deterministic dynamical systems. Thus, we call the methods "deterministic" nonlinear prediction.

In a history of the study of chaotic dynamical systems, Lorenz [29] is a pioneer on the application of predicting chaotic time series. Lorenz applied the method of "analogues" for predicting weather data. Unfortunately this trial did not succeed because of high dimensionality of weather dynamics and existence of noise. However, it was the first attempt to apply nonlinear prediction to possibly chaotic data on the basis of a local linear model.

Since a chaotic time series comes from a nonlinear system, the time series cannot be well predicted by a conventional linear model, such as ARMA model [4, 3]. Then, in the statistical literature, various nonlinear modeling methods have been studied. For example, Tong and Lim proposed a threshold autoregressive (TAR) model where a space is divided into several regions by a threshold, and an AR model is allocated to each region [44, 45]. This method is a very simple example of piece-wise linear modeling, and it should be noted that piece-wise linear modeling is one of the typical nonlinear modeling.

M. B. Priestley proposed state-dependent model (SDM) which include TAR, the bilinear model and the exponential AR model as special cases [35]. Okabe et al. also proposed a nonlinear prediction method based on the KM_2O-Langevin equation theory [33].

In the study of deterministic nonlinear dynamical systems, Sano and Sawada [37], Eckmann, Ruelle [12, 11], et al. used local linear models to estimate Lyapunov exponents. Farmer and Sidorowich proposed a prediction method with a local linear model and studied scaling properties of prediction errors [13, 14].

M. Casdagli applied radial basis functions (RBF) [34] to nonlinear prediction in 1989 [10]. D. Broomhead et al. also studied general functional interpolation problems including the prediction of chaotic time series [6]. In addition, Mees proposed an improvement in interpolation performance by adding a linear term to the RBF [32]. Recently, Judd and Mees discussed how to optimize the selection of the radial basis functions using an MDL criterion [25, 30].

The RBF method interpolates given data, on the basis of a global function obtained by superimposing many radial basic functions. Usually, localized functions are used as the radial basic functions. On the other hand, Lapedes et al. proposed a prediction method based on a neural network with back-propagation (BP) learning, in which data are interpolated by superposition of sigmoid functions [27]. Weigend et al. and Wolpert et al. also proposed a neural network method to solve this issue [49, 50].

Sugihara and May used a method called simplex projection in 1990 [40]; a state space is triangulated by simplexes with vertices defined by points on a reconstructed attractor. In 1991, Mees proposed a prediction method constructing a Voronoi tessellation which is dual to triangulation [31].

Other methods have also been proposed so far which include a local nonlinear approximation that uses higher order terms of Jacobian series [14, 5, 8, 7], a method with locally orthogonal coordinates [28, 23], an application of association memory [24], wavelets [9], a low pass embedding [38], and a local fuzzy reconstruction method [22]. It is also possible to use methods such as RBF and BP networks, not globally but locally. In Table 11.1, we summarize the above discussions from the view points of global versus local modelings, and liner versus nonlinear modelings.

When predicting for p-steps ahead, we have basically two options. That is, we can either predict a value p steps ahead directly, or repeat predicting a value one step ahead for p times. The orbital instability means that an error is gradually increased at each iteration, then, this feature can be extracted by the latter approach of recursive predictions. Which prediction strategy gives higher accuracy depends on a trade-off between the intrinsic orbital instability of a dynamical system f that generated the time series and the complexity of the function f^p of p-time iterated transformation of f.

11.4   Analysis of Complicated Time Series by Deterministic Nonlinear Prediction

As long as time series is deterministic chaos, a short-term prediction is possible although a long-term one is fundamentally not. Then, if we analyze the relation between prediction steps p and the prediction performance such as correlation coefficients, the decreasing property of prediction performance with increasing prediction steps may indicate possible existence of deterministic chaos. Noiseless periodic data can be predicted completely, then the predictive performance does not depend on prediction steps p. If the periodic data include noise, the noise has an effect of decreasing the overall prediction accuracy, so the relationship between prediction steps p and correlation coefficients is still almost independent of p and remains nearly constant with increasing p. If data are white noise, there exists no deterministic dynamics, so the prediction accuracy is 0 for any p (Figure 11.2).

In addition, since orbital instability leads long-term unpredictability of deterministic chaos, the largest Lyapunov exponent can be estimated by the slope of the relation between prediction accuracy and prediction steps [48].

Based on these principles, Sugihara and May analyzed time series data of the number of patients of infectious diseases using a nonlinear prediction method, and revealed that measles shows chaotic behavior and chickenpox has the characteristics of noisy periodicity [40].

Although the method of nonlinear modeling and prediction is effective to characterize complicated time series data from the view point of chaotic dynamical systems, the application should be careful for colored noise. When a nonlinear prediction method is applied to colored noise of fractional Brownian motion, the prediction accuracy shows decreasing trends with increasing of predicting steps, like deterministic chaos. Tsonis *et al.* revealed that a difference can be detected by plotting and comparing the prediction

TABLE 11.1. Classification of prediction schemes.

	Linear	
Global	ARMA etc	Box & Jenkins, 1970 Akaike, 1972 (AIC)
Local	Method of analogue	Lorenz, 1969
	TAR	Tong, 1978 Tong & Lim, 1980
	SDM	Priestly, 1980
	Jacobian	Sano & Sawada, 1985 Eckmann et al.,1985,86 Farmer & Sidorowich, 1987
	triangulation	Sugihara & May, 1990 Mees, 1991
	tessellation	Mees, 1991
	low pass embedding	Sauer, 1993
	Nonlinear	
Global	RBF	Casdagli, 1989 Broomhead et al., 1988 Mees et al.,1991 (affine terms) Mees, 1993 (BIC) Judd & Mees, 1995 (MDL)
	BP	Lapedes et al., 1990 Weigend et al., 1990 Wolpert et al., 1990
	KM_2O-Langevin eqs.	Okabe et al., 1995
Local	higher order Jacobian	Brown et al., 1990 Bryant et al, 1990 Abarbanel et al., 1991 Briggs, 1992
	RBF	Smith, 1993
	local fuzzy reconstruction	Iokibe et al, 1995

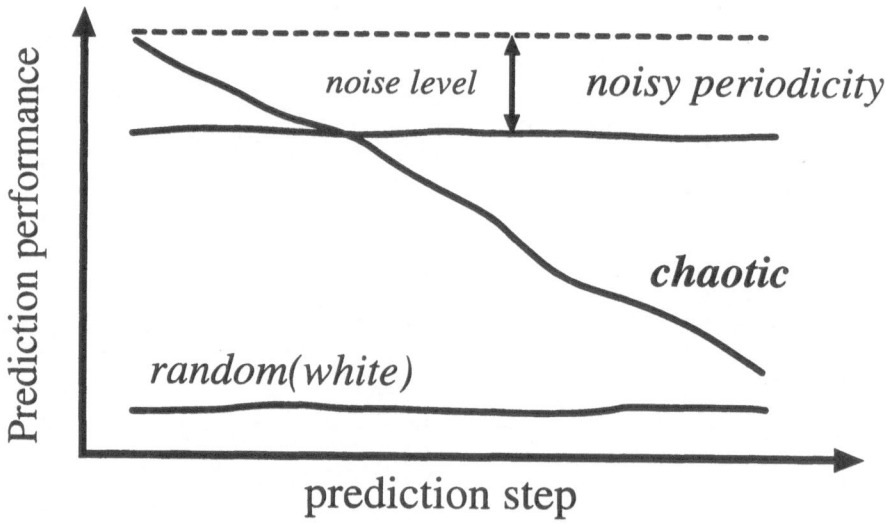

FIGURE 11.2. Discrimination of deterministic chaos, random noise and noisy periodicity by nonlinear prediction [40].

accuracy on log-log and semi-log scales with prediction steps [47]. In the case of deterministic chaos, prediction errors increase exponentially with prediction time. In the case of fractional Brownian motion (fBm), on the other hand, the property becomes a function of power of time.

However, it should be noted that deciding whether or not a slope is straight might become subjective to some extent, and also that no solution has been obtained for colored noise in general. It has also been shown that the above problem may be efficiently solved by difference correlation coefficients as a measure of evaluating prediction accuracy, as well as conventional correlation coefficients [17].

11.5   Engineering Applications of Deterministic Nonlinear Prediction

Short-term prediction is an important topic in many practical fields of engineering such as demand forecasts for electric power, gas, water, etc., forecasts of local weather, prediction of financial matters such as stock prices and foreign currency exchange rates, and sales of commodities.

Various methods have been developed to increase prediction accuracy of such data, as described before. A typical model is the ARMA model that is also applicable to time series data with seasonal variations. However, conventional linear prediction methods including this model are not appropriate for nonlinear data such as deterministic chaos.

An engineering application of the nonlinear time series analysis is how to predict time series data that have a chaotic behavior, in the short term, as described before. This is based on the principle that if the behavior of observed time series data is chaotic and its dynamics can be identified, the behavior of the data can be predicted up to the time at which its determinism is lost.

For example, the local fuzzy reconstruction and other methods [21, 22] are studied for applications to practical cases as follows:

1. Traffic volume on a highway (Figure 11.3)

2. Peak electric power demand (Figure 11.4)

3. Local weather information (Figure 11.5)

4. Inflow to a hydro power station (Figure 11.6)

These figures show results of one step ahead prediction.

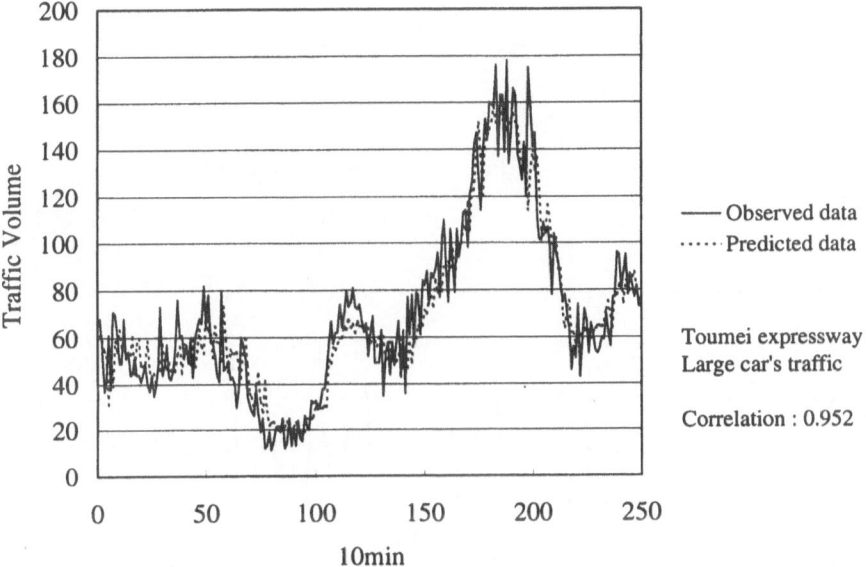

FIGURE 11.3. Traffic volume on a highway.

The nonlinear prediction method uses no other information than the observed time series data. Nevertheless, the method performs as well as or even better in some case than conventional prediction methods based on a linear stochastic process. Moreover, every new data observed, attractors in reconstructed state spaces are updated, which fact means that the dynamics of the time series data can be learned automatically. If the dynamics change considerably due to for example environmental fluctuation, prediction accuracy may become low. But, this can be compensated to some extent by updating the model.

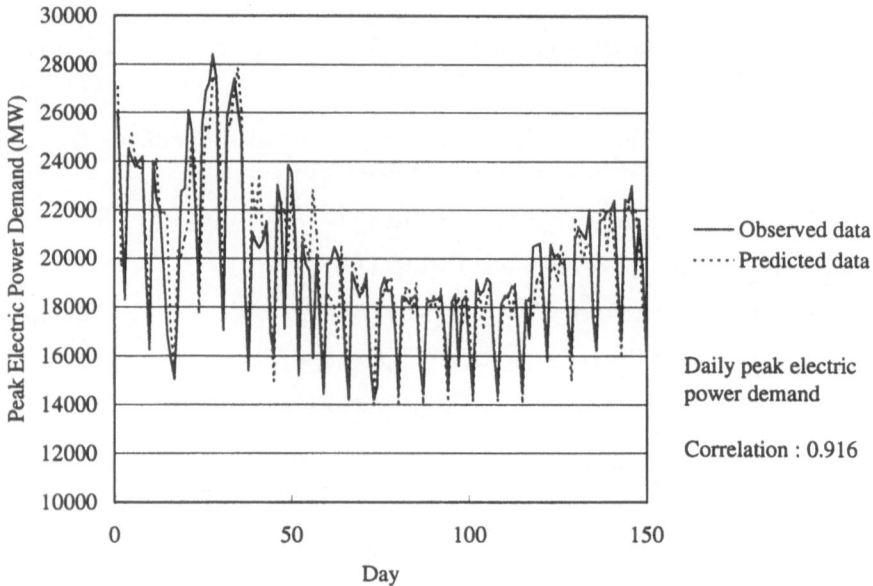

FIGURE 11.4. Peak electric power demand.

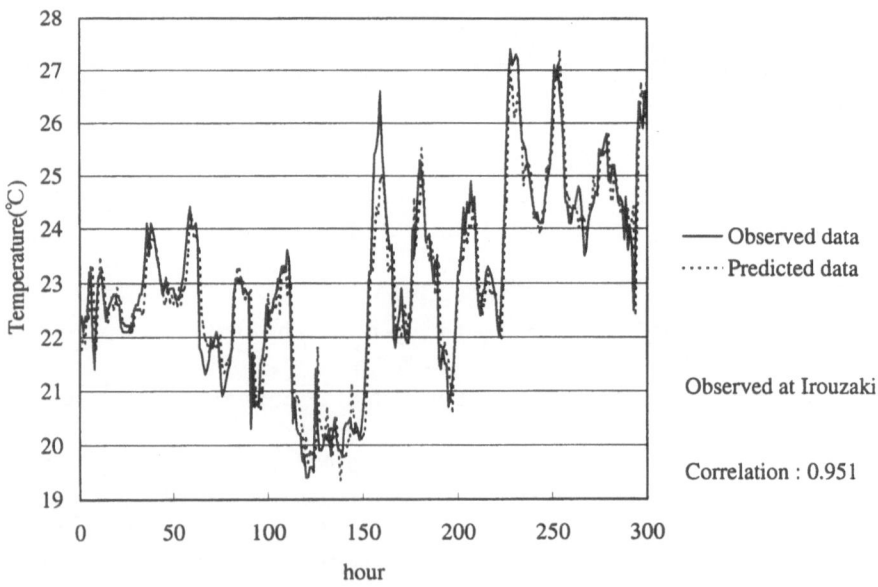

FIGURE 11.5. Local weather information.

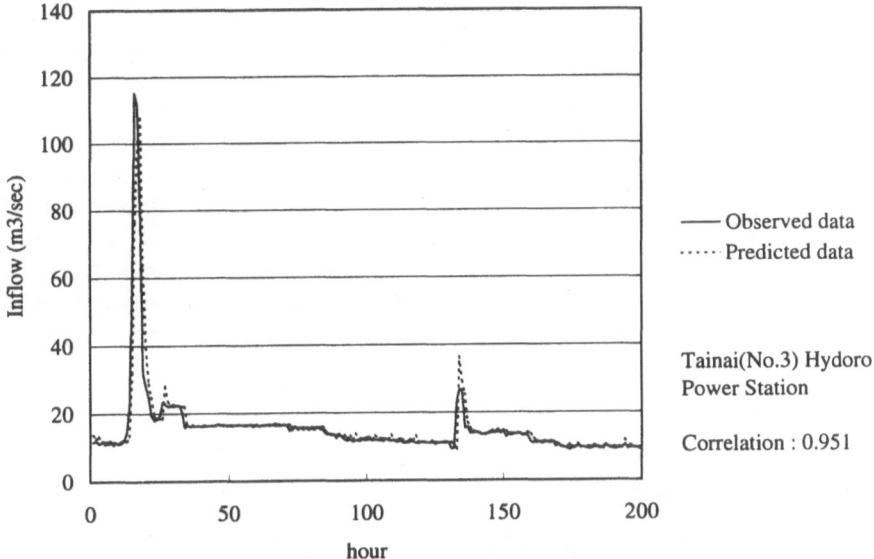

FIGURE 11.6. Inflow to a hydro power station.

There must be deterministic characteristics behind time series data if they are predictable by a deterministic model. The time period over which prediction is useful depends considerably on the dynamics behind the time series data.

11.6   Chaotic Time Series Analysis and Statistical Hypothesis Testing

Chaotic time series analysis obviously indicates new possibilities, however it is very dangerous to blindly apply the analysis [36, 15]. In the real world, phenomena cannot be purely deterministic chaos, that is, the assumption of $\eta_t = 0$ and $\xi_t = 0$ cannot be held. Actually, "real world" chaos necessarily includes such noise. In this sense, it is indispensable to take into account the effects of dynamic noise and observation noise [46].

Theiler *et al.* proposed the method of surrogate data, which is a statistical hypothesis testing [42, 43]; the method has contributed remarkably to analyzing practical data [15, 16, 18].

The method of surrogate data tests a null hypothesis in statistical way. One of the basic background of chaos is nonlinearity, and it is easier to imply that time series data are nonlinear than to show that they are chaotic. Therefore, a null hypothesis is adopted usually for the linearity of time series data; if the hypothesis is rejected, then nonlinearity is implied.

For both the original data and surrogates generated by a null hypothesis, statistical indicators Q_0 and $Q_{H_i}(i = 1, 2, \ldots, B)$ such as fractal dimensions, Lyapunov exponents and nonlinear prediction errors are estimated, respectively. If Q_0 is sufficiently separated from the distribution of Q_{H_i}, the corresponding null hypothesis can be rejected [43, 18, 19].

Although the method of surrogate data is very effective and important tool to analyze real data, it should be noted again that it is dangerous to blindly apply the method of surrogate data as well as another methods of chaotic time series analysis, as we had learned from the abuse of the GP method in 1980s. If we use the nonlinear analysis methods very carefully and can reveal that a process cannot be understood satisfactorily only with linearity, then such an indication can be an important milestone towards the goal of modeling unknown data in the real world.

11.7   Conclusions

Many researches on deterministic chaos have clarified ubiquity of low-dimensional chaos in many nonlinear dynamical systems both theoretically and numerically and various practical data have also been analyzed from the view point of nonlinear dynamics.

Although such nonlinear analysis of experimental data in the real world have apparently achieved remarkable results, these directions might also have discovered phenomena that could be only explained by low-dimensional.

In fact, there are many other phenomena that are more complicated and cannot be explained only by low-dimensional nonlinearity, and these were discarded in the past due to difficulties of analysis.

Considering this fact, we have to develop a novel time series analysis method that can deal with more complex phenomena, for example, higher-dimensional chaos that cannot be reduced into low-dimensional chaotic dynamics. Constructing such an analysis method seems to be an very important future problem for practical applications of chaos engineering [2].

11.8   References

[1] K. Aihara, *Neural computers - learning from the brain and neurons*, Tokyo Denki University Press, April 1988.

[2] K. Aihara and R. Katayama, "Chaos Engineering in Japan," *Communications of the ACM*, Vol.38, No. 11, pp.103-107, November 1995.

[3] H. Akaike and T. Nakagawa, *Statistical analysis and control of dynamical systems*, Science-sha, 1972.

[4] G.E.P. Box and G.M. Jenkins, *Time Series Analysis, Forecasting and Control*, Holden-Day, San Francisco, 1970.

[5] K. Briggs, "An improved method for estimating Liapunov exponents of chaotic time series," *Physics Letters A*, Vol.151, No.1, 2, pp.27-32, November, 1990.

[6] D.S. Broomhead and D. Lowe, "Multivariable Functional Interpolation and Adaptive Networks," *Complex Systems*, Vol.2, pp.321-355, 1988.

[7] R. Brown, P. Bryant, and H.D.I. Abarbanel, "Computing the Lyapunov spectrum of a dynamical system from an observed time series," *Physical Review A*, Vol.43, No.6, pp.2787-2806, March 1991.

[8] P. Bryant, R. Brown, and H. D.I. Abarbanel, "Lyapunov Exponents from observed time series," *Physical Review Letters*, Vol.65, No.13, pp.1523-1526, September 1990.

[9] L. Cao, Y. Hong, H.P. Fang, and G.W. He, "Predicting chaotic time series with wavelet networks," *Physica D*, Vol.85, pp.225-238, 1995.

[10] M. Casdagli, "Nonlinear Prediction of Chaotic Time Series," *Physica D*, Vol.35, pp.335-356, 1989.

[11] J.P. Eckmann, S.O. Kamphorst, D. Ruelle, and S. Ciliberto, "Liapunov exponents from time series," *Physical Review A*, Vol.34, No.6, pp.4971-4979, December 1986.

[12] J.P. Eckmann and D. Ruelle, "Ergodic theory of chaos and strange attractors," *Reviews of Modern Physics*, Vol.57, No.3, Part. 1, pp.617-656, July 1985.

[13] J.D. Farmer and J.J. Sidorowich, "Predicting chaotic time series," *Physical Review Letters*, Vol.59, No.8, pp.845-848, August 1987.

[14] J.D. Farmer and J.J. Sidorowich, "Exploiting chaos to predict the future and reduce noise," in *Evolution Learning and Cognition*, Y.C. Lee, (Ed.), pp.277-330, Singapore, 1988. World Scientific.

[15] T. Ikeguchi, "EEG and chaos," *Mathematical Sciences*, No.381, pp.36-43, 1995.

[16] T. Ikeguchi and K. Aihara, "Estimating correlation dimension of biological time series with a reliable method", *International Journal of Intelligent and Fuzzy Systems*, Vol.5, No.1, pp.33-52, 1997.

[17] T. Ikeguchi and K. Aihara, "Difference correlation can distinguish deterministic chaos from $1/f^\alpha$ type colored noise," *Physical Review E*, Vol.55, No.3, pp.2530-2538, 1997.

[18] T. Ikeguchi and K. Aihara, "Lyapunov Spectral Analysis on Random Data," *International Journal of Bifurcation and Chaos*, Vol.7, No.6, 1997.

[19] T. Ikeguchi and K. Aihara, "On dimension estimates with surrogate data sets," *IEICE Transactions on Fundamentals of Electronics, Communications and Computer Sciences*, Vol.E80-A, No.5, pp.859-868, 1997.

[20] T. Ikeguchi, M. Takakuwa, and K. Aihara, "Reconstructing attractors with higher order correlation," *Proceedings of International Symposium on Nonlinear Theory and Its Applications*, Vol.2, pp.1149-1152, December 1997.

[21] T. Iokibe, Takashi Kimura, and Kazuyuki Aihara, "An application of deterministic nonlinear short-term prediction to timeseries data of water demand," Trans. IEE of Japan, Vol.114-D, No.4, pp.409-414, 1994.

[22] T. Iokibe, M. Kanke, Y. Fujimoto, and S. Suzuki, "Local fuzzy reconstruction method for short-term prediction on chaotic timeseries," *Journal of Japan Society for Fuzzy Theory and Systems*, Vol.7, No.1, pp.186-194, 1995.

[23] J. Jiménez, J.A. Moreno, and G.J. Ruggeri, "Forecasting on chaotic time series : A local optimal linear-reconstruction method," *Physical Review A*, Vol.45, No.6, pp.3553-3557, March 1992.

[24] J. Jiménez, J. Moreno, and R.J. Ruggeri, "Detecting chaos with local associative memories," *Physics Letters A*, Vol.169, pp.25-30, September 1992.

[25] K. Judd and A. Mees, "On selecting models for nonlinear time series," *Physica D*, Vol.82, No.2, pp.426-444, 1995.

[26] K. Judd and A. Mees, "Embedding as a modeling problem," *Physica D*, Vol.120, Nos.3&4, pp.273-286, 1998.

[27] A.S. Lapedes and R.Farber, "Nonlinear signal processing using neural networks : prediction and system modeling," Technical report, Los Alamos National Laboratory, 1987.

[28] P.S. Linsay, "An efficient method of forecasting chaotic time series using linear interpolation," *Physics Letters A*, Vol.153, No.6&7, pp.353-356, 1991.

[29] E.N. Lorenz, "Atmospheric predictability as revealed by naturally occurring analogues," *Journal of the Atmospheric Sciences*, Vol.26, pp.636-646, July 1969.

[30] A.I. Mees, "Parsimonious Dynamical Reconstruction," *International Journal of Bifurcation and Chaos*, Vol.3, No.3, pp.669-675, 1993.

[31] A.I. Mees, "Dynamical systems and tessellations: detecting determinism in data," *International Journal of Bifurcation and Chaos*, Vol.1, No.4, pp.777-794, 1991.

[32] A.I. Mees, M.F. Jackson, and L.O. Chua, "Device modeling by radial basis functions," *IEEE Transactions on Circuits and Systems I:Fundamental Theory and Applications*, Vol.39, No.1, pp.19-27, January 1992.

[33] Y. Okabe and T. Ootsuka, "Application of the theory of KM_2O-Langevin equations to the non-linear prediction problem for the one-dimensional strictly stationary time series," *Journal of Mathematical Society Japan*, Vol.47, No.2, 1995.

[34] M.J.D. Powell, "Radial basis function for multivariable interpolation: a review," In *Algorithm for Approximation*, Clarendon, Oxford, 1987.

[35] M.B. Priestley, "State dependent models : a general approach to non-linear time series analysis," *Journal of Time Series Analysis*, Vol.1, No.1, pp.47-71, 1980.

[36] D. Ruelle, "Deterministic chaos : the science and the fiction," *Proc. R. Soc. Lond. A*, Vol.427, pp.241-248, 1990.

[37] M. Sano and Y. Sawada, "Measurement of the Lyapunov spectrum from a chaotic time series," *Physical Review Letters*, Vol.55, No.10, pp.1082-1085, September 1985.

[38] T. Sauer, "Time Series Prediction by Using Delay Coordinate Embedding," in *Time Series Prediction: Forecasting the Future and Understanding the Past*, A.S. Weigend and N.A. Gershenfeld, (Eds.), Vol.XV, pp.175-193, Addison Wesley, 1993.

[39] T. Sauer, J.A. Yorke, and M. Casdagli, "Embedology," *Journal of Statistical Physics*, Vol.65, No.3/4, pp.579-616, 1991.

[40] G. Sugihara and R.M. May, "Nonlinear forecasting as a way of distinguishing chaos from measurement error in time series," *Nature*, Vol.344, pp.734-741, April 1990.

[41] F. Takens, "Detecting strange attractors in turbulence," in *Dynamical Systems of Turbulence*, D.A. Rand and B.S. Young, (Eds.), Vol.898 of *Lecture Notes in Mathematics*, Springer-Verlag , Berlin, pp.366-381, 1981.

[42] J. Theiler, S. Eubank, A. Lonting, B. Galdrikian, and J.D. Farmer, "Testing for nonlinearity in time series : the method of surrogate data," *Physica D*, Vol.58, pp.77-94, 1992.

[43] J. Theiler and D. Prichard, "Constrained-realization Monte-Carlo method for hypothesis testing," *Physica D*, Vol.94, pp.221-235, 1996.

[44] H. Tong, "On a threshold model," in *Pattern Recognition and Signal-Processing*, C.H. Chen, (Ed.), 1978.

[45] H. Tong and K.S. Lim, "Threshold autoregression, limit cycles and cyclical data," *Journal of the Royal Statistical Society B*, Vol.42, No.3, pp.245-292, 1980.

[46] H. Tong, (Ed) *Chaos and Forecasting*, World Scientific, Singapore, 1995.

[47] A.A. Tsonis and J. B. Elsner, "Nonlinear prediction as a way of distinguishing chaos from random fractal sequences," *Nature*, Vol.358, pp.217-220, July 1992.

[48] D.J. Wales, "Calculating the rate of loss of information from chaotic time series by forecasting," *Nature*, Vol.350, pp.485-488, April 1991.

[49] A.S. Weigend, B.A. Huberman, and D.E. Rumelhart, "Predicting The Future : A Connectionist Approach," *International Journal of Neural Systems*, Vol.1, No.32, pp.193-209, 1990.

[50] D.M. Wolpert and R.C. Miall, "Detecting chaos with neural networks," *Proc. R. Soc. London B*, Vol.242, pp.82-86, 1990.

12

A Short Course for Fuzzy Set Theory

Sadaaki Miyamoto

12.1 Classical Sets

Intuitively, a set is any collection of objects. Many examples are found in the real world and in mathematical theories. There are two main ways of defining a set. One is to list up objects that constitutes a set. For example, $\{John, Mary, Thomas\}$ is a set of the three objects: $John$, $Mary$, and $Thomas$. We can think that these three objects are mere words or names indicating three real persons. Usually we give a name, say X, to the collection: $X = \{John, Mary, Thomas\}$, and we refer to X instead of listing up all the objects after we have given the name. Thus, in an abstract manner a set $X = \{x_1, x_2, \cdots, x_n\}$ is the collection of $x_1, x_2, \cdots, x_n$.

An object that constitutes a set is called an element or a member of that set. Thus, $John$ is a member of X and x_1 is an element of X. If x is an element of X, we write $x \in X$; if x is not an element of X, we write $x \notin X$. The relation $x \in X$ is also read as "x belongs to X" or "x is a member of X."

The other way of defining a set is to provide a criterion to distinguish membership from nonmembership. Such a criterion is usually expressed by a sentence or sentences of a definite meaning. For example, let X be the set of positive integers less than 100. Then, $90 \in X$ and $0 \notin X$. In this way, for an arbitrarily given object a, the criterion definitely decides if a belongs to the set or not.

Infinite sets should be considered as well as finite sets. The way of listing up elements of infinite sets uses dots like $X = \{a_1, a_2, \cdots\}$. For example, the set N of positive integers is $N = \{1, 2, 3, \cdots\}$. The second way of employing a criterion is used for both finite sets and infinite sets.

Remark. Although axiomatic definitions of sets have been studied in symbolic logic, many of us understands basic sets in mathematics in an intuitive way. We know what the set of integers denoted by Z is, and in general we do not care about detailed discussion on the set of real numbers such as the completion of rational numbers. Thus, in this text we understand these basic sets intuitively: the set R of real numbers is understood as a straight line.

When all elements in a set are included as members of another set, the former set is called a subset of the latter set, and the latter set is called a superset of the former. For example, $A = \{John, Mary\}$ is a subset of $X = \{John, Mary, Thomas\}$, and X is a superset of A. When a set C is a subset of another set D, we write $C \subseteq D$. $C \subseteq D$ means that if $x \in C$ then $x \in D$. The converse is also true: if $x \in C$ then $x \in D$ implies $C \subseteq D$.

If two sets A and B consist of the same elements, this means that A is equal to B, and we write $A = B$. Thus, if $A = B$, then $x \in A$ implies $x \in B$ and $x \in B$ implies $x \in A$. The converse is also true: if $x \in A$ implies $x \in B$ and $x \in B$ implies $x \in A$, then $A = B$. If A and B do not consist of the same members, we write $A \neq B$. Notice that $C \subseteq D$ does not imply $C \neq D$ in general. When $C \subseteq D$ and $C \neq D$, we write $C \subset D$ and C is called a proper subset of D.

Remark. Logical symbols $\neg$, $\wedge$, $\vee$, $\Rightarrow$, and $\Leftrightarrow$ are sometimes used here. Assume that we have two sentences P and Q. Then, $\neg P$ means not P: $P \wedge Q$ means P and Q; $P \vee Q$ means P or Q; $P \Rightarrow Q$ means P implies Q, or if P then Q; $P \Leftrightarrow Q$ means P is equivalent to Q, in other words, if P then Q and the converse is also true. Using $\Leftrightarrow$, we can write

$$\forall x \, (x \in A \Rightarrow x \in B) \Leftrightarrow A \subseteq B,$$

$$A = B \quad \Leftrightarrow \quad (A \subseteq B) \wedge (B \subseteq A).$$

It is necessary to introduce a particular set $\{\}$ of no members at all that is called the empty set and denoted by $\emptyset$ (i.e., $\emptyset = \{\}$). The empty set $\emptyset$ is a subset of any set.

There is always possibility of finding a superset of a given set. Namely, given a set A, there is a superset B such that $A \subset B$, and there is another superset C: $B \subset C$, and so on. In such an open universe we can enlarge our sets in an endless manner, but frequently it is more convenient to assume a closed universe and limit ourselves within the given universe. In the latter case we first consider a set, say X, and only discuss subsets of the closed universe X. The set X is called a universe of discourse or a universal set. Other names may be used such as a set of reference or an entire set.

Discussion of fuzzy sets uses a universe of discourse in most cases. We mainly use a finite set or the set of real numbers as the universal set for discussing illustrative examples, as we will see later.

Three well-known operations are the union, the intersection, and the complementation. Given two sets A and B, their union $A \cup B$ is the set of all elements that belong either to A or to B, or both.

The intersection of A and B is the set of all elements that are common to both A and B.

Given a universal set X and a subset A of X, the complement of A, denoted by $\bar{A}$, is the set of all elements in X that do not belong to A.

In this way, we need not assume a universe of discourse in discussing the union and the intersection, whereas the complementation requires a universal set.

Let us consider an example. Given the universal set

$$X = \{John, Mary, Thomas, Philip\}$$

and subsets

$$
\begin{aligned}
A &= \{John, Mary\}, \\
B &= \{Mary, Philip\}, \\
C &= \{Thomas\}.
\end{aligned}
$$

It is immediate to see

$$
\begin{aligned}
A \cup B &= \{John, Mary, Philip\}, \\
A \cap B &= \{Mary\}, \\
A \cup C &= \{John, Mary, Thomas\}, \\
A \cap C &= \emptyset, \\
\bar{A} &= \{Thomas, Philip\}.
\end{aligned}
$$

Given a universal set X, a subset X is characterized by a binary function which is called the characteristic function of a subset. Let A be a subset of X, then the characteristic function $f_A(x)$ of A is a 0/1-valued function which take the value 1 if x belongs to A, and the value 0 if x does not belong to A. In mathematical expression

$$
f_A(x) = \begin{cases} 1 & (x \in A) \\ 0 & (x \notin A) \end{cases}
$$

Many examples for discussing characteristic functions assume one dimensional universe. Namely, as in Figure 12.1, the universal set is the set (or a subset) of real numbers. Notice that the set of real numbers is denoted by R. In Figure 12.1, $A = [1, 2)$, i.e., the set of real numbers that are greater than or equal to 1 and less than 2. Thus, $f_A(x) = 1$ if $1 \leq x < 2$, and $f_A(x) = 0$ otherwise. In particular $f_A(1) = 1$ and $f_A(2) = 0$, which are distinguished by the black bullets and by the white circles in this figure.

When we discuss a set, we do not care about the order of its members. Thus,

$$\{John, Mary, Thomas, Philip\} = \{John, Philip, Thomas, Mary\}.$$

Generally, we use the symbol $\{\cdots\}$ when the order of the elements surrounded by braces can be changed without altering the meaning. On the

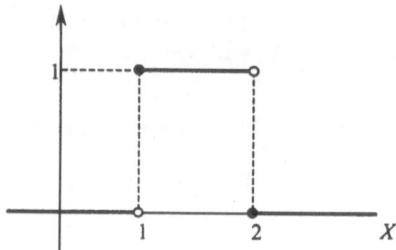

FIGURE 12.1. A characteristic function on a universe of one dimension

other hand, if we write, $(\cdots)$, the order of elements in parentheses cannot be changed: the typical example of the latter is a vector.

It should be noted that when we discuss fuzzy sets introduced from next section, classical sets are also called ordinary sets, nonfuzzy sets, or crisp sets.

12.2   Fuzzy Sets

The adjective *fuzzy* indicates indistinctness in outline in general. Thus, Zadeh [7] attempted to define a class without a clear boundary using the name of a *fuzzy set*.

In general, there is no unique way to provide an appropriate mathematical representation for such a 'vague set'. Apart from fuzzy sets, there have been other attempts such as rough sets [5]and flou sets (see Dubois and Prade [1]), but fuzzy sets are the simplest in applications and have sufficient theoretical contents.

To represent an indistinct boundary of a set, a generalization of the characteristic function is introduced in order to define a fuzzy set.

Let us consider a function $\mu(x)$ that is used to represent such a 'vague' set instead of a characteristic function for a classical set. For $\mu(x)$, suppose that the region where $\mu(x) = 1$ is inside of the set and the region where $\mu(x) = 0$ is outside of the set. On the indistinct boundary, however, $\mu(x) \neq 1$ and $\mu(x) \neq 0$.

A fuzzy set is introduced by specifying values of $\mu(x)$ such that

$$0 < \mu(x) < 1 \tag{12.1}$$

on the vague boundary. In other words, when an element $x' \in X$ satisfies $0 < \mu(x') < 1$, then the element is considered to be in an ambiguous status: it is unclear if x' belongs to the 'vague set' represented by $\mu(x)$ or not, whereas $\mu(x') = 1$ means that x' belongs to that set and $\mu(x') = 0$ implies that x' does not belong to that set.

Now, we can define a fuzzy set using this function. A fuzzy set A of X is characterized by the membership function denoted by $\mu_A(x)$. The function is defined on X and the values are in the unit interval $[0,1]$. The value $\mu_A(x)$ shows the degree of membership of x to A. The membership value is also called the grade of membership.

If $\mu_A(x) = 1$, x is sure to belong to A, and $\mu_A(x) = 0$ means that x does not belong to A at all. If $0 < \mu_A(x) < 1$, it is uncertain whether x belongs to A or not. Nevertheless, when $\mu_A(x)$ increases, the certainty of belongingness of x to A also increases, and vice versa. An example of the membership function of a fuzzy set is shown in Figure 12.2.

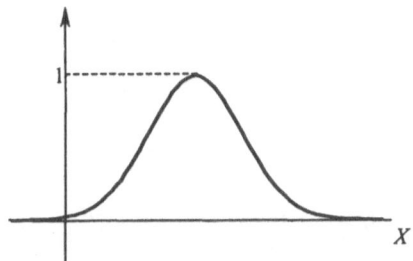

FIGURE 12.2. A membership function for a fuzzy set A

This function $\mu_A(x)$ is simply a bounded function with a label A, nothing other than this. Nevertheless, we call the label a fuzzy set A, because we consider such labels as a generalization of classical sets by providing set operations to the labels $A, B, \cdots$ of fuzzy sets.

In applications, such labels, i.e., fuzzy sets refer to concepts or properties described by words or phrases. In particular, adjectives such as $YOUNG$, $TALL$, and $HEAVY$ are represented by fuzzy sets. Let us consider a fuzzy set A for $YOUNG$. (We will informally write $A = YOUNG$ afterwards.) It is natural to assume that this concept is described in terms of ages. Hence the fuzzy set A should be defined on the universal set of all ages described as a line. Thus, we can specify a membership $\mu_A(x)$ in Figure 12.3. In this figure ages less than 20 are defined to be young ($\mu_A(x) = 1$), and ages greater than 50 are not young at all ($\mu_A(x) = 0$). For ages between 20 and 50, whether they are young or not are uncertain ($0 < \mu_A(x) < 1$), but it is true that if the age of a person becomes greater (x increases), he becomes less young ($\mu_A(x)$ decreases). Notice also that the concept $YOUNG$ cannot be represented by a nonfuzzy set. In other words, it is impossible to specify a particular figure of the age, say a, and decide that if one's age is less than a then he is young, and if his age is greater than a then he is not young.

When we discuss a set of *tall* persons, a fuzzy set $B = TALL$ should be defined on the universal set of height. In Figure 12.4, the height below 170cm is not tall at all, the height above 185cm is definitely tall, and the

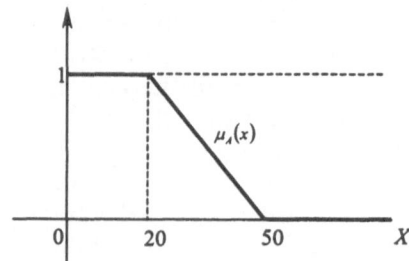

FIGURE 12.3. A fuzzy set $A = YOUNG$

height between these two numbers is in a fuzzy status.

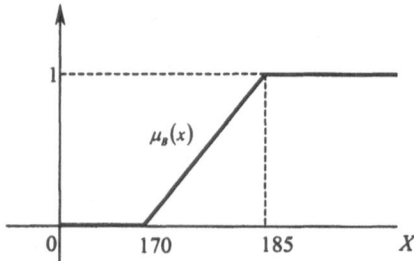

FIGURE 12.4. A fuzzy set $B = TALL$

These examples show a subjective characteristic of fuzzy sets. Determination of young ages and tall heights are subjective. There is a large degree of freedom in determining membership functions, even in these simple examples. In general membership functions need not be linear, not be continuous.

In spite of the subjectiveness, we discuss fuzzy sets mathematically and objectively after they have been defined. This is the standpoint of fuzzy systems. In practice, a general rule of determining fuzzy sets is *to take the simplest way*, unless an established or widely accepted method is available. There are many studies on the determination of membership functions, but in real applications of fuzzy control systems, simple fuzzy sets of triangular or trapezoidal membership functions have mainly been used, since they are simpler to define and require less computational effort. Moreover we rarely encounter discontinuous membership functions in real applications, unless the universal set is finite or discrete, or the set is crisp. (When the universal set is finite or discrete, we cannot, of course, define continuous membership functions.)

An information item in general can be expressed as a triplet

$$(object, \quad attribute, \quad value)$$

which means that an *attribute* of an *object* has a *value*. This type of information item has been extended to a quadruple in order to include a confidence measure: *(object, attribute, value, confidence)* (see Dubois and Prade [3]). Seeing that a fuzzy set refers to a property of an attribute, we are considering a quintuple

$$(object, \quad attribute, \quad property, \quad value, \quad confidence).$$

Let us consider an example: *(John, age, YOUNG, 30, 0.67)*. This example means that the *age* of *John* is 30 and relevance to the property $YOUNG$ is 0.67. Thus, fuzzy set $A = YOUNG$, $x = 30$, $\mu_A(x) = 0.67$, and the universal set X represents possible ages. Sometimes data of the *value* part is missing. Consider an example *(John, programming skill, good, −, 0.9)*, which means that programming skill of *John* is good, and the confidence for the goodness is 0.9. Thus, we may take a fuzzy set $B = $ "programming skill is good", $x = John$, and $\mu_B(x) = 0.9$. In the latter example, the universal set X is a collection of people including *John*. Namely, the two universal sets are different in nature, although these two fuzzy sets refer to particular attributes of objects. The former example uses a universal set represented as a line in Figure 12.4, but in the latter example we cannot draw such a figure of a fuzzy set, since no scale on the goodness of programming skill is given.

Now, we have two different methods of specifying fuzzy sets:

(a) When an attribute of an object has a scale, describe a fuzzy set on the universal set of that scale like the graphs in Figures 12.2 – 12.4.

(b) When such a scale is not provided, describe a fuzzy set by directly specifying a membership value on each object.

For example, the fuzzy set $B = $ "programming skill is good", the membership values are specified parentheses: $\mu_B(John) = 0.9$, $\mu_B(Mary) = 0.8$, $\mu_B(Philip) = 1.0$, and $\mu_B(Thomas) = 0$.

There are different notations for a fuzzy set. Membership functions are in general denoted by $\mu_A(x)$, but sometimes it is more convenient to write $A(x)$ instead of $\mu_A(x)$. Using $A(x)$, the membership function in Figure 12.3 is

$$A(x) = \begin{cases} 1 & (0 \le x \le 20) \\ \dfrac{50 - x}{30} & (20 < x \le 50), \\ 0 & (x > 50) \end{cases}$$

In the case of a finite universe of discourse $X = \{x_1, x_2, \cdots, x_n\}$, A fuzzy set A with $\mu_A(x_1) = a_1$, $\mu_A(x_2) = a_2$, $\cdots, \mu_A(x_n) = a_n$ is simply written as

$$A = a_1/x_1 + a_2/x_2 + \cdots + a_n/x_n = \sum_i \mu_A(x_i)/x_i.$$

Accordingly, the fuzzy set B is

$$B = 0.9/John + 0.8/Mary + 1.0/Philip.$$

Generally, the element with the zero membership is omitted, hence we do not write $0/Thomas$ in the above example.

Remark. Some authors prefer to use $\{(x_i, \mu_A(x_i))\}$ or $\{\mu_A(x_i)/x_i\}$ instead of $\sum \mu_A(x_i)/x_i$, such as

$$B = \{(John, 0.9), (Mary, 0.8), (Philip, 1.0)\}$$

or

$$B = \{0.9/John, 0.8/Mary, 1.0/Philip\}.$$

In the same way, a fuzzy set of a discrete universe $X = \{x_1, x_2, \cdots, x_n, \cdots\}$ is written as

$$A = a_1/x_1 + a_2/x_2 + \cdots + a_n/x_n + \cdots$$

On the other hand, a fuzzy set of a continuous universe is sometimes written as

$$A = \int_X \mu_A(x)/x.$$

Thus, the fuzzy set in Figure 12.3 is

$$A = \int_0^{20} 1/x + \int_{20}^{50} \frac{50 - x}{30}/x.$$

Note that $\int_{50}^{\infty} 0/x$ is omitted.

12.3   Basic Operations on Fuzzy Sets

Let us review basic operations for nonfuzzy sets: union, intersection, complementation, and direct product using mathematical symbols.

Let X be a universal set, and consider two nonfuzzy sets $A, B \subseteq X$.

1. [union] The union $A \cup B$ is expressed as

$$A \cup B = \{x : x \in A \quad \text{or} \quad x \in B\}.$$

The above expression is read as "the union of A and B is the set of elements x such that $x \in A$ or $x \in B$."

2. [intersection] The intersection $A \cap B$ is expressed as

$$A \cap B = \{x : x \in A \quad \text{and} \quad x \in B\}.$$

The above expression is read as "the intersection of A and B is the set of elements x such that $x \in A$ and $x \in B$."

3. [complement] The complement $\bar{A}$ is written as

$$\bar{A} = \{x : x \notin A, \quad x \in X\}.$$

4. [direct product] The coordinate of a point in a plane is denoted by (x, y), such as (1,2) or (0,-1). The point is an element of the universe $R \times R$ which is the direct product of R, the set of real numbers. In general, given two universal sets X and Y, the direct product (sometimes called the Cartesian product) is denoted by $X \times Y$ that consists of all ordered pairs (x, y), $x \in X$, $y \in Y$. In the same way, given $A \subseteq X$ and $B \subseteq Y$, the direct product $A \times B$ is the set of (x, y), where x is any element in X, and y is any element of B:

$$A \times B = \{(x, y) : x \in A \quad \text{and} \quad y \in B\}.$$

Let us turn to fuzzy sets. First, notice that a nonfuzzy set is at the same time a fuzzy set. To see this, let A be a crisp set in X with the characteristic function $f_A(x)$. Then A is a fuzzy set by the natural definition of its membership function

$$\mu_A(x) = f_A(x) \quad \text{for all} \quad x \in X.$$

In other words, a characteristic function can be taken to be the membership function of the same set. In particular, the universal set and the empty set are fuzzy sets, since we can take

$$\begin{aligned} \mu_X(x) &= 1 \\ \mu_\emptyset(x) &= 0 \end{aligned}$$

for all $x \in X$.

5. [α - cut] An α - cut, which is also called an α - level set, is an operation that maps a fuzzy set A to a nonfuzzy set A_α with the parameter α, $0 \leq \alpha \leq 1$: A_α is the set of elements on which the value of membership functions is greater than or equal to α. In other words,

$$A_\alpha = \{x : \mu_A(x) \geq \alpha \quad x \in X\}.$$

By the α - cut, a fuzzy set defined by its membership function, is regarded as a collection of nonfuzzy sets $\{A_\alpha\}$, $0 \le \alpha \le 1$. Notice that for a nonfuzzy set C, all α - cuts are identical:

$$C_{\alpha_1} = C_{\alpha_2} \quad \text{for all } 0 \le \alpha_1, \alpha_2 \le 1.$$

Many properties of fuzzy sets are described in terms of the α - cuts, as we will see below.

It is easy to see that if $\alpha_1 \le \alpha_2$, then $A_{\alpha_2} \subseteq A_{\alpha_1}$.

The following operations (or properties) of fuzzy sets are generalizations of the corresponding operations (or properties) of nonfuzzy sets in the sense that when crisp sets are regarded as fuzzy sets as described above, the results of applying fuzzy operations to the crisp sets coincide with those of the corresponding crisp operations on those nonfuzzy sets.

Equality and inclusion of fuzzy sets are defined as follows.

6. [equality] Given two fuzzy sets A and B of the universe X, we say $A = B$ if, for all $x \in X$, $\mu_A(x) = \mu_B(x)$

7. [inclusion] Given two fuzzy sets A and B of X, we say A is included in B, which is denoted by $A \subseteq B$, if, for all $x \in X$, $\mu_A(x) \le \mu_B(x)$.

It is easy to see that the equality and the inclusion are characterized by α - cuts. Namely,

(i) $A = B$ if for all $0 \le \alpha \le 1$, $A_\alpha = B_\alpha$, and the converse is also true.

(ii) $A \subseteq B$ if, for all $0 \le \alpha \le 1$, $A_\alpha \subseteq B_\alpha$, and the converse is also true.

8. [union of fuzzy sets] Given two fuzzy sets A and B of X, the union $A \cup B$ is a fuzzy set whose membership function is given by

$$\mu_{A \cup B}(x) = \max\{\mu_A(x), \mu_B(x)\}.$$

9. [intersection of fuzzy sets] Given two fuzzy sets A and B of X, the intersection $A \cap B$ is a fuzzy set whose membership function is given by

$$\mu_{A \cap B}(x) = \min\{\mu_A(x), \mu_B(x)\}.$$

It is obvious to see that the union and intersection of fuzzy sets generalize the corresponding operations for crisp sets, since if A and B are crisp,

$$\{x : \max\{\mu_A(x), \mu_B(x)\} = 1\} = \{x : x \in A \quad \text{or} \quad x \in B\},$$
$$\{x : \min\{\mu_A(x), \mu_B(x)\} = 1\} = \{x : x \in A \quad \text{and} \quad x \in B\}.$$

An interesting property holds between the union or the intersection and an α - cut. Namely,

$$(A \cup B)_\alpha = A_\alpha \cup B_\alpha, \tag{12.2}$$
$$(A \cap B)_\alpha = A_\alpha \cap B_\alpha. \tag{12.3}$$

Thus, an α - cut of the union of two fuzzy sets coincides with the union of α -cuts of the same sets. This property is useful in proving basic set-theoretical properties of fuzzy sets.

To prove the above property is true, it is sufficient to show that $(A \cup B)_\alpha \subseteq A_\alpha \cup B_\alpha$ and $A_\alpha \cup B_\alpha \subseteq (A \cup B)_\alpha$ for all $0 \leq \alpha \leq 1$ (We omit the proof for the intersection for simplicity.)

It is clear that $A \subseteq A \cup B$ and $B \subseteq A \cup B$, from which it follows that $A_\alpha \subseteq (A \cup B)_\alpha$ and $B_\alpha \subseteq (A \cup B)_\alpha$. Taking the union of the both sides, we have and $A_\alpha \cup B_\alpha \subseteq (A \cup B)_\alpha$.

On the other hand, let $x \in (A \cup B)_\alpha$, then

$$\alpha \leq \mu_{A \cup B}(x) = \max\{\mu_A(x), \mu_B(x)\}$$

which means that $\mu_A(x) \geq \alpha$ or $\mu_B(x) \geq \alpha$, that is, $x \in A_\alpha \cup B_\alpha$. Thus, $(A \cup B)_\alpha \subseteq A_\alpha \cup B_\alpha$ is proved, and therefore the desired equation is valid.

Owing to (12.2) and (12.3), it is easy to see that for fuzzy sets A, B, and C, the following properties hold:

(i) [commutativity]

$$A \cup B = B \cup A, \quad A \cap B = B \cap A.$$

(ii) [associativity]

$$(A \cup B) \cup C = A \cup (B \cup C), \quad (A \cap B) \cap C = A \cap (B \cap C).$$

(iii) [distributivity]

$$
\begin{aligned}
(A \cup B) \cap C &= (A \cap C) \cup (B \cap C), & (12.4)\\
(A \cap B) \cup C &= (A \cup C) \cap (B \cup C). & (12.5)
\end{aligned}
$$

Let us prove the distributivity. Notice that (12.4) is equivalent to

$$[(A \cup B) \cap C]_\alpha = [(A \cap C) \cup (B \cap C)]_\alpha \quad \text{for all } 0 \leq \alpha \leq 1. \quad (12.6)$$

Now, using (12.2) and (12.3), the distributive law for crisp sets, we have, for arbitrary α in the unit interval,

$$
\begin{aligned}
[(A \cup B) \cap C]_\alpha &= (A \cup B)_\alpha \cap C_\alpha\\
&= (A_\alpha \cup B_\alpha) \cap C_\alpha\\
&= (A_\alpha \cap C_\alpha) \cup (B_\alpha \cap C_\alpha)\\
&= (A \cap C)_\alpha \cup (B \cap C)_\alpha\\
&= [(A \cap C) \cup (B \cap C)]_\alpha
\end{aligned}
$$

therefore (12.6) is valid. The identity (12.5) is proved in the same way. This proof is syntactic and needs no analysis of the meaning of membership functions.

It should also be noted that since associativity holds, we can define $A \cup B \cup C$ which means either $(A \cup B) \cup C$ or $A \cup (B \cup C)$. In the same way, we can define $A \cap B \cap C$. More generally, we can define

$$A_1 \cup A_2 \cup \cdots \cup A_n \quad \text{and} \quad A_1 \cap A_2 \cap \cdots \cap A_n.$$

They are also written as

$$\bigcup_{i=1}^{n} A_i \quad \text{and} \quad \bigcap_{i=1}^{n} A_i$$

respectively.

The symbols $\vee$ and $\wedge$ are frequently used for logical OR and AND, respectively. In fuzzy sets, $\vee$ is used for max, and $\wedge$ is used for min. Thus,

$$\max\{\mu_A(x), \mu_B(x)\} = \mu_A(x) \vee \mu_B(x),$$
$$\min\{\mu_A(x), \mu_B(x)\} = \mu_A(x) \wedge \mu_B(x),$$

or

$$\max\{A(x), B(x)\} = A(x) \vee B(x),$$
$$\min\{A(x), B(x)\} = A(x) \wedge B(x).$$

Remark. The symbols $\vee$ and $\wedge$ are used in infix notations, whereas max and min are for prefix notations. It is more convenient to use infix notations in manipulations of mathematical expressions in fuzzy sets.

10. [complement] Given a fuzzy set A of X, its complement $\bar{A}$ is defined by

$$\mu_{\bar{A}}(x) = 1 - \mu_A(x), \quad \text{for all } x \in X.$$

Unlike the union and intersection,

$$\overline{(A_\alpha)} \neq (\bar{A})_\alpha$$

in general. When A is nonfuzzy,

$$\{x : \mu_{\bar{A}}(x) = 1\} = \{x : x \notin A, x \in X\}$$

that is, the fuzzy complement coincides the ordinary complement of a crisp set. The following properties holds for the complementation of fuzzy sets.

(i) $\overline{(\bar{A})} = A$

(ii) [De Morgan's law]

$$\overline{(A \cup B)} = \bar{A} \cap \bar{B}, \qquad \overline{(A \cap B)} = \bar{A} \cup \bar{B}.$$

To see De Morgan's law is valid, note the following.

$$
\begin{aligned}
\mu_{\overline{(A \cup B)}}(x) &= 1 - \mu_{A \cup B}(x) \\
&= 1 - \mu_A(x) \vee \mu_B(x) \\
&= (1 - \mu_A(x)) \wedge (1 - \mu_B(x)) \\
&= \mu_{\bar{A}}(x) \wedge \mu_{\bar{B}}(x) \\
&= \mu_{\bar{A} \cap \bar{B}}(x).
\end{aligned}
$$

It is well known that for a nonfuzzy set A,

$$
A \cap \bar{A} = \emptyset, \qquad A \cup \bar{A} = X.
$$

These properties do not hold for fuzzy sets. Consider $\mu_A(x) \equiv 0.5$, then $A = \bar{A}$; therefore $A \cap \bar{A} = A$ and $A \cup \bar{A} = A$.

11. [direct product] Let X and Y are two universal sets and A and B are fuzzy sets of X and Y, respectively. The direct product is defined by

$$
\mu_{A \times B}(x, y) = \min\{\mu_A(x), \mu_B(y)\}.
$$

It is obvious to see

$$
(A \times B)_\alpha = A_\alpha \times B_\alpha \quad \text{for all } 0 \le \alpha \le 1.
$$

12.4   Extension Principle

We will consider images and inverse images of fuzzy sets in this section. Let us first review mappings for crisp sets.

Consider a simple example. Let $X = \{x_1, x_2, x_3\}$, $Y = \{y_1, y_2, y_3\}$, and assume that f is a mapping defined on X with the values in Y, given by

$$
f(x_1) = y_1, \quad f(x_2) = y_1, \quad f(x_3) = y_3.
$$

Figure 12.5 shows the correspondence of elements by the function f using the arrows. Assume that $A = \{x_1, x_2\}$ and $B = \{y_2, y_3\}$ and suppose that we wish to have all elements that is mapped from elements in A. Immediately we see from Figure 12.5 that these elements are on the arrowheads that stem from x_1 and x_2. We denote this set by $f(A)$ and call the image of A by f. When we wish to have all elements in X that is mapped onto B, we see from the same figure that they are those elements that have arrows with the arrowheads on B. This set is called the inverse image of B by f, and is denoted by $f^{-1}(B)$. Thus,

$$
\begin{aligned}
f(A) &= \{y_1, y_3\}, \\
f^{-1}(B) &= \{x_3\}.
\end{aligned}
$$

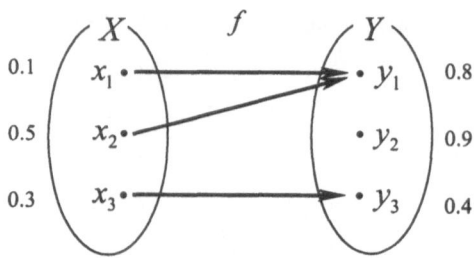

FIGURE 12.5. Two sets X, Y, and a function f

If $B_1 = \{y_2\}$, then $f^{-1}(B) = \emptyset$, which means that there are no elements that are mapped onto y_2.

Generally, image and inverse image are defined as follows.

1. [image] Let X and Y be two universal sets, and A be a subset of X. Assume that a function defined on X with the values in Y is given. The image of A by f is defined to be all the elements in Y that are mapped from elements in A by f:

$$f(A) = \{f(x) : x \in A\}.$$

2. [inverse image] With the same X, Y, and f as in the definition of the image, assume that B is a subset of Y. Then, the inverse image of B by f is those elements in X that are mapped onto elements in B by f:

$$f^{-1}(B) = \{x : f(x) \in B, x \in X\}.$$

Now, consider image and inverse image of fuzzy sets. consider the above universal set and assume

$$
\begin{aligned}
A &= 0.1/x_1 + 0.5/x_2 + 0.3/x_3, \\
B &= 0.8/y_1 + 0.9/y_2 + 0.4/y_3.
\end{aligned}
$$

(Readers may remember that the above expression means $\mu_B(y_1) = 0.8$, $\mu_B(y_2) = 0.9$, and so on.) It is natural to take

$$\mu_{f^{-1}(B)}(x_1) = \mu_{f^{-1}(B)}(x_2) = 0.8$$

but when $f(A)$ is considered for y_1, two values 0.1 and 0.5 are mapped from x_1 and x_2, respectively. Extension principle states that the maximum of the possible grades for x is defined to be the grade of the corresponding y. Thus,

$$\mu_{f(A)}(y_1) = \max\{\mu_A(x_1), \mu_A(x_2)\} = 0.5.$$

Generally, let A be a fuzzy set of X, then the image $f(A)$ is defined by

$$\mu_{f(A)}(y) = \begin{cases} \max\limits_{x \in f^{-1}(y)} \mu_A(x) & (f^{-1}(y) \neq \emptyset), \\ 0 & (f^{-1}(y) = \emptyset). \end{cases}$$

Extension principle in many textbooks are given in a more general form. Namely, let universal sets be $X_1, X_2, \cdots, X_n$ and fuzzy sets be $A_1, A_2, \cdots, A_n$ in the respective universes. Assume $f \colon X_1 \times X_2 \times \cdots \times X_n \to Y$ Then,

$$\mu_{f(A_1, A_2, \dots, A_n)}(y) = \max_{y = f(x_1, x_2, \dots, x_n)} \mu_{A_1 \times A_2 \times \cdots \times A_n}(x_1, x_2, \dots, x_n)$$

$$= \max_{y = f(x_1, x_2, \dots, x_n)} \min\{\mu_{A_1}(x_1), \mu_{A_2}(x_2), \dots, \mu_{A_n}(x_n)\}.$$

Remark. In general, max in the right hand sides are replaced by sup in advanced textbooks. We assume here, however, that sup can always be replaced by max for simplicity, as noted earlier. This assumption for the simplification is satisfied in most of practical or applicational examples.

12.5   Fuzzy Relations

Crisp or ordinary relations include equality ($=$), inequality($>$, $<$), equivalence relations for generating classes, and so on. As fuzzy sets are generalizations of crisp sets, fuzzy relations generalize crisp relations.

A relation (or more precisely, binary relation) R defined on X means that for any pairs of elements $x, y \in X$, one and only one of xRy (x and y have the relation R) or $x\not\!\!R y$ (x and y do not have the relation R) holds. This form of the infix notation can be transformed into an prefix notation $R(x, y)$ of the same meaning. Thus, using R as a 0/1- valued function:

$$xRy \iff R(x, y) = 1,$$
$$x\not\!\!R y \iff R(x, y) = 0.$$

For example, assume that $R_=(x, y)$ and $R_>(x, y)$ are the functions corresponding to equality $=$ and inequality $>$, respectively,

$$R_=(x, y) = 1 \iff x = y,$$
$$R_=(x, y) = 0 \iff x \neq y,$$
$$R_>(x, y) = 1 \iff x > y,$$
$$R_>(x, y) = 0 \iff x \leq y.$$

The latter form $R(x, y)$ of the binary relation is a 0/1 valued function, and furthermore, $R(x, y)$ is the characteristic function of the set of elements (x, y) such that xRy :

$$R = \{(x, y) : R(x, y) = 1\} = \{(x, y) : xRy\}.$$

In this way, a relation R on X is equivalent to the set R of the same symbol in the product space $X \times X$, by abuse of terminology.

Fuzzy relations have been considered for representing relations of "approximately equal" ($\approx$) or "far greater than" ($\gg$), and so on. Since a crisp relation is equivalent to a set in $X \times X$, a fuzzy relation should be defined as a fuzzy set in $X \times X$. Equivalently, $R(x,y)$ as a characteristic function of the set R should be generalized to a membership function defined on $X \times X$.

Thus, we define a fuzzy relation R on X to be a fuzzy set of the same symbol R of $X \times X$. The membership function for R is denoted by $\mu_R(x,y)$ or more simply $R(x,y)$ for all $x, y \in X$.

More generally, given universes $X_1, X_2, \cdots, X_n$, n-ary fuzzy relation R is defined to be a fuzzy set R of $X_1 \times X_2 \times \cdots \times X_n$. Its membership function is denoted by $\mu_R(x_1, x_2, \cdots, x_n)$ or $R(x_1, x_2, \cdots, x_n)$.

Let us consider a fuzzy relation $R_\approx$ for "approximately equal" relation. When $x = y$, $R_\approx(x,y) = 1$. According as $|x - y|$ becomes greater, $R_\approx(x,y)$ decreases. There are many functions to satisfy these conditions, one of which is $\exp(-\frac{|x-y|}{C})$. Thus we can take

$$R_\approx(x,y) = \exp\left(-\frac{|x - y|}{C}\right)$$

where C is an appropriate positive constant.

Well-known crisp relations are thus generalized by specifying appropriate membership functions. Another important class of fuzzy relations are expressed using fuzzy graphs, which visualize fuzzy relations on finite sets.

Consider a finite universe $X = \{John, Mary, Thomas, Philip\}$ which is simplified as $X = \{J, M, T, P\}$ hereafter. In section 12.2, we have considered a similar universe and fuzzy set of that universe. Let us try to quantize "strength of friendship" between a pair of individuals in this universe. The strongest friendship is coded by 1.0. On the other hand, the weakest friendship is given the value 0.0. Other strength of friendship are given membership between these extreme values. Assume that the data are given by the following matrix.

$$R = \begin{array}{c} \\ J \\ M \\ T \\ P \end{array} \begin{array}{cccc} J & M & T & P \\ \begin{pmatrix} 1 & 0.5 & 0.9 & 0 \\ 0.5 & 1 & 0.4 & 0.3 \\ 0.9 & 0.4 & 1 & 0.7 \\ 0 & 0.3 & 0.7 & 1 \end{pmatrix} \end{array}.$$

Friendship between an individual and himself is set to unity and the above matrix is symmetric. Now, assume that this matrix is a fuzzy relation of the same symbol: $R(J,J) = 1$, $R(M,T) = R(T,M) = 0.4$, and so on. As we define later, if the matrix of a fuzzy relation is symmetric and has unity as its diagonal elements, the relation is called symmetric and reflexive. Such a fuzzy relation can be expressed as an undirected graph with weights on the edges in Figure 12.6. Notice that an undirected graph is simply

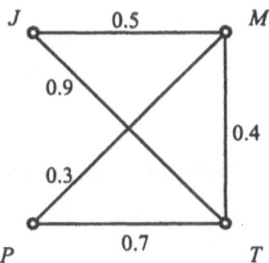

FIGURE 12.6. Undirected fuzzy graph which expresses the relation R on $X = \{J, M, T, P\}$

called a graph. When we regard such a weighted graph as a visualization of a fuzzy relation, the graph is called a fuzzy graph. Note that an edge with the grade 0 can be eliminated from the graph. Moreover the edges of the reflexive parts like $R(J, J) = 1$ are also omitted from the graph for simplicity.

Let us consider another fuzzy relation of "much older than" on the same universe. Assume that this relation Q is given by the following matrix.

$$
Q = \begin{array}{c} \\ J \\ M \\ T \\ P \end{array}
\begin{array}{c} \begin{array}{cccc} J & M & T & P \end{array} \\
\left(\begin{array}{cccc}
0 & 0.6 & 0.3 & 0 \\
0 & 0 & 0 & 0 \\
0 & 0.4 & 0 & 0 \\
0.7 & 1 & 0.9 & 0
\end{array} \right)
\end{array}.
$$

$Q(J, M) = 0.6$ means *John* is considerably "much older than" *Mary*, and $Q(T, J) = 0$ implies that *Thomas* is not at all "much older than" *John*. The matrix for this relation is not symmetric at all. Such a nonsymmetric fuzzy relation is visualized by a directed graph with weights as in Figure 12.7, and is called a fuzzy digraph, although the name *fuzzy graph* in the broad sense refers to both fuzzy undirected graph and fuzzy digraph.

Generally, given a finite universal set $X = \{x_1, x_2, \cdots, x_n\}$ and a fuzzy relation R on X, the fuzzy graph is a digraph on X with weight $R(x_i, x_j)$ on the edge (x_i, x_j). If $R(x_i, x_j) = 0$, we do not draw the edge (x_i, x_j) in the figure. A fuzzy digraph may have two edges between a pair of vertices,

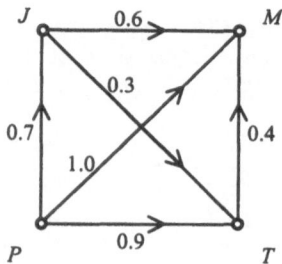

FIGURE 12.7. Fuzzy digraph which expresses the relation Q on $X = \{J, M, T, P\}$

and a loop on a vertex. Consider an example of $X = \{x_1, x_2\}$ and

$$
R = \begin{array}{c} \\ x_1 \\ x_2 \end{array} \begin{pmatrix} x_1 & x_2 \\ 0.5 & 1 \\ 0.8 & 0 \end{pmatrix}.
$$

which is visualized as Figure 12.8.

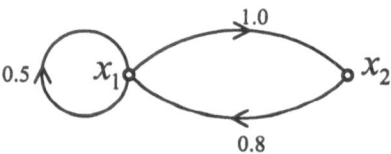

FIGURE 12.8. Fuzzy digraph R with a loop and two edges between x_1 and x_2

Let X be finite or infinite for the moment. A fuzzy relation R is called reflexive if

$$R(x, x) = 1, \quad \forall x \in X.$$

(Note that the symbol $\forall$ is read as *for all*.)
 R is called symmetric if

$$R(x, y) = R(y, x) \quad \forall x, y \in X.$$

A reflexive and symmetric fuzzy relation on a finite universe can be visualized, not by a fuzzy digraph, but by a fuzzy (undirected) graph.

Another type of a fuzzy graph is used for visualizing a fuzzy relation, say R, on $X \times Y$, where X and Y are considered to be two different universes.

Consider an example in which $X = \{x_1, x_2\}$ and $Y = \{y_1, y_2\}$. The relation R on $X \times Y$ is given by

$$R = \begin{array}{c} \\ x_1 \\ x_2 \end{array} \begin{array}{c} y_1 \quad y_2 \\ \left(\begin{array}{cc} 1 & 0.3 \\ 0 & 0.4 \end{array} \right). \end{array}$$

Then the corresponding fuzzy graph is shown in Figure 12.9.

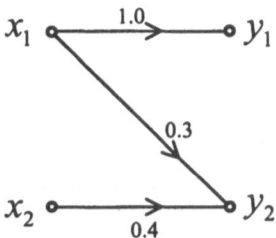

FIGURE 12.9. Fuzzy graph R on $X \times Y$

Since a fuzzy relation is a fuzzy set of the product space, all fuzzy set operations are applied to fuzzy relations. Furthermore, there are other operations specific to fuzzy relations, the most important of which is the max-min composition. The max-min composition is frequently abbreviated as the composition. To discuss the composition, let us review briefly the composition of crisp relations. Let P and Q are crisp relations, for the moment, on $X \times Y$ and $Y \times Z$, respectively. The composition of P and Q, denoted by $P \circ Q$, is defined as follows. For $x \in X$ and $z \in Z$, $P \circ Q(x, z) = 1$ if for some $y \in Y$, $P(x, y) = 1$ and $Q(y, z) = 1$; otherwise $P \circ Q(x, z) = 0$.

Let us turn to fuzzy case and assume now that P and Q are fuzzy relations on the same universes. In view of the crisp composition, it is natural to consider the fuzzy composition in terms of α-cuts. Thus, we should define $P \circ Q$ as:

"For all $\alpha \in [0, 1]$, $(P \circ Q)_\alpha(x, z) = 1$ if $P_\alpha \circ Q_\alpha(x, z) = 1$; otherwise $(P \circ Q)_\alpha(x, z) = 0$."

where $R_\alpha(x, y) = 1 \iff R(x, y) \geq \alpha$.

Most of textbooks in fuzzy sets define the composition without the use of α-cuts. Namely, $P \circ Q$ is defined to be

$$(P \circ Q)(x, z) = \max_{y \in Y} \min\{P(x, y), Q(y, z)\}.$$

Using $\vee$ and $\wedge$, the same equation is written as

$$(P \circ Q)(x, z) = \bigvee_{y \in Y} (P(x, y) \wedge Q(y, z)).$$

(Since this definition uses max and min operations, the composition is called the max-min composition.) The latter definition without an α-cut is more elegant than the former with α-cuts; it is shown that these definitions are equivalent.

To see this, it is sufficient to note that $P \circ Q$ by the latter definition satisfies

$$(P \circ Q)_\alpha = P_\alpha \circ Q_\alpha$$

for all $\alpha \in [0, 1]$.

The last equation is proved as follows. For all $\alpha \in [0, 1]$,

$$
\begin{aligned}
(P \circ Q)_\alpha(x, z) = 1 \quad &\Longleftrightarrow \quad \exists y \in Y, \min\{P(x, y), Q(y, z)\} \geq \alpha \\
&\Longleftrightarrow \quad \exists y \in Y, P(x, y) \geq \alpha \text{ and } Q(y, z) \geq \alpha \\
&\Longleftrightarrow \quad \exists y \in Y, P_\alpha(x, y) = 1 \text{ and } Q_\alpha(y, z) = 1 \\
&\Longleftrightarrow \quad \exists y \in Y, (P_\alpha \circ Q_\alpha)(x, z) = 1.
\end{aligned}
$$

Let P, Q, R be fuzzy relations on appropriately defined universes. Notice that for crisp relations $P_\alpha, Q_\alpha, R_\alpha$, the associative property is valid:

$$(P_\alpha \circ Q_\alpha) \circ R_\alpha = P_\alpha \circ (Q_\alpha \circ R_\alpha).$$

Now, for an arbitrary $\alpha \in [0, 1]$,

$$
\begin{aligned}
[(P \circ Q) \circ R]_\alpha &= (P \circ Q)_\alpha \circ R_\alpha = (P_\alpha \circ Q_\alpha) \circ R_\alpha \\
&= P_\alpha \circ (Q_\alpha \circ R_\alpha) = P_\alpha \circ (Q \circ R)_\alpha \\
&= [P \circ (Q \circ R)]_\alpha.
\end{aligned}
$$

Seeing that two fuzzy sets are equal if and only if all α-cut of them are equal, we have

$$(P \circ Q) \circ R = P \circ (Q \circ R).$$

Thus we have established the associativity of the composition. It should be noted that the above property is proved in the same way as the argument by which the distributivity of fuzzy sets has been proved.

In this way, α-cut is essential in consideration of fuzzy sets and fuzzy relations. Consequently, the meaning of fuzzy graphs (and digraphs) should also be described in terms of α-cuts. Let us consider the example in Figure 12.9 again. Unlike ordinary graphs with weights in which the weight means the amount of a flow, the grade in a fuzzy graph is degree of attainability. If we cut the fuzzy graph in Figure 12.9 by various α, we have a collection of crisp graphs. Let $\alpha = 0.3$, then x_1 and y_2 are connected, but if $\alpha = 0.5$, then there is only one edge in the α-cut, that is, (x_1, y_1). The degree of attainability between x_i and x_j means the maximum value of α such that x_i and x_j are connected in that α-cut of the fuzzy graph.

Since the grade in a fuzzy graph is degree of attainability, the composition of two fuzzy graphs should be interpreted in the same manner. Indeed, the simplest way to see the meaning of the max-min composition is to observe fuzzy graphs. Let us consider an example of two relations P on $X \times Y$ and Q on $Y \times Z$, where $X = \{x_1, x_2\}$, $Y = \{y_1, y_2\}$, $Z = \{z_1, z_2\}$,

$$P = \begin{array}{c} \\ x_1 \\ x_2 \end{array} \begin{array}{c} y_1 \quad\ y_2 \\ \left(\begin{array}{cc} 0.5 & 0.7 \\ 0.3 & 1.0 \end{array} \right) \end{array}.$$

$$Q = \begin{array}{c} \\ y_1 \\ y_2 \end{array} \begin{array}{c} z_1 \quad\ z_2 \\ \left(\begin{array}{cc} 0.4 & 0.7 \\ 0.8 & 0.6 \end{array} \right) \end{array}.$$

As in Figure 12.10, the two fuzzy graphs P and Q are combined into one fuzzy graph.

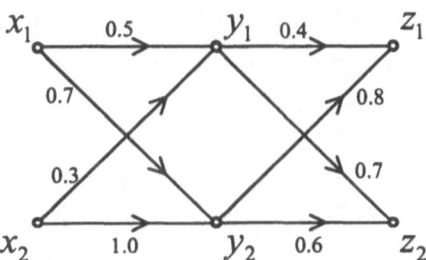

FIGURE 12.10. Fuzzy graphs P and Q are combined.

By the analogy of the attainability in a fuzzy relation, the grade $(P \circ Q)(x_i, z_k)$ should imply the degree of attainability between x_i and z_k through y_j, $j = 1, 2$.

Take x_1 and z_1 to consider $(P \circ Q)(x_1, z_1)$. Remember that the degree of attainability means the maximum value of α such that x_1 and z_1 are connected in that α-cut. There are two paths $x_1 - y_1 - z_1$ and $x_1 - y_2 - z_1$. The attainability using $x_1 - y_1 - z_1$ is $\min\{R(x_1, y_1), Q(y_1, z_1)\} = \min\{0.5, 0.4\} = 0.4$, since if $\alpha > 0.4$, one of edges (x_1, y_1) and (y_1, z_1) disappears in that α-cut of the fuzzy graph. The degree using $x_1 - y_2 - z_1$ is $\min\{R(x_1, y_2), Q(y_2, z_1)\} = \min\{0.7, 0.8\} = 0.7$ by the same reason. Now, to reach from x_1 to z_1, we can use either of the two paths, and consequently the latter path $x_1 - y_2 - z_1$ is more favorable, since the latter path has the larger degree of attainability. Namely, the overall degree of attainability is

$$\max\{\min(R(x_1, y_1), Q(y_1, z_1)), \min(R(x_1, y_2), Q(y_2, z_1))\}$$
$$= \max\{0.4, 0.7\} = 0.7.$$

The last equation is just the expression of the max-min composition defined before.

Generally, given $X = \{x_1, x_2, \cdots, x_\ell\}$, $Y = \{y_1, y_2, \cdots, y_m\}$, and $Z = \{z_1, z_2, \cdots, z_n\}$, P on $X \times Y$, and Q on $Y \times Z$, we have m paths of $x_i - y_j - z_k$, $j = 1, 2, \cdots, m$, to connect x_i and z_k. The attainability through $x_i - y_j - z_k$ is $\min\{P(x_i, y_j), Q(y_j, z_k)\}$, and the overall attainability is

$$\max_{y_j \in Y} \min\{P(x_i, y_j), Q(y_j, z_k)\}$$

since we can use any of these m paths. Thus, $\min\{P(x_i, y_j), Q(y_j, z_k)\}$ describes attainability of a path, and $\max_{y_j \in Y}$ implies that we can choose the path of the maximum attainability out of possible routes.

Apart from the graphical interpretation, algebraic calculation of the composition should use $\wedge$ and $\vee$ instead of min and max. Thus, in the former example,

$$P \circ Q = \begin{array}{c} \\ x_1 \\ x_2 \end{array} \begin{pmatrix} \overset{z_1}{(0.5 \wedge 0.4) \vee (0.7 \wedge 0.8)} & \overset{z_2}{(0.5 \wedge 0.7) \vee (0.7 \wedge 0.6)} \\ (0.3 \wedge 0.4) \vee (1.0 \wedge 0.8) & (0.3 \wedge 0.7) \vee (1.0 \wedge 0.6) \end{pmatrix}$$

$$= \begin{array}{c} \\ x_1 \\ x_2 \end{array} \begin{pmatrix} \overset{z_1}{0.7} & \overset{z_2}{0.6} \\ 0.8 & 0.6 \end{pmatrix}.$$

Sometimes $\wedge$ is replaced by a dot $(\cdot)$ and $\vee$ is replaced by plus $(+)$. Then the same calculation is written as

$$P \circ Q = \begin{array}{c} \\ x_1 \\ x_2 \end{array} \begin{pmatrix} \overset{z_1}{0.5 \cdot 0.4 + 0.7 \cdot 0.8} & \overset{z_2}{0.5 \cdot 0.7 + 0.7 \cdot 0.6} \\ 0.3 \cdot 0.4 + 1.0 \cdot 0.8 & 0.3 \cdot 0.7 + 1.0 \cdot 0.6 \end{pmatrix} = \begin{array}{c} \\ x_1 \\ x_2 \end{array} \begin{pmatrix} \overset{z_1}{0.7} & \overset{z_2}{0.6} \\ 0.8 & 0.6 \end{pmatrix}.$$

It is interesting to note that the way of using the symbols $+$ and $\cdot$ in the last calculation is the same as the ordinary matrix calculation, although the result is different from the ordinary matrix product.

Given a fuzzy set A of X and a fuzzy relation R on $X \times Y$, the max-min composition

$$(A \circ R)(y) = \max_{x \in X} \min\{A(x), R(x, y)\}$$

is used in fuzzy inference. Figure 12.11 is the fuzzy graph representing $A \circ R$ for

$$A = \begin{array}{c} \\ \end{array} \begin{pmatrix} \overset{x_1}{0.4} & \overset{x_2}{0.5} \end{pmatrix} \qquad R = \begin{array}{c} \\ x_1 \\ x_2 \end{array} \begin{pmatrix} \overset{y_1}{0.2} & \overset{y_2}{0.7} \\ 0.3 & 0.6 \end{pmatrix}.$$

The fuzzy set $A(x)$ is represented by weights on two edges from a dummy vertex to x_1 and x_2. The resulting membership values of $A \circ R$ is written in parentheses by the vertices y_1 and y_2.

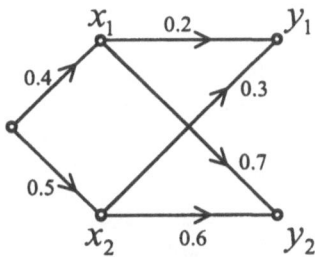

FIGURE 12.11. Fuzzy graph for calculating $A \circ R$, where A is a fuzzy set of X

Algebraically,

$$
\begin{aligned}
A \circ R &= [0.4 \quad 0.5] \begin{bmatrix} 0.2 & 0.7 \\ 0.3 & 0.6 \end{bmatrix} \\
&= [(0.4 \wedge 0.2) \vee (0.5 \wedge 0.3) \quad (0.4 \wedge 0.7) \vee (0.5 \wedge 0.6)] \\
&= [0.3 \quad 0.5]
\end{aligned}
$$

If A is a fuzzy set on real numbers $X = \mathbf{R}$ and R is a fuzzy relation on $X \times Y = \mathbf{R}^2$, the composition is described by such a figure as Figure 12.12. In this figure the triangle on the X-axis in the left side is the fuzzy set $A(x)$. The large pyramid is the relation R. Then the trapezoid with the shadow (shaded area) on Y-axis is $A \circ R$.

To obtain $A \circ R$, we consider the cylindrical extension

$$ A'(x, y) = A(x), \quad \forall y \in Y $$

and take $A' \cap R$. Notice that

$$ (A' \cap R)(x, y) = A'(x, y) \wedge R(x, y) = A(x) \wedge R(x, y). $$

In Figure 12.12, $A' \cap R$ is a polyhedron within the pyramid R. Then the shadow projected onto the plane of $Y \times [0, 1]$ shows the membership function of $A \circ R$. Namely, the projection of $A' \cap R$ is given by

$$ \max_{x \in X}(A' \cap R)(x, y) = \max_{x \in X} A(x) \wedge R(x, y) $$

which is just the max-min composition.

12.6 Possibility and Necessity Measures

Possibility theory has been proposed by Zadeh [8] and extensively discussed by Dubois and Prade [3].

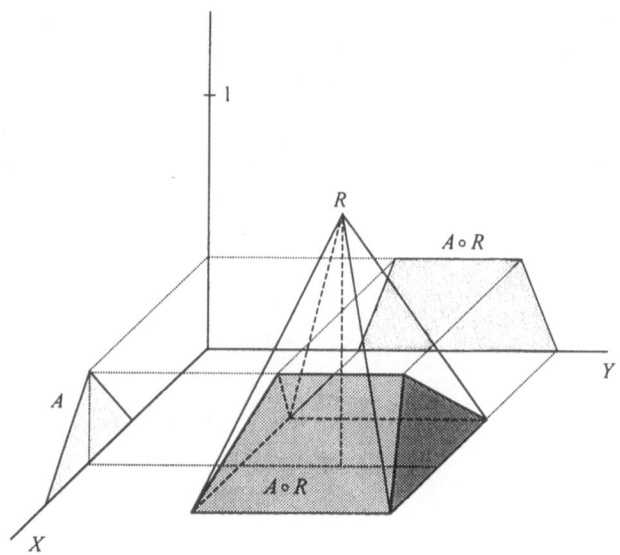

FIGURE 12.12. Geometrical calculation of $A \circ R$ when A is a fuzzy set of real numbers and R is a fuzzy relation on $\mathbf{R}^2$

Let X be a universal set. In most cases X is supposed to be finite, $X = \{x_1, \cdots, x_n\}$, for simplicity, and we briefly refer to infinite sets of universe.

We begin with the possibility measure which is firstly defined on the set 2^X of the collection of crisp sets.

Assume $A \in 2^X$. A measure, denoted by $\Pi(A)$, $(\Pi(\cdot) \colon 2^X \to [0, \infty))$ is called a *possibility measure* if it satisfies the following (**I–III**):

(**I**)

$$\Pi(\emptyset) = 0, \qquad \Pi(X) = 1,$$

(**II**)

$$\Pi(A) \leq \Pi(B) \quad \text{if} \quad A \subseteq B,$$

(**III**)

$$\Pi(A \cup B) = \Pi(A) \vee \Pi(B).$$

Remark. A real-valued measure that satisfies (**I–II**) and a continuity condition (in case of an infinite universe) is called a *fuzzy measure*, of which details are omitted here.

A measure, denoted by $N(A)$, $(N(\cdot) \colon 2^X \to [0, \infty))$ is called a *necessity measure* if it satisfies the following (**I'–III'**):

(I')

$$N(\emptyset) = 0, \qquad N(X) = 1,$$

(II')

$$N(A) \le N(B) \quad \text{if} \quad A \subseteq B,$$

(III')

$$N(A \cap B) = N(A) \wedge N(B).$$

These nonadditive measures can be contrasted with the ordinary additive measures such as those in probability theory.

Assume that a possibility measure $\Pi(\cdot)$ is given, we define

$$N(A) = 1 - \Pi(\bar{A})$$

where $\bar{A}$ is the complement of A. We see that $N(\cdot)$ is a necessity measure, since

$$
\begin{aligned}
N(A \cap B) &= 1 - \Pi(\overline{(A \cap B)}) = 1 - \Pi(\bar{A} \cup \bar{B}) \\
&= 1 - \max\{\Pi(\bar{A}), \Pi(\bar{B})\} = \min\{1 - \Pi(\bar{A}), 1 - \Pi(\bar{B})\} \\
&= \min\{N(A), N(B)\}
\end{aligned}
$$

In the same way, given a necessity measure $N(\cdot)$, the measure defined by

$$\Pi(A) = 1 - N(\bar{A}) \qquad (12.7)$$

is proved to be a possibility measure.

A possibility measure is expressed in terms of a possibility distribution. To see this, suppose X is finite and $A = \{x_i, \cdots, x_\ell\}$, then

$$\Pi(A) = \max\{\Pi(\{x_i\}), \cdots, \Pi(\{x_\ell\})\}.$$

Define

$$\pi(x) = \Pi(\{x\}). \qquad (12.8)$$

$\pi(\cdot)$ is called a possibility distribution associated with the possibility measure $\Pi(\cdot)$. We have

$$\Pi(A) = \bigvee_{x \in A} \pi(x). \qquad (12.9)$$

A possibility measure is thus represented by the corresponding possibility distribution. Suppose moreover that a necessity measure $N(A)$ is given, the possibility distribution $\pi(x)$ is defined for the possibility measure defined by (12.7), then we have

$$N(A) = 1 - \bigvee_{x \in \bar{A}} \pi(x) = \bigwedge_{x \in \bar{A}} \{1 - \pi(x)\}. \qquad (12.10)$$

Remark. $\bigvee_{x \in \bar{A}} \pi(x)$, which means the maximum of $\pi(x)$ for all $x \in A$ is generalized to the *supremum* in case of an infinite set. In the latter case, $\bigvee_{x \in \bar{A}} \pi(x) = \sup_{x \in \bar{A}} \pi(x)$.

We now proceed to define possibility and necessity measures for fuzzy sets. Assume that A is a fuzzy set of X. (The collection of all fuzzy sets of X is sometimes denoted by $[0,1]^X$; thus $A \in [0,1]^X$.) Suppose that a possibility distribution for crisp sets is given. (If a possibility measure is given, we can define the corresponding possibility distribution by (12.8).

We define $\Pi(A)$ as follows.

$$\Pi(A) = \sup_{x \in X} \min\{\pi(x), \mu_A(x)\} = \bigvee_{x \in X} \{\pi(x) \wedge \mu_A(x)\}. \qquad (12.11)$$

The above measure reduces to the ordinary possibility measure if A is crisp, hence it generalizes the previous measure (12.9).

It is immediate to generalize the necessity measure. We define

$$N(A) = 1 - \sup_{x \in X} \min\{\pi(x), \mu_{\bar{A}}(x)\} = \bigwedge_{x \in X} \{1 - \pi(x), \mu_A(x)\}. \qquad (12.12)$$

12.7   Fuzzy Numbers

Fuzzy numbers [2] have frequently been used for data analysis including fuzzy quantities; fuzzy regression analysis is a typical example of their applications.

A fuzzy interval [3] is a fuzzy set in $\mathbf{R}$, such that every α-cut of it is a closed interval and its height is unity: suppose $J \in [0,1]^{\mathbf{R}}$ is a fuzzy interval, then

$$J_\alpha = [j_1(\alpha), j_2(\alpha)], \qquad \forall \alpha \in (0,1]$$

and

$$\mathrm{hgt}(J) = \max_{x \in \mathbf{R}} \mu_J(x) = 1.$$

We assume hereafter that membership functions of all fuzzy intervals are upper semi-continuous.

A fuzzy number is a special type of fuzzy intervals for which the α-cut with $\alpha = 1$ is a singleton. Suppose M is a fuzzy number, then $M_1 = \{x_M\}$ for some $x_M \in \mathbf{R}$.

The most frequent form of fuzzy numbers is the triangular fuzzy number: its membership function is given by

$$\mu_M(x) = \begin{cases} 0 & (x \leq a_1) \\[2mm] \dfrac{x - a_1}{a_2 - a_1} & (a_1 < x \leq a_2) \\[2mm] \dfrac{a_3 - x}{a_3 - a_2} & (a_2 < x \leq a_3) \\[2mm] 0 & (x > a_3) \end{cases}$$

where a_1, a_2, and a_3 are appropriately chosen real parameters.

Fuzzy numbers are used for representing a phrase such as 'about n' by a fuzzy set. For example, 'about 3' can be represented by the above triangular fuzzy number with $a_2 = 3$ (a_1 and a_3 are somewhat arbitrary, say $a_1 = 2.5$ and $a_3 = 3.5$).

Algebraic calculations between fuzzy numbers (and fuzzy intervals) such as $M + N$ and $M - N$ are possible. Before the discussion of fuzzy numbers, let us return to the crisp case.

Given two intervals $C_a = [a_1, a_2]$ and $C_b = [b_1, b_2]$, we define the addition, subtraction, multiplication, and division as follows.

$$C_a * C_b = \{x * y \in \mathbf{R} : x \in C_a, \text{ and } y \in C_b\} \tag{12.13}$$

where $*$ stands for each of $+$, $-$, $\times$, and $/$. We can immediately see that $C_a * C_b$ is a closed interval.

Let us proceed to consideration of fuzzy numbers. given two fuzzy numbers M and N, we define the addition, subtraction, multiplication, and division by the following.

$$\mu_{M*N}(z) = \sup_{z = x * y} \min\{\mu_M(x), \mu_N(y)\} \tag{12.14}$$

where $*$ represents $+$, $-$, $\times$, and $/$.

It is not difficult to see that the α-cut commutes with the $*$ operation:

$$(M * N)_\alpha = M_\alpha * N_\alpha, \qquad \forall \alpha \in (0, 1]. \tag{12.15}$$

Notice that the right hand side uses the definition of crisp intervals.

Remark. Frequently the symbols $\oplus$, $\ominus$, $\otimes$, and $\oslash$ are used for addition, subtraction, multiplication, and division, respectively in literature. The symbol $*$ can represent the minimum $\wedge$ and maximum $\vee$ in addition.

Moreover, the symbol $\otimes$ is used for the scalar product of two vectors of fuzzy numbers. Let $M = (M_1, \cdots, M_p)^T$ and $N = (N_1, \cdots, N_p)^T$ be vectors in which M_i and N_i $(i = 1, \cdots, p)$ are fuzzy numbers. Then, the scalar product is a fuzzy number defined by

$$M \otimes N = M_1 \times N_1 + \cdots + M_p \times N_p.$$

Possibility of equality and inequality between fuzzy numbers

We consider relational operators $=$ and $\leq$ between fuzzy numbers. Notice that a principle in fuzzy sets is to define a quantity that represents the maximum value of α-cut such that a condition is true.

Let us for example consider the condition 'M equals N' for two fuzzy numbers. A trivial way is to use the equality of fuzzy sets. However, the equality of fuzzy sets is too strict, i.e., we rarely encounter the strict equality $M = N$ in real situations and therefore the equality of fuzzy sets is not useful. We therefore consider a weaker condition: the maximum value of α-cut such that there exists an $x \in M$ and $x \in N$ at the same time. According to Dubois and Prade [2], this value is called the possibility by which M equals N denoted by $Pos(M = N)$, which is given by the following.

$$Pos(M = N) = \sup_{x=y} \mu_M(x) \wedge \mu_N(y). \tag{12.16}$$

It is immediate to see that $(M \cap N)_\alpha \neq \emptyset$ if and only if $\alpha \leq Pos(M = N)$; in other words, there exists $x \in M_\alpha$ and $y \in N_\alpha$ such that $x = y$ for all $\alpha \leq Pos(M = N)$ and such x and y do not exist for all $\alpha > Pos(M = N)$.

In the same way, $Pos(M \leq N)$, the maximum value of α-cut by which 'M is less than or equal to N' is introduced:

$$Pos(M \leq N) = \sup_{x \leq y} \mu_M(x) \wedge \mu_N(y). \tag{12.17}$$

Let $\lambda = Pos(M \leq N)$. Then it is easily seen that there are $x \in M_\alpha$ and $y \in N_\alpha$ such that $x \leq y$ for all $\alpha \leq \lambda$ and such x and y do not exist for all $\alpha > \lambda$.

12.8   Discussion and Remarks

Several chapters in this book need basic knowledge of fuzzy sets, which have been described in the present chapter.

As set-theoretical discussions are based on the crisp logic, fuzzy sets are on the basis of fuzzy logic. Typical consideration is found in Takeuti and Titani [6], in which fuzzy logic is dealt with within the scope of intuitionistic logic. There are other types of studies in fuzzy logic (e.g., Novák [4]) but we omit the details, since they use sophisticated and advanced mathematical arguments beyond the scope of this book.

Nevertheless, Chapter 3 in this book is devoted to modal logic and rough sets which are also mathematical. Modal logic is now known to be an important fundamental theory in information systems. Rough sets represent a set-theoretical aspect of modal logic, and hence is considered to be more useful in applications when compared with modal logic itself, in view of the fact that fuzzy logic has been put into practice after fuzzy sets had been proposed. We therefore have included rough sets and modal logic in this book.

Basic operations have been shown in Section 12.3 whereas Chapter 7 discusses advanced operations of aggregating fuzzy sets such as t-norms, conorms, mean operators, and OWA operators. Combination of the discussion in that chapter and that in Chapter 2 should be considered.

Basic concept of fuzzy sets is sufficient in understanding fuzzy clustering using c-means, while fuzzy relations in Section 12.5 are used when hierarchical clustering in Chapter 5 should be studied.

In Chapter 9, advanced features of truth qualifications are employed, which we have omitted in this chapter, since Chapter 9 is self-contained in this aspect. Readers will find details the truth qualifications in [1].

It should also be remarked that fuzzy numbers and the possibility discussed in Sections 12.6 and 12.7 are employed in discussing regression of fuzzy data in Chapter 10.

12.9 REFERENCES

[1] D. Dubois and H. Prade, *Fuzzy Sets and Systems: Theory and Applications*, Academic Press, New York, 1980.

[2] D. Dubois and H. Prade, "Fuzzy numbers: an overview," in *Analysis of Fuzzy Information*, Vol.1, J. C. Bezdek, (Ed.), CRC Press, pp.3-39, 1987.

[3] D. Dubois and H. Prade, *Possibility Theory*, Plenum, New York, 1988.

[4] V. Novák, *Fuzzy Sets and Their Applications*, Adam Hilger, Bristol, 1989.

[5] Z. Pawlak, *Rough Sets*, Kluwer, Dordrecht, 1991.

[6] G. Takeuti and S. Titani, "Intuitionistic fuzzy logic and intuitionistic fuzzy set theory," *J. Symbolic Logic*, Vol.49, pp.851-866, 1984.

[7] L. A. Zadeh, Fuzzy sets, *Information and Control*, Vol.8, pp.338-353, 1965.

[8] L. A. Zadeh, "Fuzzy sets as a basis for a theory of possibility," *Fuzzy Sets and Systems*, Vol.1, pp.3-28, 1978.